NANOSCIENCE AND TECHNOLOGY

NanoScience and Technology

Series Editors:
P. Avouris B. Bhushan D. Bimberg K. von Klitzing H. Sakaki R. Wiesendanger

The series NanoScience and Technology is focused on the fascinating nano-world, mesoscopic physics, analysis with atomic resolution, nano and quantum-effect devices, nanomechanics and atomic-scale processes. All the basic aspects and technology-oriented developments in this emerging discipline are covered by comprehensive and timely books. The series constitutes a survey of the relevant special topics, which are presented by leading experts in the field. These books will appeal to researchers, engineers, and advanced students.

Bharat Bhushan
Harald Fuchs
Satoshi Kawata (Eds.)

Applied Scanning Probe Methods V

Scanning Probe Microscopy Techniques

With 194 Figures and 12 Tables
Including 5 Color Figures

 Springer

Editors:

Professor Bharat Bhushan
Nanotribology Laboratory for Information
Storage and MEMS/NEMS (NLIM)
W 390 Scott Laboratory, 201 W. 19th Avenue
The Ohio State University, Columbus
Ohio 43210-1142, USA
e-mail: Bhushan.2@osu.edu

Satoshi Kawata
Osaka City University, Graduate School
of Science, Department of Mathematics
Sugimoto 3-3-138, 558-8585 Osaka, Japan
e-mail: skawata@skawata.com

Professor Dr. Harald Fuchs
Center for Nanotechnology (CeNTech)
and Institute of Physics
University of Münster
Gievenbecker Weg 11, 48149 Münster, Germany
e-mail: fuchsh@uni-muenster.de

Series Editors:

Professor Dr. Phaedon Avouris
IBM Research Division
Nanometer Scale Science & Technology
Thomas J. Watson Research Center, P.O. Box 218
Yorktown Heights, NY 10598, USA

Professor Bharat Bhushan
Nanotribology Laboratory for Information
Storage and MEMS/NEMS (NLIM)
W 390 Scott Laboratory, 201 W. 19th Avenue
The Ohio State University, Columbus
Ohio 43210-1142, USA

Professor Dr. Dieter Bimberg
TU Berlin, Fakutät Mathematik,
Naturwissenschaften,
Institut für Festkörperphysik
Hardenbergstr. 36, 10623 Berlin, Germany

Professor Dr., Dres. h. c. Klaus von Klitzing
Max-Planck-Institut für Festkörperforschung
Heisenbergstrasse 1, 70569 Stuttgart, Germany

Professor Hiroyuki Sakaki
University of Tokyo
Institute of Industrial Science,
4-6-1 Komaba, Meguro-ku, Tokyo 153-8505, Japan

Professor Dr. Roland Wiesendanger
Institut für Angewandte Physik
Universität Hamburg
Jungiusstrasse 11, 20355 Hamburg, Germany

DOI 10.1007/b136626
ISSN 1434-4904
ISBN-10 3-540-37315-2 Springer Berlin Heidelberg New York
ISBN-13 978-3-540-37315-5 Springer Berlin Heidelberg New York
Library of Congress Control Number: 2006932714

Springer is a part of Springer Science+Business Media

springer.com

© Springer-Verlag Berlin Heidelberg 2007

Typesetting and production: LE-TEX Jelonek, Schmidt & Vöckler GbR, Leipzig
Cover: WMX Design, Heidelberg
Printed on acid-free paper   2/3100/YL - 5 4 3 2 1 0

Preface

The scanning probe microscopy field has been rapidly expanding. It is a demanding task to collect a timely overview of this field with an emphasis on technical developments and industrial applications. It became evident while editing Vols. I–IV that a large number of technical and applicational aspects are present and rapidly developing worldwide. Considering the success of Vols. I–IV and the fact that further colleagues from leading laboratories were ready to contribute their latest achievements, we decided to expand the series with articles touching fields not covered in the previous volumes. The response and support of our colleagues were excellent, making it possible to edit another three volumes of the series. In contrast to topical conference proceedings, the applied scanning probe methods intend to give an overview of recent developments as a compendium for both practical applications and recent basic research results, and novel technical developments with respect to instrumentation and probes.

The present volumes cover three main areas: novel probes and techniques (Vol. V), charactarization (Vol. VI), and biomimetics and industrial applications (Vol. VII).

Volume V includes an overview of probe and sensor technologies including integrated cantilever concepts, electrostatic microscanners, low-noise methods and improved dynamic force microscopy techniques, high-resonance dynamic force microscopy and the torsional resonance method, modelling of tip cantilever systems, scanning probe methods, approaches for elasticity and adhesion measurements on the nanometer scale as well as optical applications of scanning probe techniques based on nearfield Raman spectroscopy and imaging.

Volume VI is dedicated to the application and characterization of surfaces including STM on monolayers, chemical analysis of single molecules, STM studies on molecular systems at the solid–liquid interface, single-molecule studies on cells and membranes with AFM, investigation of DNA structure and interactions, direct detection of ligand protein interaction by AFM, dynamic force microscopy as applied to organic/biological materials in various environments with high resolution, noncontact force microscopy, tip-enhanced spectroscopy for investigation of molecular vibrational excitations, and investigation of individual carbon nanotube polymer interfaces.

Volume VII is dedicated to the area of biomimetics and industrial applications. It includes studies on the lotus effect, the adhesion phenomena as occurs in gecko feet, nanoelectromechanical systems (NEMS) in experiment and modelling, application of STM in catalysis, nanostructuring and nanoimaging of biomolecules for biosensors, application of scanning electrochemical microscopy, nanomechanical investigation of pressure sensitive adhesives, and development of MOEMS devices.

As in the previous volumes a distinction between basic research fields and industrial scanning probe techniques cannot be made, which is in fact a unique factor in nanotechnology in general. It also shows that these fields are extremely active and that the novel methods and techniques developed in nanoprobe basic research are rapidly being transferred to applications and industrial development.

We are very grateful to our colleagues who provided in a timely manner their manuscripts presenting state-of-the-art research and technology in their respective fields. This will help keep research and development scientists both in academia and industry well informed about the latest achievements in scanning probe methods. Finally, we would like to cordially thank Dr. Marion Hertel, senior editor chemistry, and Mrs. Beate Siek of Springer for their continuous support and advice without which these volumes could have never made it to market on time.

July, 2006
Prof. Bharat Bhushan, USA
Prof. Harald Fuchs, Germany
Prof. Satoshi Kawata, Japan

Contents – Volume V

Contents – Volume VI

Contents – Volume VII

Contents – Volume I

Part III　Industrial Applications

Contents – Volume II

Contents – Volume III

Contents – Volume IV

List of Contributors – Volume V

Maria Allegrini
Dipartimento di Fisica "Enrico Fermi", Università di Pisa
Largo Bruno Pontecorvo, 3, 56127 Pisa, Italy
e-mail: maria.allegrini@df.unipi.it

Yasuhisa Ando
Tribology Group, National Institute of Advanced Industrial Science and Technology
1-2 Namiki, Tsukuba, Ibaraki, 305-8564, Japan
e-mail: yas.ando@aist.go.jp

Bharat Bhushan
Nanotribology Laboratory for Information Storage and MEMS/NEMS (NLIM)
W 390 Scott Laboratory, 201 W. 19th Avenue, Ohio State University
Columbus, Ohio 43210-1142, USA
e-mail: bhushan.2@osu.edu

Daniel Ebeling
Center for NanoTechnology (CeNTech), Heisenbergstr. 11, 48149 Münster
e-mail: Daniel.Ebeling@uni-muenster.de

Pietro Guiseppe Gucciardi
CNR-Istituto per i Processi Chimico-Fisici, Sezione di Messina
Via La Farina 237, I-98123 Messina, Italy
e-mail: gucciardi@its.me.cnr.it

Sadik Hafizovic
ETH-Zurich, PEL, Hoenggerberg HPT H6, 8093 Zurich, Switzerland
e-mail: hafizovi@phys.ethz.ch

Andreas Hierlemann
ETH-Zurich, PEL, Hoenggerberg HPT H6, 8093 Zurich, Switzerland
e-mail: hierlema@phys.ethz.ch

Hendrik Hölscher
Center for NanoTechnology (CeNTech), Heisenbergstr. 11, 48149 Münster
e-mail: Hendrik.Hoelscher@uni-muenster.de

Lin Huang
Veeco Instruments, 112 Robin Hill Road, Santa Barbara, CA 93117, USA
e-mail: lhuang@veeco.com

Hideki Kawakatsu
Institute of Industrial Science, University of Tokyo
Komaba 4-6-1, Meguro-Ku, Tokyo 153-8505, Japan
e-mail: kawakatu@iis.u-tokyo.ac.jp

Kay-Uwe Kirstein
ETH-Zurich, PEL, Hoenggerberg HPT H6, 8093 Zurich, Switzerland
e-mail: kirstein@phys.ethz.ch

Christine Kranz
School of Chemistry and Biochemistry, Georgia Institute of Technology
311 Ferst Dr., Atlanta GA 30332-0400, USA
e-mail: Christine.Kranz@chemistry.gatech.edu

Boris Mizaikoff
School of Chemistry and Biochemistry, Georgia Institute of Technology
311 Ferst Dr., Atlanta GA 30332-0400, USA
e-mail: Boris.Mizaikoff@chemistry.gatech.edu

Salvatore Patanè
Dipartimento di Fisica della Materia e Tecnologie Fisiche Avanzate
Università di Messina, Salita Sperone 31, I-98166 Messina, Italy
e-mail: patanes@unime.it

Craig Prater
Veeco Instruments, 112 Robin Hill Road, Santa Barbara, CA 93117, USA
e-mail: cprater@veeco.com

Elisa Riedo
Georgia Institute of Technology, School of Physics
837 State Street, Atlanta, GA 30332-0430, USA
e-mail: elisa.riedo@physics.gatech.edu

Tilman E. Schäffer
Institute of Physics and Center for Nanotechnology, University of Münster
Heisenbergstr. 11, 48149 Münster, Germany
e-mail: tilman.schaeffer@uni-muenster.de

Udo D. Schwarz
Department of Mechanical Engineering, Yale University
P.O. Box 208284, New Haven, CT 06520-8284, USA
e-mail: Udo.Schwarz@yale.edu

Yaxin Song
Nanotribology Lab for Information Storage and MEMS/NEMS (NLIM)
The Ohio State University
650 Ackerman Road, Suite 255, Columbus, Ohio 43202, USA
e-mail: Song.220@osu.edu

Chanmin Su
Veeco Instruments, 112 Robin Hill Road, Santa Barbara, CA 93117, USA
e-mail: csu@veeco.com

Robert Szoszkiewicz
Georgia Institute of Technology, School of Physics
837 State Street, Atlanta, GA 30332-0430, USA
e-mail: robert.szoszkiewicz@physics.gatech.edu

Sebastiano Trusso
CNR-Istituto per i Processi Chimico-Fisici, Sezione di Messina
Via La Farina 237, I-98123 Messina, Italy
e-mail: trusso@its.me.cnr.it

Cirino Vasi
CNR-Istituto per i Processi Chimico-Fisici, Sezione di Messina
Via La Farina 237, I-98123 Messina, Italy
e-mail: vasi@its.me.cnr.it

Justyna Wiedemair
School of Chemistry and Biochemistry, Georgia Institute of Technology
311 Ferst Dr., Atlanta GA 30332-0400, USA
e-mail: Justyna.Wiedemair@chemistry.gatech.edu

List of Contributors – Volume VI

Asa H. Barber
Queen Mary, University of London, Department of Materials
Mile End Road, London E1 4NS, UK
e-mail: a.h.barber@qmul.ac.uk

Lilia A. Chtcheglova
Institute for Biophysics, Johannes Kepler University of Linz
Altenbergerstr. 69, A-4040 Linz, Austria
e-mail: lilia.chtcheglova@jku.at

Sidney R. Cohen
Chemical Research Support, Weizmann Institute of Science
Rehovot 76100, Israel
e-mail: Sidney.cohen@weizmann.ac.il

Andreas Ebner
Institute for Biophysics, Johannes Kepler University of Linz
Altenbergerstr. 69, A-4040 Linz, Austria
e-mail: andreas.ebner@jku.at

Hermann J. Gruber
Institute for Biophysics, Johannes Kepler University of Linz
Altenbergerstr. 69, A-4040 Linz, Austria
e-mail: hermann.gruber@jku.at

Norihiko Hayazawa
Nanophotonics Laboratory
RIKEN (The Institute of Physical and Chemical Research)
2-1 Hirosawa, Wako, Saitama, 351-0198, Japan
e-mail: hayazawa@riken.jp

Peter Hinterdorfer
Institute for Biophysics, Johannes Kepler University of Linz
Altenbergerstr. 69, A-4040 Linz, Austria
e-mail: peter.hinterdorfer@jku.at

Ferry Kienberger
Institute for Biophysics, Johannes Kepler University of Linz
Altenbergerstr. 69, A-4040 Linz, Austria
e-mail: ferry.kienberger@jku.at

Kei Kobayashi
International Innovation Center, Kyoto University, Katsura, Nishikyo
Kyoto 615-8520, Japan
e-mail: keicoba@iic.kyoto-u.ac.jp

Tadahiro Komeda
Institute of Multidisciplinary Research for Advanced Materials (IMRAM)
Tohoku University
2-1-1, Katahira, Aoba, Sendai, 980-0877 Japan
e-mail: komeda@tagen.tohoku.ac.jp

Andrzej J. Kulik
Ecole Polytechnique Fédérale de Lausanne, EPFL – IPMC – NN
1015 Lausanne, Switzerland
e-mail: Andrzej.Kulik@epfl.ch

Piotr Laidler
Institute of Medical Biochemistry, Collegium Medicum Jagiellonian University
Kopernika 7, 31-034 Kraków, Poland
e-mail: mblaidle@cyf-kr.edu.pl

Małgorzata Lekka
The Henryk Niewodniczański Institute of Nuclear Physics
Polish Academy of Sciences, Radzikowskiego 152, 31–342 Kraków, Poland
e-mail: Malgorzata.Lekka@ifj.edu.pl

Thomas Mueller
Veeco Instruments, 112 Robin Hill Road, Santa Barbara, CA 93117, USA
e-mail: tmueller@veeco.com

Theeraporn Puntheeranurak
Institute for Biophysics, Johannes Kepler University of Linz
Altenbergerstr. 69, A-4040 Linz, Austria
e-mail: theeraporn.puntheeranurak@jku.at

Yuika Saito
Nanophotonics Laboratory
RIKEN (The Institute of Physical and Chemical Research)
2-1 Hirosawa, Wako, Saitama, 351-0198, Japan

Yasuhiro Sugawara
Department of Applied Physics, Graduate School of Engineering,
Osaka University, Yamada-oka 2-1, Suita, Osaka 565-0871, Japan
e-mail: sugawara@ap.eng.osaka-u.ac.jp

Neil H. Thomson
Molecular and Nanoscale Physics Group, University of Leeds
EC Stoner Building, Woodhouse Lane, Leeds, LS2 9JT, UK
e-mail: n.h.thomson@leeds.ac.uk

Kohei Uosaki
Division of Chemistry, Graduate School of Science, Hokkaido University
N10 W8, Sapporo, Hokkaido, 060-0810, Japan
e-mail: uosaki@pcl.sci.hokudai.ac.jp

H. Daniel Wagner
Dept. Materials and Interfaces, Weizmann Institute of Science
Rehovot 76100, Israel
e-mail: daniel.wagner@weizmann.ac.il

Hirofumi Yamada
Department of Electronic Science & Engineering, Kyoto University
Katsura, Nishikyo, Kyoto 615-8510, Japan
e-mail: h-yamada@kuee.kyoto-u.ac.jp

Ryo Yamada
Division of Material Physics, Graduate School of Engineering Science
Osaka University, Machikaneyama-1-3, Toyonaka, Osaka, 060-0810, Japan
e-mail:yamada@molectronics.jp

List of Contributors – Volume VII

Flemming Besenbacher
Interdisciplinary Nanoscience Center (iNANO), University of Aarhus
DK-8000 Aarhus C, Denmark
e-mail: fbe@inano.dk

François Bessueille
LSA, Université Lyon I, 43 Boulevard du 11 Novembre 1918
69622 Villeurbanne Cedex, France
e-mail: francois.bessueille@univ-lyon1.fr

Bharat Bhushan
Nanotribology Laboratory for Information Storage and MEMS/NEMS (NLIM)
W 390 Scott Laboratory, 201 W. 19th Avenue, Ohio State University
Columbus, Ohio 43210-1142, USA
E-mail: bhushan.2@osu.edu

Malte Burchardt
Faculty of Mathematics and Sciences
Department for Pure and Applied Chemistry
and Institute of Chemistry and Biology of the Marine Environment (ICBM)
Carl von Ossietzky University of Oldenburg
D-26111 Oldenburg, Germany
e-mail: malteburchardt@gmx.de

Abdelhamid Errachid]
Laboratory of NanoBioEngineering, Barcelona Science Park
Edifici Modular, C/Josep Samitier 1–5, 08028-Barcelona, Spain
e-mail: aerrachid@pcb.ub.es

Horacio D. Espinosa
Department of Mechanical Engineering, Northwestern University
2145 Sheridan Rd., Evanston, IL 60208-3111, USA
e-mail: espinosa@northwestern.edu

Yanxia Hou
Ecole Centrale de Lyon, STMS/CEGELY
36 Avenue Guy de Collongue, F-69131 Ecully Cedex, France
e-mail: yanxiahou24@yahoo.com

Nicole Jaffrezic-Renault
Ecole Centrale de Lyon, STMS/CEGELY
36 Avenue Guy de Collongue, F-69131 Ecully Cedex, France
e-mail: Nicole.Jaffrezic@ec-lyon.fr

Changhong Ke
Department of Mechanical Engineering, Northwestern University
2145 Sheridan Rd., Evanston, IL 60208-3111, USA
e-mail: c-ke@northwestern.edu

Keun-Ho Kim
Department of Mechanical Engineering, Northwestern University
2145 Sheridan Rd., Evanston, IL 60208-3111, USA
e-mail: kkim@nualumni.edu

Jeppe Vang Lauritsen
Interdisciplinary Nanoscience Center (iNANO)
Department of Physics and Astronomy
University of Aarhus, DK-8000 Aarhus C, Denmark
e-mail: jvang@phys.au.dk

Huiwen Liu
Nanotribology Laboratory for Information Storage and MEMS/NEMS (NLIM)
The Ohio State University
650 Ackerman Road, Suite 255, Columbus, Ohio 43202, USA
e-mail: Huiwen.Liu@seagate.com

Claude Martelet
Ecole Centrale de Lyon, STMS/CEGELY
36 Avenue Guy de Collongue, F-69131 Ecully Cedex, France
France
e-mail: Claude.Martelet@ec-lyon.fr

Nicolaie Moldovan
Department of Mechanical Engineering, Northwestern University
2145 Sheridan Rd., Evanston, IL 60208-3111, USA
e-mail: n-moldovan@northwestern.edu

Martin Munz
National Physcial Laboratory (NPL), Quality of Life Division
Hampton road, Teddington, Middlesex TW11 0LW, UK
e-mail: martin.munz@npl.co.uk

Michael Nosonovsky
Nanomechanical Properties Group
Materials Science and Engineering Laboratory
National Institute of Standards and Technology
100 Bureau Dr., Mail Stop 8520, Gaithersburg, MD 20899-8520, USA
e-mail: michael.nosonovsky@nist.gov

Sascha E. Pust
Faculty of Mathematics and Sciences
Department for Pure and Applied Chemistry
and Institute of Chemistry and Biology of the Marine Environment (ICBM)
Carl von Ossietzky University of Oldenburg
D-26111 Oldenburg, Germany
E-mail: sascha.pust@uni-oldenburg.de

Robert A. Sayer
Nanotribology Lab for Information Storage and MEMS/NEMS (NLIM)
The Ohio State University
650 Ackerman Road, Suite 255, Columbus, Ohio 43202, USA
e-mail: Sayer.11@osu.edu

Heinz Sturm
Federal Institute for Materials Research (BAM), VI.25
Unter den Eichen 87, D-12205 Berlin, Germany
e-mail: heinz.sturm@bam.de

Ronnie T. Vang
Interdisciplinary Nanoscience Center (iNANO)
Department of Physics and Astronomy
University of Aarhus, DK-8000 Aarhus C, Denmark
e-mail: rtv@inano.dk

Gunther Wittstock
Faculty of Mathematics and Sciences
Department for Pure and Applied Chemistry
and Institute of Chemistry and Biology of the Marine Environment (ICBM)
Carl von Ossietzky University of Oldenburg, D-26111 Oldenburg, Germany
e-mail: gunther.wittstock@uni-oldenburg.de

1 Integrated Cantilevers and Atomic Force Microscopes

Sadik Hafizovic · Kay-Uwe Kirstein · Andreas Hierlemann

1.1
Overview

Since Binnig, Quate and Gerber presented the first atomic force microscope (AFM) in 1986 [1], it has developed at a remarkable pace from a laboratory instrument to a commercial tool that is currently used in many domains in research and in industry. In research, surface and interface properties down to the molecular level are of interest, whereas typical industrial applications are related to quality control. The optical readout of a passive cantilever by means of a laser beam is the most common detection principle. The optical readout is, in most cases, combined with a piezoelectric scanning stage. The combination of passive cantilevers with an optical readout facilitated the rapid development of the AFM techniques due to the rather simple setup, but, at the same time, it entails several limitations for many modern applications. The exchange of the AFM probe, especially under vacuum or low-temperature conditions is laborious, and the standard setup does not offer the possibility of using an array of cantilevers for parallel scanning to increase the throughput.

To overcome some of these limitations, AFM probes with integrated detection schemes such as capacitive [2], piezoelectric, or piezoresistive schemes [3], and high-speed scanning systems that rely on arrays of active cantilevers featuring piezoelectric excitation and piezoresistive/piezoelectric readout [3–7] have been developed.

The use of industrial complementary metal oxide semiconductor (CMOS) processes in combination with post-CMOS microelectromechanical systems (MEMS) technology can help to advance AFM technology. In particular, it is possible to realize active cantilevers (cantilevers with integrated actuation and readout schemes) and to integrate multiple active cantilevers with the associated actuation and readout electronics into a single monolithic system.

In this chapter, the design and use of CMOS-based AFM probes is discussed in great detail. Different cantilever actuation and sensing principles suitable for integrated microelectronics-based systems are described, and the results that have been obtained from monolithic AFM chips and systems are presented.

Glossary

ADC	Analog-to-digital converter
ASIC	Application-specific integrated circuit
CMOS	Complementary metaloxide semiconductor. Most important technology for integrated circuits

Integrated AFM	Device that monolithically includes one or more active cantilevers together with the associated actuation and readout circuitry
DAC	Digital-to-analog converter
DSP	Digital signal processor
FPGA	Field programmable gate array. Reprogrammable logic device
IC	Integrated circuit
IIR	Infinite impulse response
I/O	Input-output
MAC	Multiplier accumulator
MEMS	Microelectromechanical systems
Piezoelectricity	Coupling of electrical and mechanical effects. Upon mechanical stress, charged particles are displaced generating a measurable electric charge in the crystal. In turn, mechanical deformations can be achieved by applying a voltage to the crystal
Piezoresistor	Resistor that changes its resistance upon mechanical stress. In the context of this chapter, it is a CMOS piezoresistor realized by means of a (p+) diffusion in the silicon n-well
PID	Proportional integral derivative
SNR	Signal-to-noise ratio
SAC	Sensor-actuator crosstalk
SQRT	Square root
ULSI	Ultra large-scale integration
VLSI	Very large-scale integration

1.2
Active Cantilevers

In Fig. 1.1, an example of an active cantilever, which includes actuation and readout features, is displayed. The cantilever is 5.5 to 6 μm thick with its thickness being defined mainly by the depth of the n-well of the CMOS process. There is a central section of the cantilever that contains the actuator, a heater on a bimorph layer sandwich that is deformed upon heating, and a second area closer to the cantilever base that hosts the force sensors, i.e., the piezoresistors.

The design criteria for the active CMOS-based cantilevers include:

– noise
– contact force
– readout sensitivity
– actuation efficiency

There are two major sources of noise to be taken into account, firstly the mechanical cantilever (Brownian) noise, and, secondly, the electronic noise of the force sensor. Mechanical noise considerations show that a spring constant as low as 0.01 N/m constitutes a lower limit, since it yields a noise-vibration amplitude of 0.64 nm at room temperature. An upper limit of 10 N/m for the cantilever spring constant is given by the estimated spring constant of an atomic oscillator, where an atom of mass $m = 10^{-25}$ kg oscillates with $\omega = 10^{13}$ Hz such that $k = \omega^2 m = 10$ N/m.

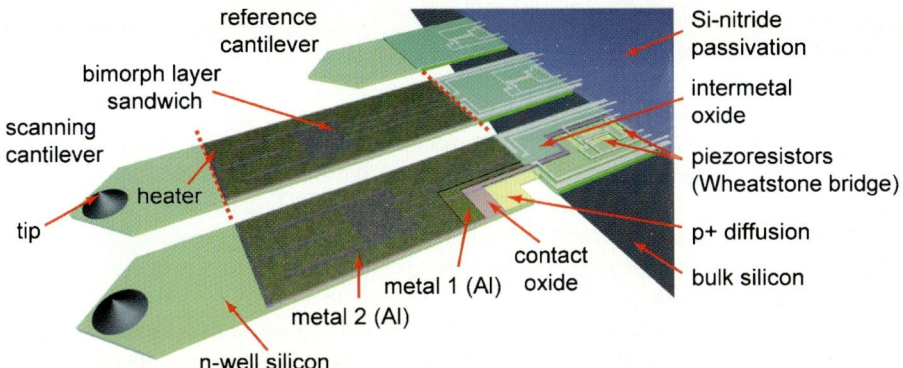

Fig. 1.1. Three-dimensional schematic of CMOS-based active cantilevers: three cantilevers (two scanning cantilevers and one shorter reference cantilever) and the different CMOS layers as they are used for the various cantilever components (actuation: bimorph and heater; detection: piezoresistors in a Wheatstone bridge configuration)

When exposed to a force gradient, the cantilever is stabilized and its vibrations are reduced. If the cantilever stiffness significantly exceeds the above-mentioned value, the cantilever spring constant dominates the net spring constant and the stabilization is reduced.

The electronic sensor noise is dominated by white or Johnson noise and depends on the size of the piezoresistors. It is further discussed in Sect. 1.2.1.1.

For contact-mode measurements, a spring constant on the order of 1 N/m is desirable to keep the tip wear and the tip-induced alterations to the sample (indentation) low while still maintaining good contact with the sample. For dynamic non-contact measurements, a spring constant of 10–100 N/m is desirable to prevent snap-in-effects of the cantilever (the pull-down of the cantilever to the surface by, e.g., van der Waals forces) and to have a high resonance frequency and, consequently, a high oscillation quality factor.

The spring constant of a uniform silicon cantilever is given by

$$k = \frac{Ebh^3}{4L^3}$$

where E is the apparent Young's modulus, and b, h, and L denote the cantilever width, thickness, and length, respectively. For the sandwich-type cantilevers that are discussed in the following, another equation has to be used, since the CMOS n-well (5.5–6 μm deep), that represents most of the cantilever material is only a bit thicker than the silicon oxide layers (2.4 μm), silicon nitride (1.1 μm), and aluminum (1.1 μm) layers. For such a composite cantilevers, the spring constant k is given by

$$k = \frac{3 \sum_i E_i I_i}{L^3}$$

with E_i and I_i denoting the apparent Young's moduli and the moments of inertia of the individual layers of the cantilever.

1.2.1
Integrated Force Sensor

Force sensors that can be realized in integrated circuit (IC) technologies like CMOS, depend, in most cases, on two different transduction principles: (1) the capacitive (displacement) sensor and (2) the piezoresistive (strain) sensor (see Fig. 1.2). The transducer provides the translation of a force applied to a body of well-defined stiffness into a mechanical displacement of the body or into strain generated in the body. While the capacitive approach offers advantages with regard to power consumption and readout bandwidth, it is often difficult to achieve a linear readout over a larger range if there is no feedback architecture with compensation features available [11]. To achieve large cantilever displacements, it is beneficial to use mechanical structures with a vertical instead of a horizontal direction of displacement. Horizontal mechanical structures are more difficult to realize in planar CMOS technology so that the devices become less reliable and/or more expensive.

Piezoresistors can be designed and used to measure the force and the related strain in the plane of the chip surface or in any other direction.

Piezoelectric materials can also be used as stress sensors [12, 13], but the respective materials are not common CMOS materials so that they have to be processed in additional steps. Recent devices are based on the deposition of thin layers, which show a good piezoelectric performance, but a reliable CMOS integration of those materials is not trivial [14–18]. The advantages of piezoelectric stress sensors and actuators in comparison to piezoresistive elements are most evident for the resonant-mode operation, where larger signal bandwidths are required. Piezoelectric sensors, however, are less suitable for measuring static deflections [19].

1.2.1.1
Piezoresistive Force Transducer

Figure 1.3 shows the basic structure of a silicon-based cantilever featuring a piezoresistive stress sensor. The bulk silicon is thinned down to a defined thickness by anisotropic silicon etching using a dedicated etch-stop technique. A local implantation and/or diffusion is used to form a doped area of silicon, which acts as the

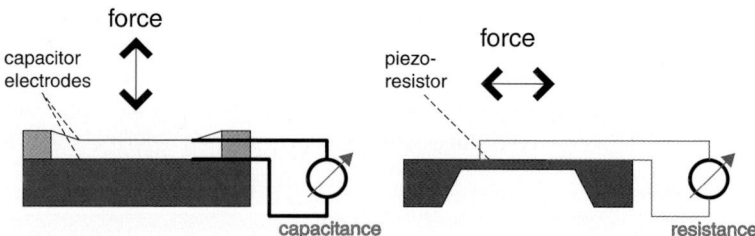

Fig. 1.2. Comparison between capacitive and piezoresistive transducers realized in integrated circuit (IC) technology

Fig. 1.3. Schematic of a cantilever with integrated piezoresistive detection (piezoresistors) in silicon technology

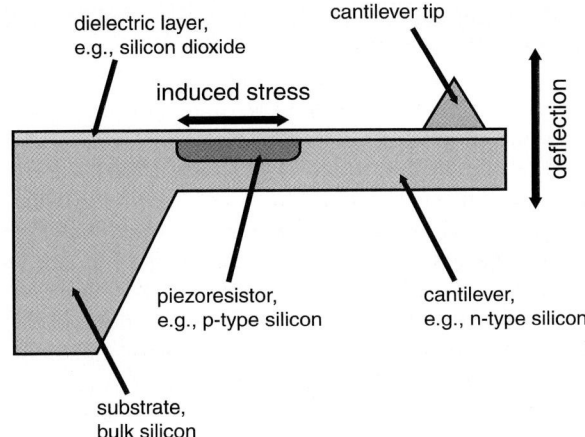

piezoresistive stress sensor. The area is usually isolated from the surrounding silicon by means of a reverse-biased pn-junction. Additional layers on top of the cantilever structure include dielectric layers, like silicon dioxide or nitride, and can be utilized as a mask for implantation. A pre-manufactured tip is either mounted or glued to the cantilever end or it can be micromachined directly on the cantilever by additional steps.

The two simplest methods to measure a resistance by means of an integrated circuit is to either apply a constant voltage and measure the current flowing through, e.g., the piezoresistor by a low-input-impedance current amplifier, or to apply a constant electrical current to the resistor and measure the voltage drop with a high-input-impedance amplifier. The latter approach can be easily realized in CMOS technology, as MOS-transistors are commonly used for high-input-impedance (instrumentation) amplifiers. Figure 1.4 shows a circuit realization of a resistance readout using a single operational amplifier. An additional gain for the voltage output can be realized by carefully dimensioning the feedback resistors R_1 and R_2.

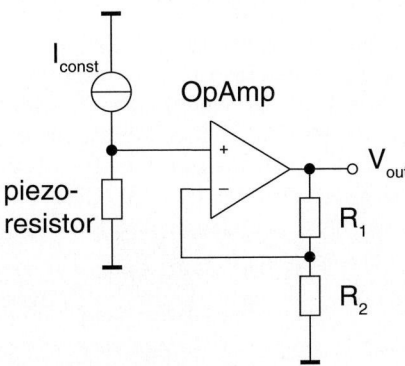

Fig. 1.4. Integrated readout circuit for a constant-current-based resistance measurement

Constant-current sources are widely used in integrated circuits, and many approaches that cover the whole range from simple to high-precision and temperature-compensated current sources, can be found in the literature [20, 21]. Potential problems in using the circuit topology of Fig. 1.4 include the signal sensitivity to temperature fluctuations and the fact that a piezoresistor fabricated in CMOS technology may feature a production-induced resistance spread of up to 20%. Temperature variations will change the measured resistance value owing to the temperature coefficient of the piezoelement. For instance, the temperature coefficient of a silicon-based piezoresistor usually shows strong non-linear characteristics and generally depends on the doping. It is therefore difficult to compensate for temperature-induced fluctuations at the signal processing stage. The production spread makes additional calibration measures necessary, as large offsets would saturate the readout electronics or would limit the maximum signal gain and, therefore, the transducer resolution. Several silicon-based cantilevers with single integrated piezoresistive force sensors have been realized by local diffusion of p-type silicon in an n-type substrate or vice versa [22–26].

The use of advanced fabrication technologies like CMOS enables the realization of more complex structures using metal lines as interconnects. The implementation of a Wheatstone bridge configuration as shown, for instance, in Fig. 1.5 helps to overcome some of the problems mentioned before. On the left-hand side a so-called half-bridge configuration is shown. If R_1 represents the active (piezoresistive) element, the actual output voltage is given by a voltage-divider equation

$$V_{out} = \frac{R_1}{R_1 + R_2} (V_{DD} - V_{SS}) \tag{1.1}$$

Bridge configurations are usually designed symmetrically to maximize their sensitivity, so that no output voltage is generated for zero sensor signal. The symmetric design can easily be achieved by using identical supply voltages ($V_{SS} = -V_{DD}$) and symmetrically designed bridge resistors ($R_1 = R_2$). The resulting bridge sensitivity upon a resistance change of R_1 can be calculated as

$$S_{R_1} = \frac{\partial}{\partial R_1} \left[\frac{R_1}{R_1 + R_2} (V_{DD} - V_{SS}) \right] \Bigg|_{R_1 = R_2 = R} \approx \frac{1}{R} \cdot (V_{DD} - V_{SS}) \tag{1.2}$$

In this configuration, the temperature influence is canceled to the first order, but high-frequency fluctuations on the supply voltages directly affect the output voltages through capacitive signal coupling so that a co-integrated digital circuit (strong, clocked signals) would interfere with the weak sensor signal.

An effective temperature compensation can be achieved using the full-bridge configuration, as shown in Fig. 1.5b. In case of a symmetric configuration the sensitivity is given by (1.3). Here, both branches of the Wheatstone bridge contain a piezoresistive element (R_1 and R_4), as represented by the solid arrows in Fig. 1.5

$$S_{R_{1,4}} = \frac{\partial}{\partial R_{1,4}} \left[\left(\frac{R_1}{R_1 + R_2} + \frac{R_4}{R_3 + R_4} \right) \cdot (V_{DD} - V_{SS}) \right] \Bigg|_{R_1 = R_2 = R_3 = R_4 = R}$$

$$\approx \frac{2}{R} \cdot (V_{DD} - V_{SS}) \ . \tag{1.3}$$

Fig. 1.5. Possible configurations of measurement bridges. (**a**) A half-bridge configuration with either one or two piezotransducers (*solid* or *solid* and *dotted arrows*). (**b**) A full-bridge configuration with two or four active transducers (*solid* or *solid* and *dotted arrows*)

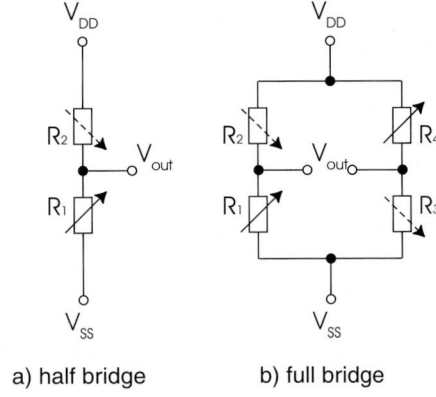

a) half bridge   b) full bridge

One advantage of the full-bridge configuration is its double sensitivity in comparison to the half bridge, but the differential output voltage is even more important for integrated instrumentation circuits, since any effect of the supply voltages is eliminated, and the differential output is, therefore, much less sensitive to interferences. In contrast to discrete electronics, IC-based instrumentation and sensor readout electronics often feature a fully differential signal path for weak analog signals to overcome the problems of interference with clocked circuit blocks, of power supply sensitivity, and those of limited signal swing in low-power, low-voltage applications. As will be shown later in Sect. 1.3.1 a differential implementation of the analog signal processing is crucial for the complete integration of an AFM deflection control on a single silicon die. Without differential implementation, the weak voltage signals of the deflection sensors would be strongly distorted by the driving signals of the deflection actuators and the clock signals of the on-chip digital signal processor (DSP).

The sensitivity of the full Wheatstone bridge configuration can be further increased by also realizing the resistors R_2 and R_3 as piezoresistors. As indicated by the dotted arrows in Fig. 1.5b their piezoresistivity must feature a sign opposite to that of R_1 and R_4. This can be achieved by placing the piezoresistors on the opposite faces of a cantilever, e.g., R_1 and R_4 on the top and R_2 and R_3 at the bottom of the cantilever.

However, electronic units or transducers can only be placed on the top face of a cantilever structure in integrated systems. This holds particularly true for monolithic systems based on CMOS technology [13].

While the sensitivity of such a deflection sensor is mainly determined by the material properties of the piezoresistors, the overall signal resolution will be also limited by the electronic and mechanic noise introduced by the active cantilever. The main noise contribution of resistive-bridge circuits is the electronic white noise or Johnson noise, the origin and properties of which are well-known and can be analytically described by the generated noise power [13]

$$P_N = 4kTR\Delta f \qquad (1.4)$$

The signal-to-noise ratio can then be analytically calculated knowing also the power of the sensor signal

$$P_S = \frac{[(V_{DD} - V_{SS})(\pi_L - \pi_T) \cdot \sigma_L]^2}{R} \tag{1.5}$$

$$SNR = \frac{P_S}{P_N} = \frac{[(V_{DD} - V_{SS})(\pi_L - \pi_T) \cdot \sigma_L]^2}{4kT\,\Delta f \cdot R^2}\ , \tag{1.6}$$

where $V_{DD} - V_{SS}$ is the supply voltage of the Wheatstone bridge, π_L and π_T are the longitudinal and transversal piezocoefficients, σ_L represents the longitudinal stress at the Wheatstone bridge, and R the resistivity of the bridge elements. Therefore, for a given supply voltage, the SNR can be increased by lowering the resistivity of the Wheatstone bridge, which leads to an increased current consumption on the cantilever.

In some cases, additional noise sources like (mechanical) Brownian noise or electronic flicker noise have also to be considered. Particularly for lightly-doped piezoresistive devices $1/f$-noise may prevail over Johnson noise, which is usually not the case for the CMOS-based realizations described in this chapter.

1.2.2
Integrated Actuation

Integrated cantilever actuation schemes have several principal advantages over external actuation by either the sample holder or the probe holder. The most important advantage is the possibility to perform parallel scanning, which also requires an integrated detection scheme, i.e., integrated force sensors. Even when only one cantilever is used at a time, it is useful to have a set of several, eventually differently coated or modified cantilevers at hand and to use one of them while having the non-used cantilevers retracted from the sample surface. Another benefit of integrated actuation and detection per cantilever is the possibility to have a closed feedback loop on the chip. When the whole feedback loop for both, dynamic and static operation modes is integrated with the cantilever, the individual feedback loop components can be better adapted to optimize the overall system performance.

The key system criteria include the overall force and, correspondingly, displacement measurement range, the measurement precision, accuracy, and speed.

The two predominantly used actuation principles for integrated cantilevers include:

1. Electrothermal actuation
2. Piezoelectric actuation

As has been previously explained, the deposition of piezoelectric materials on CMOS-based devices requires extra processing so that we will focus here on thermal actuation.

1.2.2.1
Thermal Actuation Principle

Thermal cantilever actuation makes use of the bimorph effect, i.e., the different thermal expansion coefficients of silicon, silicon oxide and silicon nitride, and the aluminum layers (metal 1 and metal 2 of the CMOS process; see [27,28]). Upon heating the layer sandwich the cantilever bends, and the bending can be precisely controlled via the heating current passing through diffused heating resistors embedded in the cantilever. Figure 1.6 [27] shows that 0.5 μm static deflection are attained for a heating power of 20 mW. This cantilever features only short heating resistors close to the cantilever support. Also the dynamic characteristics of the cantilever deflection are visible. The thermal time constant of the cantilever is approximately 1 ms. Figure 1.7 shows the actuator efficiency for a cantilever as shown in Fig. 1.1 featuring a large heater covering half of the cantilever length. The efficiency here is 0.3 μm/mW [29].

An unwanted side-effect of heating the cantilever is the sensor-actuator crosstalk (SAC), a thermally-induced signal in the Wheatstone bridge. The underlying phenomenon is the thermal stress generated in the Wheatstone bridge. The interference signal has to be suppressed in order to be able to scan in, e.g., the constant-force mode at low contact force. The SAC is defined as the ratio of the unwanted Wheatstone bridge output voltage, $V_{actuator}$, upon a defined electrothermally produced cantilever displacement in the z-direction and the Wheatstone bridge output signal, V_{sample}, for a topographical feature of a height equal to the defined cantilever displacement

$$SAC = \frac{V_{actuator}(z)}{V_{sample}(z)} \tag{1.7}$$

The SAC can be compensated for in an integrated AFM system as will be described in Sect. 1.4.

It has to be noted that systems relying on piezoelectric actuation and piezoresistive readout also suffer from SAC. The respective cantilevers feature a stiffer actuator region close to the cantilever base and a softer sensing section [3,30]. The aluminum lines in these designs have to cross the actuation section where they are

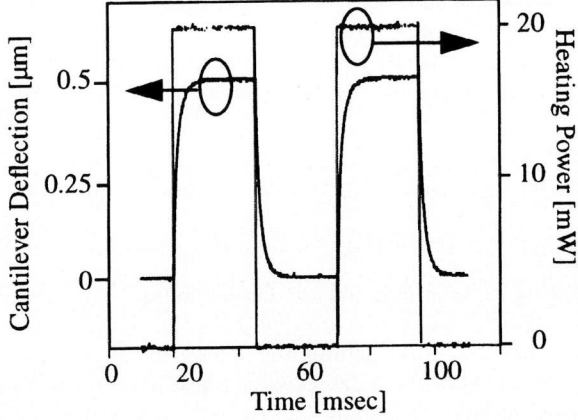

Fig. 1.6. Deflection of a 500 μm-long cantilever upon electrothermal actuation. A square-wave heating pulse of 20 mW amplitude and 20 Hz frequency has been applied. The deflection was measured using the piezoresistive Wheatstone bridge [27]

Fig. 1.7. Cantilever deflection, z, as a function of the heating power, P. Actuator efficiency is $0.3\,\mu\text{m/mW}$, and the deflection is linearly proportional to the heating power [29]

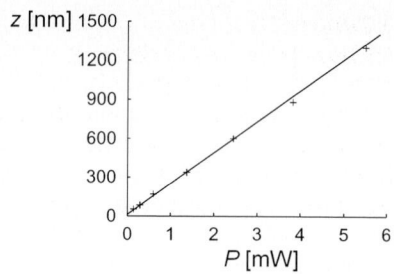

subjected to varying mechanical stress, so that the SAC originates from the gauge effect of the aluminum line bending in the actuation section.

1.3
System Integration

The different system components include (see Fig. 1.8):

1. Cantilever or MEMS structures including actuation and sensor features
2. Analog signal processing circuitry
3. Digital signal processing circuitry
4. Communication interface

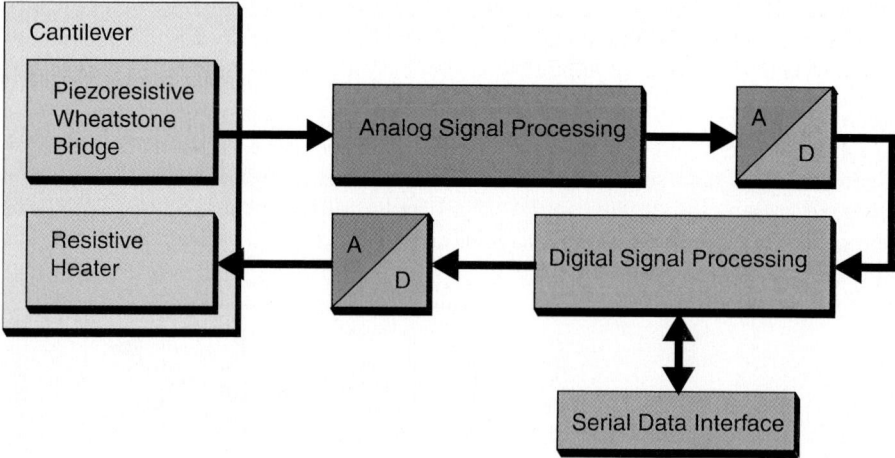

Fig. 1.8. Block diagram showing the main building blocks of the integrated AFM system

1.3.1
Analog Signal Processing and Conditioning

The key issue in designing the analog signal processing block is to match the electrical signal of the piezoresistive transducers to the input range of the implemented ADC in order to obtain the desired signal resolution and an optimal SNR for the measured force signal. Typical ADCs realized in CMOS technology feature a voltage input range similar to the supply voltage to relax noise and accuracy requirements. As CMOS technology with feature sizes below 0.25 μm (ULSI) is not well-suited for analog instrumentation applications, and since such high-end technology is too expensive to be used for large-area MEMS devices, a CMOS technology with a medium feature size is commonly used. This medium feature size includes a range from 1 μm down to 0.35 μm, and the respective chips are capable of handling supply voltages between 3 and 5 V, which entails ADC input ranges of the same magnitude.

As a consequence of these considerations, the required signal gain of the analog circuitry, which transfers the Wheatstone bridge output to the input of the ADC for further digital processing is also defined.

Figure 1.9 shows the integrated analog processing circuitry for the minute Wheatstone bridge output signals [31]. A first pre-amplifier mainly acts as an impedance converter in order to provide a high-impedance output for the successive stages. A first-order offset compensation is implemented by subtracting the output signal of a reference cantilever from the signal of the actual measurement cantilever. Stress-induced offset from the CMOS fabrication and the subsequent post-processing can thus be compensated for, since such stress is usually very similar for the cantilevers on the same die. An additional offset can arise from a mismatch of the diffused resistors of the Wheatstone bridge on the cantilevers. This offset has to be individually compensated for each cantilever and each readout channel. The two-stage approach as selected for the system here offers several advantages. A coarse offset compensation is provided by the first differential stage using the reference cantilever structure, which relaxes the requirements on the input range of the succeeding signal processing stages. The fine compensation can be controlled by the digital subsystem and can, therefore, be applied in a flexible way or can be applied repeatedly

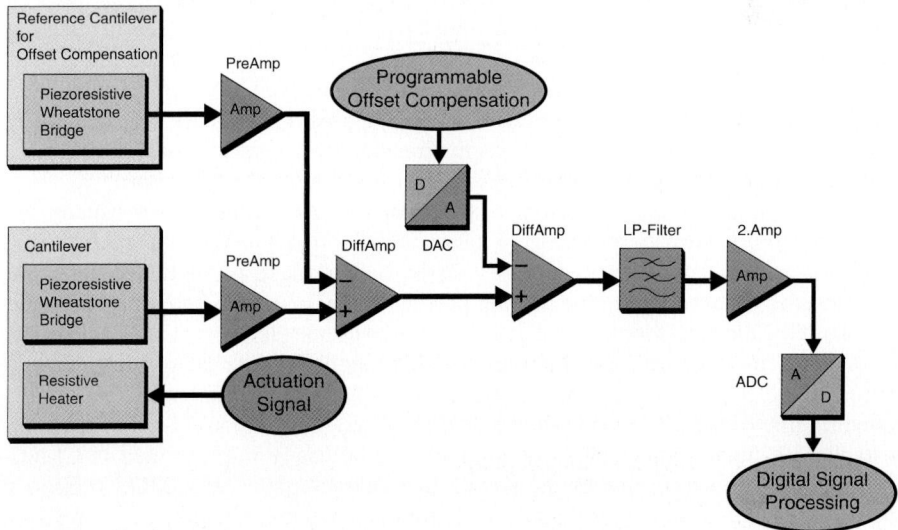

Fig. 1.9. Block diagram of the integrated analog signal processing unit [31]

depending on the system operating conditions like temperature, measurement range, or scanning speed. The segmentation in coarse and fine compensation stages also relaxes the requirements on the resolution and accuracy of the digital programmable second offset compensation stage, which is very important with respect to area and power consumption of the circuitry implementation. This holds particularly true for realizing a whole cantilever array on the same chip.

To improve the signal-to-noise ratio, the out-of-band noise is suppressed by adding a low-pass filter in the signal path. Though additional higher-order filtering will be performed in the DSP, the inclusion of a low-pass filter is useful to meet the requirements for the succeeding ADC imposed by the sampling theorem. In order to prevent aliasing effects, the bandwidth of the analog input signal must not exceed the Nyquist rate of the ADC.

As already described in Sect. 1.2.1.1, a fully differential signal path as provided by the full-bridge configuration makes the complete analog signal processing more robust with regard to distortion and interference by neighboring readout channels and digital processing blocks. The differential architecture is hence, besides the possibility to perform on-chip digital signal processing, a pre-requisite for a monolithic integration of a cantilever array.

The analog-to-digital converter (ADC) forms the interface between the analog and the digital domain. There are many possible ADC architectures that have been optimized with regard to different criteria, like resolution, conversion speed, area and power consumption [32]. For sensing applications requiring medium resolution and conversion rates, the successive-approximation architecture features good area and power efficiency. Both, power consumption and area constraints are important aspects in integrating multiple readout channels as required in the case of an array of AFM cantilevers.

The advantages of the digital signal processing include the possibility to realize signal filter functions of higher order at better configurability as compared to an analog implementation. Moreover, a digital communication interface to external hardware is more robust and easier to implement. More details on the integration and use of digital signal processing techniques will be given in the next section.

Figure 1.10 shows the analog part of an integrated cantilever relying on the thermal actuation principle [31]. The actuation signal is, in most cases, provided by a digital controller and has to be translated to the analog domain by a DAC. This DAC determines the characteristics of the cantilever actuation, as it is usually the limiting component with respect to signal resolution, signal bandwidth and linearity. Using thermal actuation by means of a resistive heater, the resulting deflection of the cantilever is proportional to the heating power, which, in turn, is proportional to the square of the electrical input signal, i.e., the input voltage (see Sect. 1.2.2.1). In order to obtain a linear relationship between actuation signal and cantilever deflection to simplify the implementation and configuration of the deflection controller, a pre-conditioning of the analog actuation signal by a square root block (SQRT) has been established. Such a signal processing block can be easily implemented in CMOS technology by making use of the square law characteristics of a MOS transistor operated in saturation [33]. An additional reference structure for the heating resistor as shown in Fig. 1.10 can be used to compensate for the fabrication spread of integrated heaters in CMOS-MEMS technology.

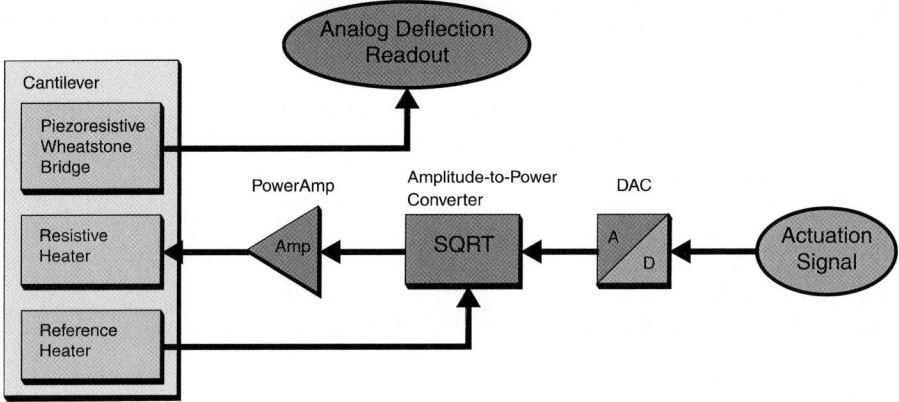

Fig. 1.10. Block diagram showing the analog actuation signal chain. The input signal is generated by a digital deflection controller

1.3.2
Digital Signal Processing

The advantages of digital signal processing have their foundations in the predictability of the respective performance. The fabrication spread of electrical components does not affect the performance of digital circuitry. The systems are inherently interference-free and long-term stable. The tuning of components, which is often necessary for analog circuits, becomes obsolete. In addition, the processing of extremely-low-frequency signals does not entail any problems in comparison to analog filters.

Owing to these facts, there is a strong trend to shift circuit functions from the analog to the digital domain. DACs are implemented using pulse-modulation techniques, and the analog anti-aliasing filters can be designed with plenty of slack if a high sampling rate is used and the signal is afterwards digitally decimated.

There is a variety of possible digital signal processing configurations. Depending on the overall level of system integration of the AFM, the digital signal processing can be implemented directly on chip to yield an application-specific circuit (ASIC), on an FPGA, or on a programmable digital signal processor. The rationale for full integration of the feedback loop on the chip is that the input-output (I/O) data volume is drastically reduced, which entails a smaller number of interconnections and relaxes electromagnetic compatibility constraints.

1.3.2.1
Universal Filter Architecture

As can be seen in Fig. 1.8, there is at least one digital unit per cantilever required for realizing a closed-loop mode. Later, in Sect. 1.4, we will show how the sensor-actuator crosstalk can be removed by additional filters. The feedback filter should be versatile with a large variability of possible configurations. An infinite-impulse-response (IIR) filter structure based on a bi-quadratic function is a universal and

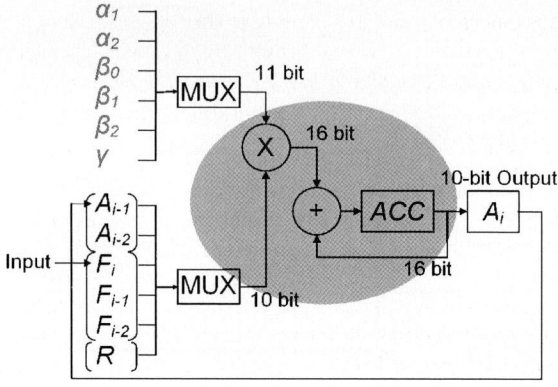

Fig. 1.11. Equation (1.8) implemented in a multiplier-accumulator architecture. The coefficients are 11-bit, fixed-point numbers in the range of −1 to 1

straightforward filter implementation. This filter structure is at the basis of many VLSI filters and is used in a large variety of filter applications similar to a standard digital signal processing chip [34]. Possible functions include proportional-integral-derivative (PID) controlling, averaging, low- and high-pass filtering and any combination thereof. The filter algorithm is given by

$$A_i = A_{i-1} + \alpha_1 A_{i-1} + \alpha_2 A_{i-2} + \beta_0 A_i + \beta_1 A_{i-1} + \beta_2 A_{i-2} + \gamma R \qquad (1.8)$$

where α, β, γ are coefficients in the range of −1 to 1, F_i are the force (input) values, A_i are the actuation (output) values, and R is a constant, programmable reference value.

Figure 1.11 depicts a filter structure that can be implemented in IC technology. The controller stores two precedent force and actuation values. The calculation is performed on a 16 bit multiplier accumulator (MAC), and the result is clipped to 10 bits. One multiplication and addition operation are performed per clock cycle.

The filter coefficients reflect thermal and mechanical time constants [29], the cantilever heat dissipation, and the selected averaging method. As an example, the most simple tunable controller, the integral controller, is implemented with $\gamma = 0.1$, $\beta_0 = -0.1$ and 0 for all remaining coefficients to yield $A_i = A_{i-1} - 0.1 \times (R - F_i)$. Using these parameters, the actuation will be such that the contact force becomes equal to R.

1.3.2.2
Variable-Deflection Mode

No actuation is performed, only the cantilever deflection is measured. This mode enables high scanning speed. The interpretation of the images, however, can be difficult, since the interaction forces depend on the cantilever deflection.

1.3.2.3
Constant-Force Mode

The constant-force mode is the most widely used mode, and the results are easy to interpret. A feedback loop keeps the force on the cantilever constant by regulating the cantilever deflection while the sample is scanned. For passive cantilevers, the sample is moved up and down by means of, e.g., a piezoelectric scanner stage. The active cantilevers are actuated or deformed in a way that the contact force is kept constant. An example of a suitable filter configuration for this operation mode has already been given in Sect. 1.3.2.1.

1.3.2.4
Dynamic Mode

In the dynamic mode, the cantilever is excited to resonate at its fundamental frequency, which is given by the cantilever mechanical properties like stiffness and mass. For the CMOS-integrated cantilevers the resonance is on the order of tens to hundreds of kHz. The thermally actuated cantilevers are hence operated at non-optimal conditions, as the thermal time constant of the cantilever is typically on the order of milliseconds. Expressed in terms of system theory it can be stated that the actuation is considerably attenuated by the low-pass of the thermal time constant. As a consequence, the supplied energy is not only dissipated in the fundamental resonance mode but also in a DC component, which contributes to heating the cantilever.

The measurement principle of any dynamic mode relies on the fact that the interaction of a resonating cantilever with a substrate or surface causes small shifts in the cantilever resonance frequency. This shift can either be directly measured, or it is indirectly accessible via amplitude or phase measurements.

The readout and actuation boundary conditions for dynamic mode operation are very different from those for the static mode. To achieve a good bandwidth of the overall scanning probe system, a cantilever featuring a resonance frequency of 50 kHz requires readout and actuation rates of at least 500 kHz. Higher sampling rates allow for a shorter integration time of the lock-in detection units and for a reduction of the signal contribution of harmonics, which result from the actuation.

The following paragraph describes different types of dynamic modes. For all these dynamic modes, the sample topology is contained in the DC part of the actuation signal, i.e., the DC signal is recorded and processed to produce the AFM image:

- Controlling the cantilever **actuation phase**. In a first frequency sweep, the resonance frequency is identified, and the actuation is adjusted to the measured resonance frequency. A change in the resonance frequency will then affect the phase relation of cantilever and actuation. The phase is controlled by the DC part of the actuation.
- Controlling the cantilever **resonance amplitude**. The advantage in locking in on the cantilever amplitude is the simplicity of the setup, especially in analog circuit implementations. If the controller is to be digitally implemented the disadvantages outweigh the advantages since the amplitude is a noisy signal, even when a Shannon-compliant digital signal is available.
- Controlling the cantilever **resonance frequency**. This measurement method requires a phase-locked loop (PLL) that maintains a pre-set phase relation between actuation and cantilever signal. The DC actuation is controlled so that the resonance frequency is kept at a preset value.

1.4
Single-Chip CMOS AFM

In this section, a standalone single-chip AFM unit that has been developed at ETH Zurich [35] is presented. Figure 1.12 shows a photograph of the AFM chip mounted onto a PCB. It includes a fully integrated array of cantilevers, each of which has its individual actuation, detection, control, amplification and on-chip digital processing unit as well as its individual offset compensation so that, e.g., constant-force operation can be performed without any external controller. The cantilevers can be moved and precisely controlled within a range of 0.5 to 6 µm (at 0.5 to 6 nm resolution) so that the chip has only to be brought within a distance of maximal 6 µm to a surface and then can be used to carry out, e.g., multiple force-distance measurements using the integrated electronics via a LabVIEW™ interface. Only a x-y-scanning stage is necessary for imaging or lateral scanning by using one or several of the individually controlled cantilevers. The individual cantilever control is of paramount importance in using multicantilever arrays, since there is always a slight tilt between the chip plane determining the cantilever positions and the sample surface plane so that some cantilevers are closer to the surface than others. Parallel scanning of multiple cantilevers only provides good results when a fast and simultaneous individual force feedback for each scanning cantilever is implemented in a closed-loop configuration (no multiplexing).

Figure 1.13 shows a micrograph of the overall microsystem chip, which features a die size of $10 \times 7 \, mm^2$, and a close-up of the 12-cantilever array. The spacing between the cantilevers is 25 µm. The ten scanning cantilevers, four of which are connected to the circuitry and can be individually controlled (Fig. 1.13b), are located in the center of the array and are 500 µm long, 85 µm wide and 5 µm thick. The two reference cantilevers located at the left and right end of the array are 250 µm long, 85 µm wide and 5 µm thick. They feature only the Wheatstone bridge for deflection detection, and they are shorter than the scanning cantilevers in order not to contact the sample surface. The reference cantilevers are used for the offset compensation of the Wheatstone bridge.

The electronics covering most of the chip area include four repeated mainly analog circuitry units (Figs. 1.12, 1.13), which are connected to the four central cantilevers and a common digital block, which is used for digital signal processing, and which includes a serial digital bus interface to connect to off-chip components.

Fig. 1.12. The single-chip AFM system mounted onto a printed circuit board. The 10-cantilever array is on the *far left*. Bond wires featuring a diameter of 25 µm form the electrical connections between the chip and the circuit board

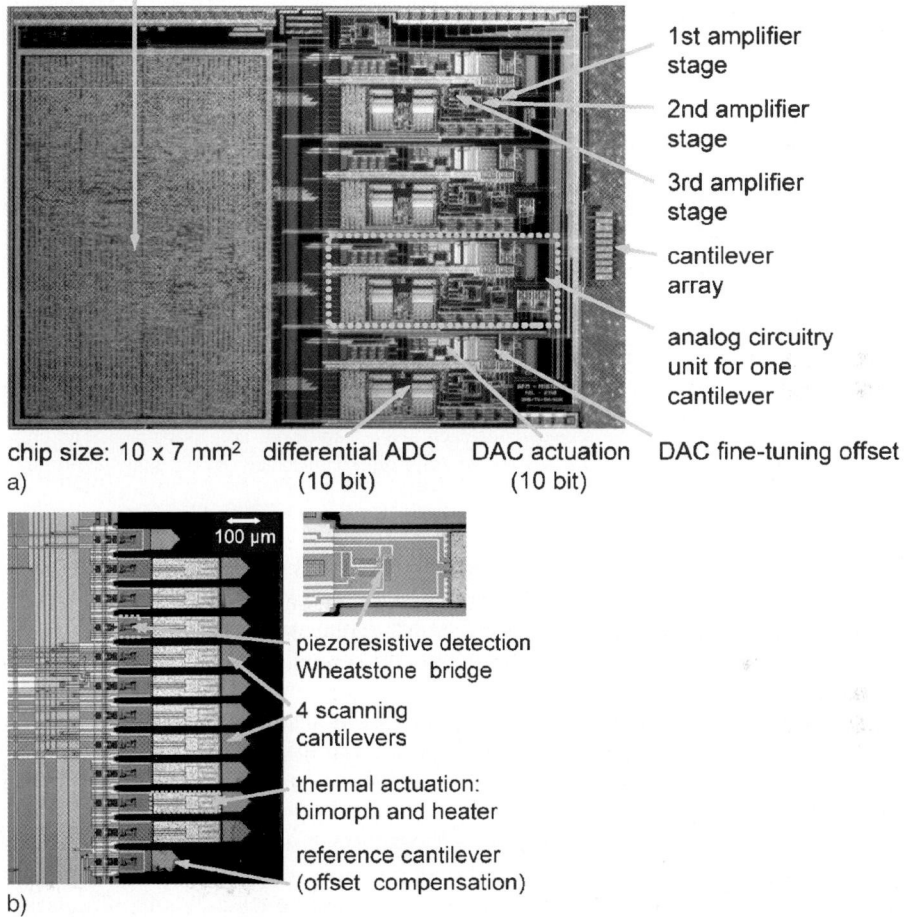

digital block including digital
signal processor (DSP), 4 serial
interfaces, 4 digital controllers,
4 filters (sensor-actuator coupling)

1st amplifier
stage

2nd amplifier
stage

3rd amplifier
stage

cantilever
array

analog circuitry
unit for one
cantilever

chip size: 10 x 7 mm²   differential ADC   DAC actuation   DAC fine-tuning offset
a)   (10 bit)   (10 bit)

100 µm

piezoresistive detection
Wheatstone bridge

4 scanning
cantilevers

thermal actuation:
bimorph and heater

reference cantilever
(offset compensation)

b)

Fig. 1.13. (**a**) Micrograph of the overall microsystem chip featuring the digital block *on the left side* and four identical analog units to control the four central cantilevers *on the right side* of the chip. The different circuitry subunits are indicated (ADC: analog-to-digital converter; DAC: digital-to-analog converter). (**b**) Close-up of the cantilever area. The chip features twelve cantilevers, ten of which can be potentially used for scanning (500 µm long, 85 µm wide). Only the four cantilevers in the *center* are connected to the circuitry. The two shorter ones at the *flanks* (250 µm long) serve as references [35]

Due to chip area constraints, only four readout and digital processing channels have been realized on-chip to provide a proof of concept. The integration of 10 readout channels can be realized by a redesign in CMOS technology with smaller feature size.

A simplified block diagram of the major circuitry units is shown in Fig. 1.14. The microsystem has an analog-digital, mixed-signal architecture. The cantilever

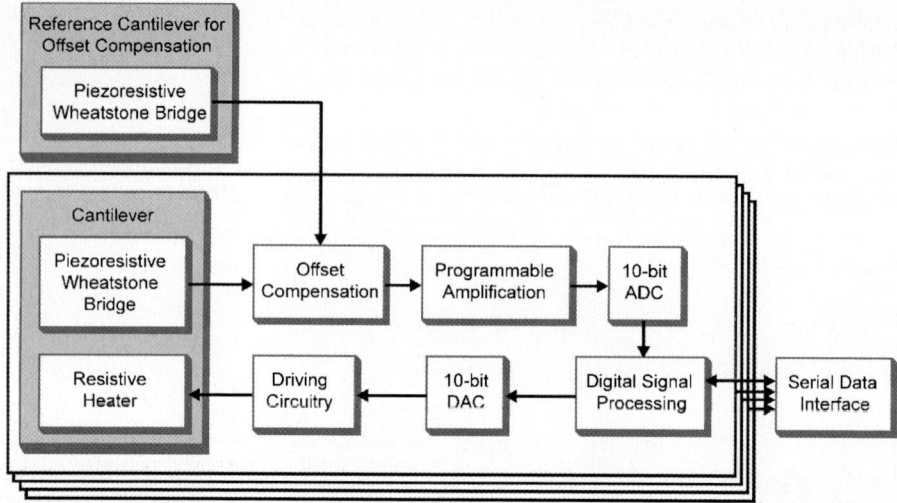

Fig. 1.14. Schematic of the chip architecture and the electronic components. The four frames indicate that these subunits are repeated for each active cantilever, i.e., four times. The chip also includes a digital signal processor and a serial digital interface (ADC: analog-to-digital converter; DAC: digital-to-analog converter) [35]

deflection signal coming from the Wheatstone bridge is amplified, filtered, conditioned and translated into the digital domain. All control operations are handled in the digital signal processor, which then issues actuation signals to the cantilever heater via a digital-to-analog converter. This way, a closed-loop operation is realized for every cantilever.

The circuitry unit repeated with each cantilever includes a fully differential analog amplification (three-stage amplification: fully differential low-noise programmable amplifier, programmable amplifier, fixed-gain amplifier) and filtering unit, which provides low-noise readout and the signal processing of the Wheatstone bridge signal. Multiple stages of offset compensation ensure the maximum possible force resolution to be achieved. After the first amplifier stage the coarse offset of the piezoresistive Wheatstone bridge is compensated by subtracting the offset signal of the reference cantilever structure (Fig. 1.13). This offset is mainly due to fabrication spread in the cantilever and the Wheatstone bridge [29] and also includes the offsets of the signal amplifiers. The imperfect matching of the diffused piezoresistive sensors from cantilever to cantilever can be compensated by another compensation stage, which enables an individual adjustment of the DC value of each readout channel with an 8-bit resolution by means of a programmable digital-to-analog converter (DAC). The overall gain of the analog signal processing is programmable from 18 dB to up to 44 dB to cover the whole range of expected forces in various applications. The amplified force signal is converted by a 10-bit successive-approximation analog-to-digital converter (ADC) and fed into the digital signal processor (DSP). The DSP unit comprises two programmable infinite-impulse-response (IIR) filters, each with six coefficients, which can be configured to act as proportional-integral-derivative (PID) controllers for the constant-force imaging mode. The controllers also provide averaging functions to further improve the force resolution and they compensate for thermal sensor-actuator crosstalk. The computing power of the DSP unit for the four cantilevers amounts to 16 million arithmetic operations per second, which is one of the highest values ever realized in CMOS-based micromechanical systems. This large value enables to issue 100,000 actuation signals per second per cantilever for repositioning and allows for precise cantilever position control even when fast force changes are to be expected. The cantilever actuation signals as coming from the DSP unit are converted to the analog domain by 10-bit flash DACs and provide together with analog square root circuits linear actuation characteristics. For a total deflection range of 1 μm, a resolution of 1 nm has been achieved, which can be further improved to 0.5 nm by lowering the overall deflection range to 0.5 μm.

1.4.1
Measurements

Two prototype applications were selected to show the performance of the monolithic AFM microsystem: (1) surface imaging and (2) force-distance measurements. Only the imaging mode will be shown here. For force-distance measurements, we refer to [35].

For surface imaging, the x-y-scanning function of a Nanoscope III (Digital Instruments, USA) was used, and the microsystem was operated in contact mode. The on-chip force controller measures the force acting on the cantilever and keeps it constant while the tip is scanned over the surface. Height information of the scanned sample is obtained from the actuation signal that is required to keep the force on the cantilever constant. Figure 1.15a shows a scanning image of a silicon grating with 18 nm steps at 3 μm distance recorded at a scanning speed of 20 μm/s and a force of 50 nN. Figure 1.15b shows a small-range line scan recorded at 3 μm/s. The maximum possible scanning speed of the monolithic system is approximately 1 mm/s. The maximum readout rate of the force signal is 100 kHz, which allows for averaging multiple values per data point in the DSP for better noise suppression. Each data point displayed in Figs. 1.15a, 1.15b represents the average of 300 values. The measured error signal indicates an excellent tracking of the surface topography. A vertical resolution of better than 1 nm has been achieved (Fig. 1.15b). A larger area scan of a biological sample at a force of 10 nN and a scanning speed of 100 μm/s is shown in Fig. 1.15c, representing a network of dried-out chicken neurites on a silicon oxide surface after fixation. A light microscope image is given for comparison (Fig. 1.15d). The force exerted on a sample can be kept as low as 5 nN so that also soft samples can be imaged. Tapping-mode operation can be also realized with the monolithic AFM system but was not tested so far.

1.5
Parallel Scanning

Parallel scanning is important when planar millimeter or centimeter-scale areas with nanometer-scale features are of interest. The most prominent samples include integrated circuitry chips. Minne et al. [36] have demonstrated 2 × 2 mm area scans with 10 cantilevers in parallel with a resolution of 0.4 μm.

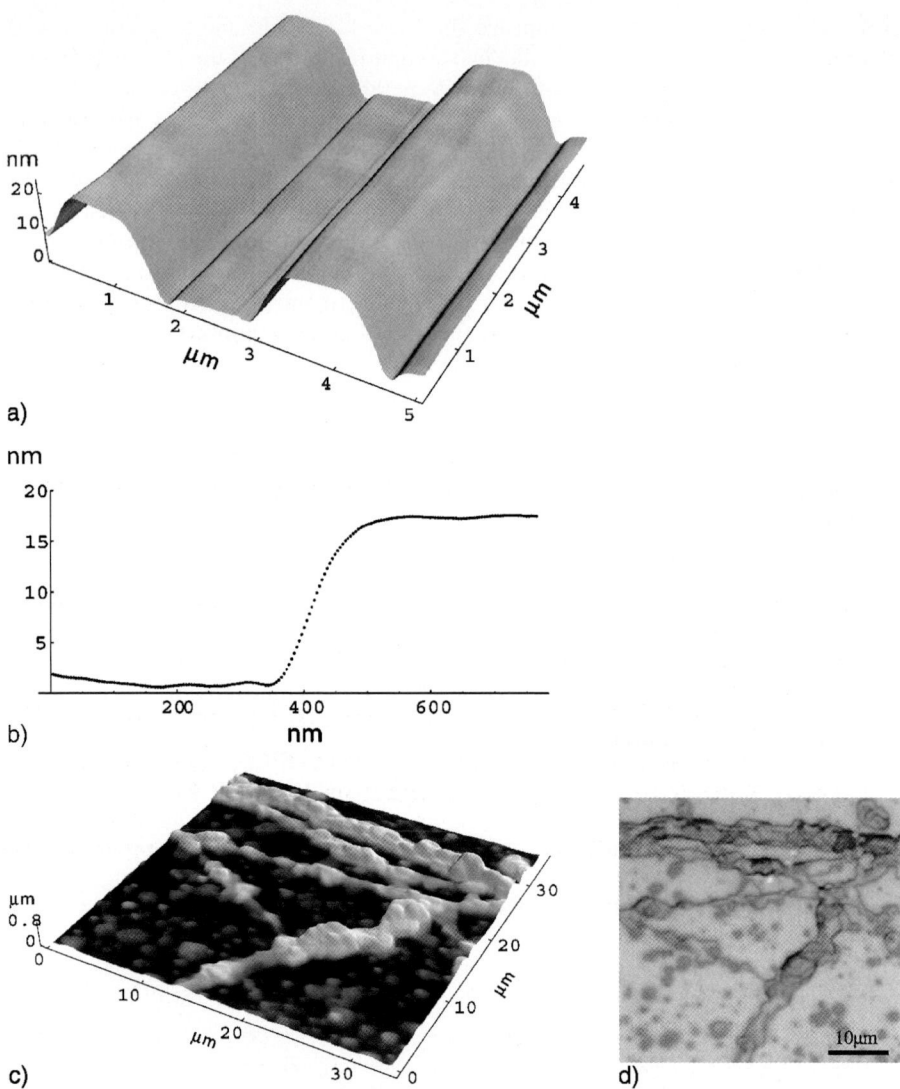

Fig. 1.15. (**a**) Scanning image (constant-force mode) of a silicon grating featuring step heights of 18 nm at 3 μm distance. A Gaussian smoothing with a radius of 3 data points was applied to yield the image [35]. (**b**) High-resolution line scan over one 18 nm step. Each displayed data point represents the average of 300 values. The vertical resolution is better than 1 nm [35]. (**c**) Scan of a biological sample at a force of 10 nN and a scanning speed of 100 μm/s representing a network of dried-out chicken neurites on a silicon oxide surface after fixation [35]. (**d**) Light microscope image of the same area of a network of dried-out chicken neurites for comparison [35]

1.6
Outlook

Future integrated AFMs will feature different floor plans with different area fractions covered by the transducer, analog and digital circuitry: the cantilever array will occupy only a small fraction of the chip. Analog circuits can be compacted since more a priori knowledge of the cantilever characteristics, such as offset etc., is available. Analog-to-digital conversion is likely to be more complex to provide larger bandwidths. The largest part will be the digital core. The focus of future development work will be on increasing the system precision, on providing larger bandwidths, on integrating digital phase-locked loops and on further reducing the number of interconnections.

New actuation schemes have to be developed that can be applied in a liquid-phase environment to satisfy the needs of the rapidly expanding field of bioanalytics. A candidate actuation scheme could be the magnetic actuation by means of Lorentz forces [37]. The packaging of an integrated AFM system for operation in liquid-phase environments will pose further challenges.

References

1. Binnig G, Quate CF, Gerber C (1986) Phys Rev Lett 56:930
2. Brugger J, de Rooij NF (1992) J Microeng 2:218
3. Minne SC, Manalis SR, Quate CF (1999) Bringing Scanning Probe Microscopy up to Speed. Kluwer Academic, Boston
4. Linnemann R, Gotszalk T, Hadjiiski L, Rangelow IW (1995) Thin Solid Films 264:159
5. Gotszalk T, Radojewski J, Grabiec PB, Dumania P, Shi F, Hudek P, Rangelow IW (1998) J Vac Sci Technol B 16:3948
6. Jumpertz R, Hart AVD, Ohlsson O, Saurenbach F, Schelten J (1998) Microelectro Eng 41/42:441
7. Thaysen J, Boisen A, Hansen O, Bouwstra S (2000) Sens Actuators A 83:47
8. Minne SC, Yaralioglu G, Manalis SR, Adams JD, Zesch J, Atalar A, Quate CF (1998) Appl Phys Lett 72:2340
9. Minne SC, Adams JD, Yaralioglu G, Mannalis SR, Atalar A, Quate CF (1998) Appl Phys Lett 73:2340
10. Kim YS, Nam HJ, Cho SM, Hong JW, Kim DC, Bu JU (2003) Sens Actuators A 103:122
11. Neumann JJ, Greve DW, Oppenheim IJ (2004) Proc SPIE 5391:230
12. Sze SM (1994) Semiconductor Sensors. Wiley, New York
13. Baltes H, Brand O, Fedder GK, Hierold C, Korvink JG, Tabata O (eds) (2005) Advanced Micro and Nanosystems, Vol. 2, p 596. Wiley, Weinheim
14. Dorey RA, Whatmore RW (2004) J Electroceramics 12:19
15. Aigner R (2003) In: Proceedings of the IEEE 2003 Custom Integrated Circuits Conference 2003, p 141–146. IEEE, New York
16. Ruby RC et al. (2002) In: IEEE ISSC Conference Digest of Technical Papers, Vol. 1, p 184. IEEE, New York
17. Callaghan LA et al. (2004) In: Proceedings of the 2004 IEEE International Frequency Control Symposium and Exposition. IEEE, New York
18. Mishin S et al. (2003) In: IEEE Symposium on Ultrasonics Proceedings. IEEE, New York
19. Lange D, Brand O, Baltes H (2002) CMOS Cantilever Sensor Systems. Springer, Berlin Heidelberg New York

20. Rincon-Mora GA (2001) Voltage References: From Diodes to Precision High-Order Bandgap Circuits. Wiley, New York
21. Allen PE, Holberg DR (2002) CMOS Analog Circuit Design, 2nd edn. Oxford University Press, New York
22. Gotszalk T, Grabiec P, Rangelow IW (2000) Piezoresistive sensors for scanning probe microscopy. Ultramicroscopy 82(1–4):39
23. King WP, Kenny TW, Goodson KE (2004) Appl Phys Lett 85:2086
24. Chow EM et al. (2002) Appl Phys Lett 80:664
25. Yu X et al. (2003) In: IEEE Conference on Electron Devices and Solid State Circuits, p 121. IEEE, New York
26. Hyun SJ et al. (2005) In: TRANSDUCERS '05. The 13th International Conference on Solid-State Sensors, Actuators and Microsystems. Digest of Technical Papers, p 1792. IEEE, New York
27. Lange D (2000) Cantilever-Based Microsystems for Gas Sensing and Atomic Force Microscopy. Dissertation (no. 13984) ETH-Zurich, Switzerland
28. Akiyama T, Staufer U, de Rooij NF (2002) Rev Scient Instr 73:2643
29. Volden T (2004) CMOS-Integrated Cantilevers for Biosensing and Probe Microscopy. Dissertation (no. 15984), ETH Zurich, Switzerland
30. Minne SC, Manalis SR, Quate CF (1995) Appl Phys Lett 68:3918
31. Barrettino D et al. (2005) IEEE JSSC 40:951
32. Van de Plassche R (2003) CMOS Integrated Analog-to-Digital and Digital-to-Analog Converters, 2nd edn. Kluwer Academic, Boston
33. Barrettino D et al. (2003) In: IEEE International Symposium on Circuits and Systems (ISCAS), Bangkok, Thailand, 25–28 May 2003, Digest of Technical Papers. IEEE, New York
34. Antoniou A (1993) Digital Filters, 2nd edn. McGraw-Hill, New York
35. Hafizovic S, Barrettino D, Volden T, Sedivy J, Kirstein K-U, Brand O, Hierlemann A (2004) PNAS 101:17011
36. Minne SC, Adams JD, Yaralioglu G, Mannalis SR, Atalar A, Quate CF (1998) Appl Phys Lett 73:1742
37. Lange D, Hagleitner C, Herzog C, Brand O, Baltes H (2002) Sens Actuators A 103:150

2 Electrostatic Microscanner

Yasuhisa Ando

2.1
Introduction

Various detecting principles for the scanning probe microscope (SPM) are used to obtain the desired atomic to nanoscale properties on the sample surface, which is done by changing the probe used for detection. For example, an scanning tunneling microscope (STM) uses tunneling current between probe and sample [1], whereas both the atomic force microscope (AFM) and lateral force microscope (LFM) detect mechanical forces from the bending [2] and twisting of the cantilever [3, 4].

The scanning device (scanner) in SPM has changed from a tripod of piezoelectric lead zirconate titanate (PZT) stack actuators [1] to a PZT tube actuator that realizes three-dimensional (3D) motion by itself. Most commercially available SPMs use a PZT tube actuator (i.e., PZT scanner), which has a rigid, simple structure and is suitable as a stage in SPM because it allows continuous motion by controlling the driving voltage [5]. PZT actuators, however, have drawbacks, such as hysteresis in motion [6] and unsuitability for use at temperatures exceeding 100 °C [7].

Electrostatic comb actuators are often used as a driving force in microelectrome-chanical systems (MEMS) for linear actuations [8, 9]. The comb actuator can be fabricated in ether polysilicon [9] or single-crystal silicon [10]. Compared with PZT actuators [11], comb actuators yield low hysteresis motion [12] and can be used at higher temperatures. Thus, electrostatic comb actuators are advantageous for use in a scanner. However, conventional electrostatic comb actuators can control the displacement only along the substrate [12, 13].

Applying electrostatic comb actuators to a scanner in SPM requires 3D motion. Although vertical comb microactuators have been developed [14], it has been difficult to combine them with lateral motion because of their complex structure. A simple displacement conversion mechanism was developed to achieve 3D motion using electrostatic comb actuators [15, 16]. A traveling table with the comb supported by a suspension system that included inclined leaf springs was used. With the inclined leaf springs, the suspension system converted the lateral displacement of the comb actuator into displacement along a line having a finite angle to the substrate.

In this chapter, microscanner technologies based on the above-mentioned displacement conversion mechanism will be described. In Sect. 2.2 the principle of the displacement conversion mechanism and methods of acquiring a vertical displacement using a combination of comb actuators will be explained. Section 2.3

describes the mechanisms of 3D microstages, fabrication based on a bulk micromachining technique, and the relationship between 3D displacement and the driving voltage on comb actuators. In Sect. 2.4, 3D microstages are applied to an AFM and used as a microscanner. The basic performances are then evaluated by measuring grating images, and the issues and features of application to SPM will be discussed.

2.2
Displacement Conversion Mechanism

2.2.1
Basic Conception

The principle for realizing vertical motion relative to the substrate is quite simple. Figure 2.1 shows the concept of the suspension for displacement conversion mechanism. When a force F_1 parallel to the substrate is applied to the end of the leaf spring inclined against the substrate at an angle of θ, components F_{1X0} and F_{1Z0} work on the end. If we ignore the displacement of the spring end caused by F_{1Z0}, the displacement in the z-direction of the end of the leaf spring (d_Z) is given by

$$d_Z = F_1 K_{X0} \cos\theta \sin\theta \,, \tag{2.1}$$

where K_{X0} is a spring constant in direction of x_0. Therefore, force parallel to the substrate can generate a vertical component in displacement by using inclined leaf springs. If a comb actuator is used to generate the applied force F_1, the vertical displacement d_Z can be continuously controlled by the applied voltage to the comb actuator, because a comb actuator is suitable for controlling lateral displacement.

Equation (2.1) shows that d_Z is a function of θ and is proportional to the applied force F_1. To calculate the inclined angle θ, which gives the maximum d_Z, (2.1) is changed to

$$\partial d_Z/\partial\theta = F_1 K_{X0} \cos 2\theta \,. \tag{2.2}$$

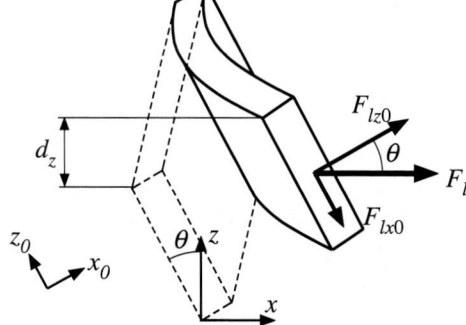

Fig. 2.1. Basic conception of the suspension. When F_1 is applied to the end of the leaf spring, the end displaces in another direction. The vertical displacement d_z is proportional to $F_1 \cos\theta \sin\theta$

Equation (2.2) shows that d_Z increases with θ when $0° \leq \theta < 45°$ and becomes the maximum at $\theta = 45°$. This indicates that the optimum angle of the inclined leaf spring is $45°$, when large displacement in the z-direction is needed for a microstage.

2.2.2
Combination with Comb Actuator

Figure 2.2 shows the design of a simple z-stage. The stage has four suspensions and each suspension incorporates a pair of leaf springs inclined at θ against the substrate. The suspensions are formed as all leaf springs have the same angle. When the electrostatic force pulls the traveling table in the lateral direction, the traveling table is lifted up with the lateral displacement. The lift-up amount is proportional to the lateral pulling force. Figure 2.3 shows the dimensions of the simple z-stage for calculating the motion of the table by using finite element method (FEM). The thickness of the leaf spring is $4\,\mu m$ and the length and width are both $20\,\mu m$. We also assumed the air gap between adjacent two fingers ($s = 3\,\mu m$), the height of a finger plane facing to another one ($h = 20\,\mu m$), and the number of fingers in the comb on the traveling table ($N = 28$). The electrostatic driving force F_1 caused by the driving voltage V_D is given by:

$$F_1 = \frac{2\varepsilon_0 Nh}{s} V_D^2 , \qquad (2.3)$$

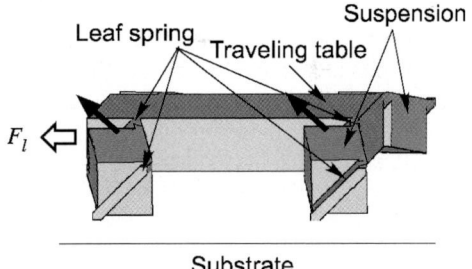

Fig. 2.2. Schematic of simple z-stage. When the lateral force acts on traveling table the suspensions lift up the table

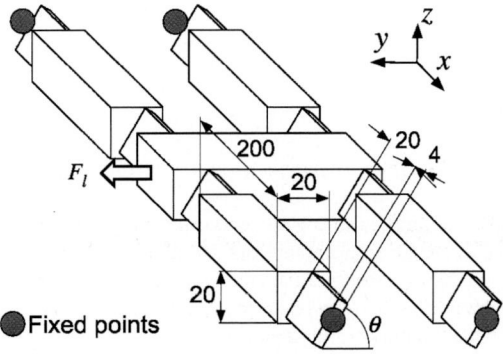

Fig. 2.3. Dimensions in simple z-stage used in FEM analyses. The 8 leaf springs have the same dimensions and each angle against the substrate is θ

where ε_0 is the permittivity constant of air (8.85×10^{-12} F/m). The driving force F_1 is then calculated to be 34 µN at $V_D = 100$ V and 137 µN at $V_D = 200$ V.

Figure 2.4a,b shows the displacements in the lateral and vertical directions calculated by FEM for the simple z-stage shown in Fig. 2.3. Young's modulus of 130 GPa and Poisson's ratio of 0.28 [17], which are in the direction of [100] in single-crystal silicon, are used for the calculation. Figure 2.4a shows the displacements as a function of the leaf spring angle θ, when a constant lateral force of 137 µN is applied to the end of the traveling table. The lateral displacement increases with increase of θ. The vertical displacement becomes the maximum value of 0.57 µm at $\theta = 45°$.

Figure 2.4b shows the displacements as a function of the thickness of the table and suspensions. The model used for the calculation is similar to Fig. 2.3, but the leaf spring angle θ is fixed to 45°. When decreasing the thickness of the microstructure, upper and lower parts of required thickness were cut off from the model shown in Fig. 2.3 and the shape of the suspension and leaf spring changes as shown in Fig. 2.4b. In the calculation, the applied force is changed according to the thickness of the microstructure by using (2.3). When the thickness decreased from 20 µm to 15 µm, the vertical displacement decreases because the applied force decreases while the height of the inclined leaf spring changes little. The vertical

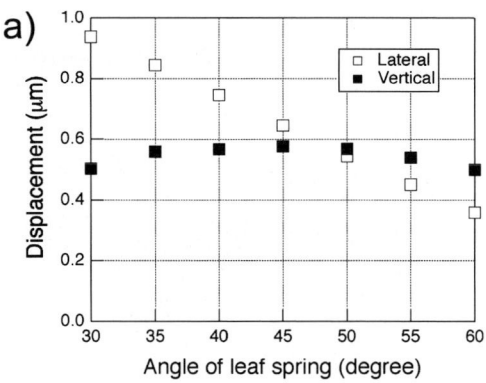

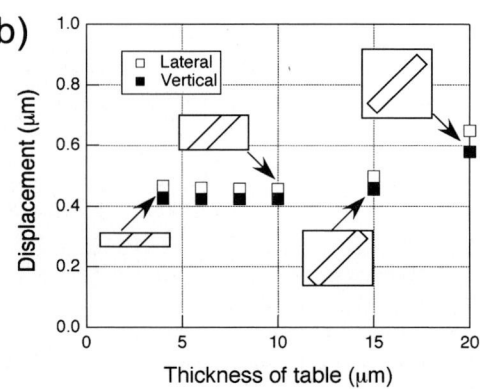

Fig. 2.4. FEM calculation results for simple z-stage. (**a**) The relationship between vertical and lateral displacements and angle of inclined leaf spring. The vertical displacement became the maximum when the angle was 45°. (**b**) The relationship between vertical and lateral displacements and thickness of the microstructure. The applied force was not constant and was proportional to the thickness to the table

displacement is almost constant at the thickness of 10 to 4 µm. Therefore, the displacements in the lateral and vertical directions are almost independent of thickness of microstructure when the height of the inclined leaf is the same thickness of microstructure.

2.2.3
Various Types of Displacement Conversion Mechanism

Figure 2.5a shows a design of the real z-stage excluding the lateral motion. The outer traveling table is supported by four outer suspensions. Each outer suspension has four leaf springs inclined at $45°$ against the substrate. The outer table supports the inner traveling table by four inner suspensions. Each inner suspension also has four leaf springs but its angle of inclination against the substrate is $-45°$. When a force of $2F_1$ is applied to the outer traveling table, the outer traveling table moves in the $+x$-direction and is lifted up by the outer suspension system (Fig. 2.5b). Simultaneously, a force of F_1 is applied to the inner traveling table, then the inner table moves in the $-x$-direction and is lifted up by the inner suspension system. Because the lateral motions in $+x$ and $-x$ compensate each other, only a real vertical motion is obtained.

Figure 2.6b shows a real z-stage fabricated in the device layer of a silicon on insulator (SOI) wafer. The microstructure with the vertical wall was fabri-

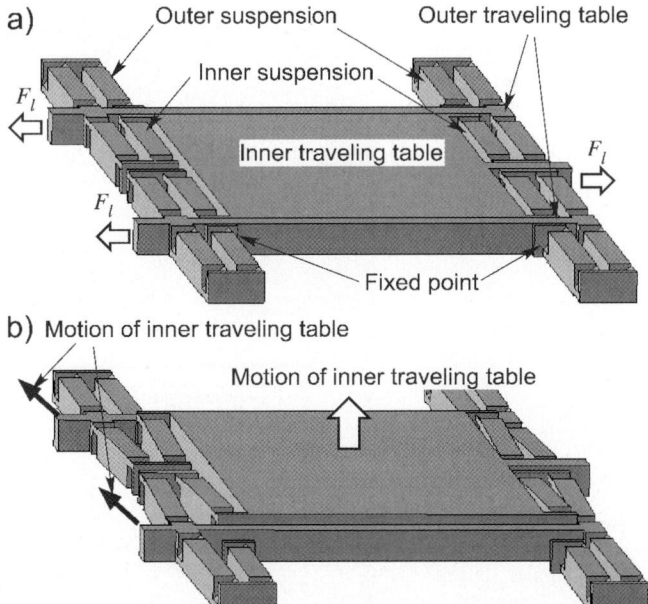

Fig. 2.5. Schematic of motion of real z-stage excluding lateral motions. (**a**) The inner traveling table installed on outer traveling table via 4 suspensions. When $2F_1$ and F_1 are applied to the outer and inner traveling tables, respectively. (**b**) The inner traveling table vertically lifted up

cated using inductively coupled plasma-reactive ion etching (ICP-RIE) and inclined leaf springs were fabricated using a focused ion beam (FIB) system. All suspensions are arranged in the area of 300 μm × 400 μm for efficient processing of FIB. The thickness of the microstructure is 20 μm and the size of the inner traveling table is approximately 120 μm × 100 μm. The number of fingers in the comb actuator on the inner and outer traveling tables is 29 and 58, respectively. Figure 2.6b shows the lift-up of the inner traveling table of the z-stage. In this figure, the maximum lift-up is 1.9 m when the driving voltage is 133 V. The vertical displacements were measured for each increasing and decreasing voltage process, and there was a negligible difference between the two processes.

It is also possible to realize a tilt motion by using inclined leaf springs. Figure 2.7a shows the design of the tilt-stage. The suspensions are formed as two pairs of leaf springs inclined at 45° and another two pairs inclined at −45° against the substrate. When the electrostatic force is applied, the traveling table moves in the lateral direction and changes its angle against the substrate (Fig. 2.7b).

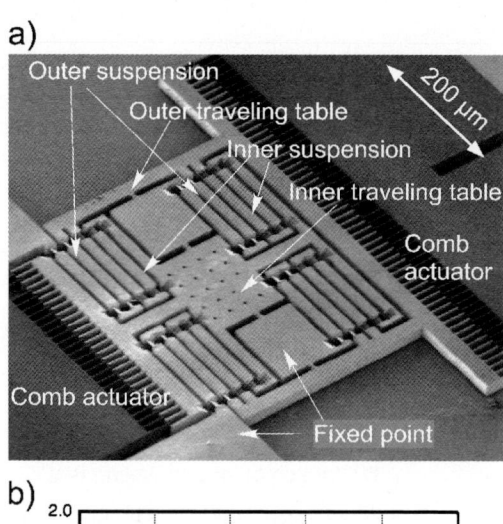

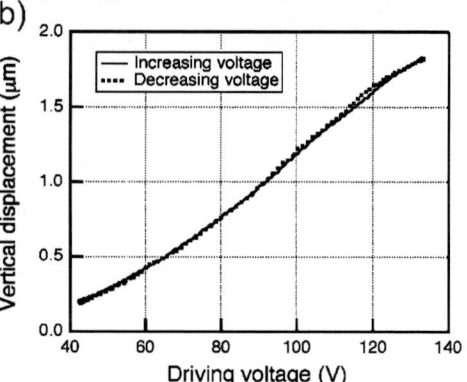

Fig. 2.6. Real z-stage was fabricated in device layer of SOI wafer and was tested by laser Doppler sensor. (**a**) Secondary electron image of a fabricated real z-stage observed using FIB. (**b**) Relationship between comb actuator driving voltage and inner traveling table lift-up

Fig. 2.7. Schematic of tilt-stage and its motion measured by laser Doppler sensor. (**a**) When the lateral force acts on the traveling table, a pair of suspensions lifts up the table while the other pair lowers the table. (**b**) Relationship between tilting angle and tilt-stage applied voltage. Distance between the suspensions on the front and rear sides was about 75 μm

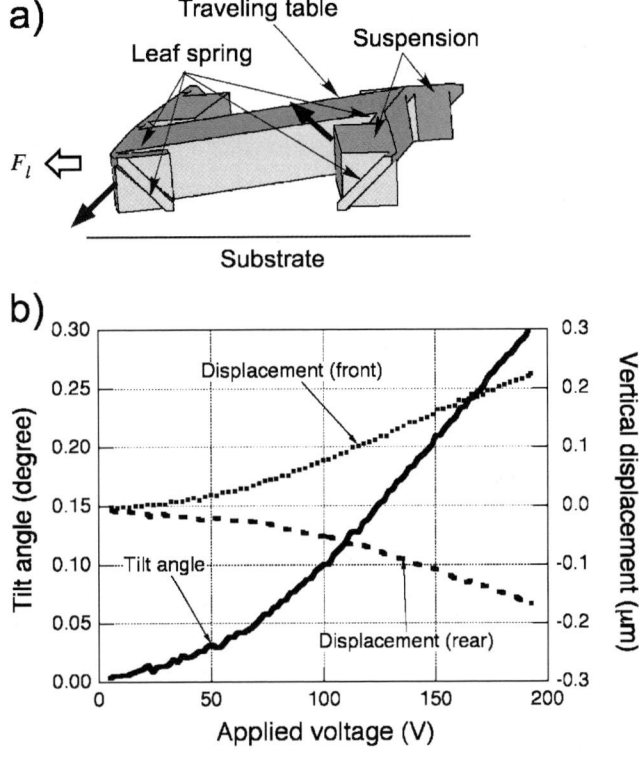

2.3
Design, Fabrication Technique, and Performance

2.3.1
Main Structure of 3D Microstage

The real z-stage shown in Fig. 2.6 includes two sets of comb actuators and suspensions including leaf springs inclined to the substrate. When the same voltage is applied to each comb actuator, the inner traveling table moves only in the z-direction (i.e., vertical to the substrate). If one controls the driving voltage to each comb actuator independently, the traveling table will move not only along a line (one dimension) but also in a plane (two dimensions) normal to the substrate. Therefore, the z-stage can work as an x-z-stage. Three-dimensional motion (i.e., additional y motion) will be achieved by adding another combination of a comb actuator and a suspension system that moves along with the substrate.

Figure 2.8 shows the mechanism for generating 3D motion. Tables A, B, and C are supported by suspension systems A, B, and C, respectively, and forces F_A, F_B, and F_C are applied to tables A, B, and C, respectively. When the corresponding force is applied, each table moves in the direction indicated by the respective white arrows in the figure. In the real z-stage in Fig. 2.6, each suspension has a hairpin

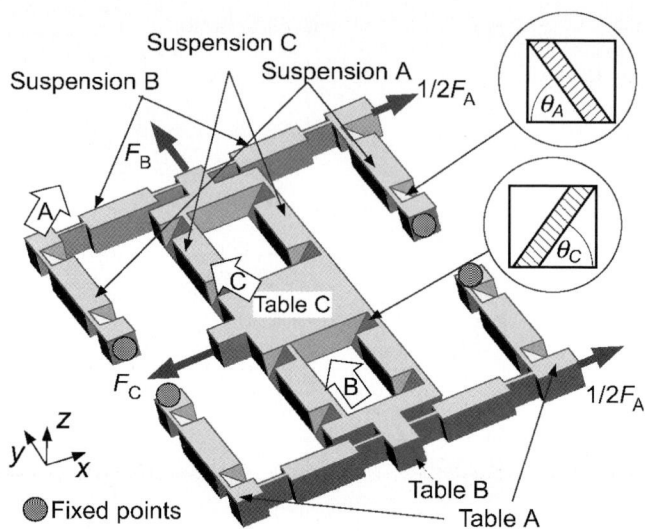

Fig. 2.8. Main structure of a 3D electrostatic stage used to convert a lateral displacement into angled motion (Fig. 2.1) and generate 3D motion. *White arrows* A, B, and C indicate the motion induced by F_A, F_B, and F_C, respectively. Inclined leaf springs are incorporated in suspension systems A and C, and their inclined angle with respect to the substrate are θ_A for suspension system A and θ_C for suspension system C

structure and includes four inclined leaf springs. In the 3D microstage, the structure of each suspension is simplified to increase rigidity in the y-direction. Figure 2.9 shows a simplified schematic of the motion for tables A and C in Fig. 2.8. In the figure, the arrows show the vector of each table motion. When the angles θ_A and θ_C between the deformation direction and the substrate were the same for suspension systems A and C ($\theta = \theta_A = \theta_C$) and the spring constants of suspension systems A and C are the same ($k = k_A = k_C$) as shown in Fig. 2.9a, the position of table C in the x-z-plane (x_{C1}, z_{C2}) is given as follows:

$$x_{C1} = (F_A - 2F_C) \cos^2 \theta / k \tag{2.4}$$

$$z_{C1} = F_A \cos \theta \sin \theta / k . \tag{2.5}$$

The displacement z_{C1} can be controlled by only F_A. If pure vertical displacement without lateral displacement ($x_{C1} = 0$) is needed, then F_A should be changed while maintaining the relation $F_A = 2F_C$. When F_A is constant, the displacement x_{C1} would be proportional to F_C while z_{C1} stays constant. Thus, this type of stage (model 1) can be controlled by using a simple algorithm. Drawbacks of model 1 are that the stage contains 16 inclined springs and therefore needs complex processes.

Figure 2.9b,c shows alternative models of combinations of vertical and inclined leaf springs to reduce the number of the inclined leaf springs. Model 2 (Fig. 2.9b) has inclined leaf springs only in suspension system A, whereas model 3 (Fig. 2.9c) has them only in suspension system C. Each model can realize motion in the x-z-plane.

Fig. 2.9. Models of 2D motion of table C and table A in the
x-z-plane based on Fig. 2.3. (**a**) Model 1, where $\theta_A = \theta_C = \theta$,
adopted in a previous study. (**b**) Model 2, where $\theta_A = \theta$,
$\theta_C = 90°$. (**c**) Model 3, where $\theta_A = 0°$, $\theta_C = \theta$

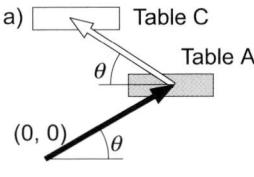

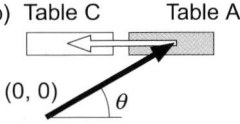

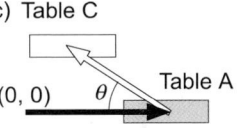

The position of table C in model 2 (x_{C2}, z_{C2}) and model 3 (x_{C3}, z_{C3}) are given as
follows:

$$x_{C2} = (F_A - F_C) \cos^2 \theta / k_A - F_C / k_C \tag{2.6}$$

$$z_{C2} = (F_A - F_C) \cos \theta \sin \theta / k_A \tag{2.7}$$

$$x_{C3} = (F_A - F_C) / k_A - F_C \cos^2 \theta / k_C \tag{2.8}$$

$$z_{C3} = F_C \cos \theta \sin \theta / k_C . \tag{2.9}$$

In model 2, larger lateral displacement (x_{C2}) can be obtained by controlling F_C.
The vertical displacement (z_{C2}), however, changes with F_C. In contrast, in model 3,
the vertical displacement (z_{C3}) is determined by only F_C and is not affected by F_A,
although changes in F_A would yield a smaller lateral displacement compared with
model 2 where F_C changes.

When maintaining the range of displacement and for ease in controlling the table
position, each model has both advantages and disadvantages. Model 1 is superior to
models 2 and 3 in its lager displacement in the z-direction. But a PZT scanner of a conventional SPM has a much smaller ($< 1/10$) z-displacement than x and y. Therefore,
the larger z-displacement is not strictly required. As to simplify the processes as described in Sect. 2.3.3, model 3 would be superior to models 1 and 2. Therefore, we
will further discuss model 3 as the basis for the design of the main structure of the
3D microstage, which is the same as shown in Fig. 2.8 where $\theta_A = 90°$.

2.3.2
Amplification Mechanism of Scanning Area

In Fig. 2.8, comb actuators are used to apply the driving force to tables A, B, and C.
When a comb actuator is directly attached to these tables and the air gap in the comb
actuator is a few microns ($\sim 2 \, \mu m$), interference between the combs is a problem.
For the comb directly attached to table B, collision between combs would occur

when F_A or F_C is applied and the displacement approaches the width of the air gap. The same problem would occur for the comb on table C when F_B is applied and a larger displacement is generated. Figure 2.10 shows an improved structure of the 3D microstage based on model 3. The mechanism of generating 3D motion is similar to Fig. 2.8 (shown enclosed by dashed lines). Although, the comb for F_A (comb A) is directly attached to table A, the combs for table B (comb D) and table C (comb E) are independently supported by support suspension systems D and E. Support suspension systems D and E restrict the actuator's motion to one axis along y and x, respectively. Link beams B and C connect tables B and C and combs D and E, respectively, and absorb the motion in x- and y-directions and transmit motions only in the y- and x-directions, respectively. In addition, torsion beams are added to table C and comb E to absorb the motion in the z-direction.

In the discussion on the main structure of the 3D microstage (Fig. 2.8), the effects of the support suspension systems and link beams was not considered. To predict the motion of table C more precisely, FEM analysis was conducted on the design shown in Fig. 2.10. The sizes of the suspensions and leaf springs used for the calculation are shown in Table 2.1. The thickness of the microstructure is 20 μm and $\theta_C = 70°$. For the applied forces F_A and F_E calculated using FEM, Fig. 2.11 shows the position of table C when F_A is changed from 0 to 2.4 mN at a constant F_E of 0.4 (solid triangles) and 0.8 mN (solid circles), and when F_E is changed from 0 to 0.8 mN while maintaining the relationship of $F_A = 2F_E$ (open squares). Although the position of table C does not change in the z-direction at a constant F_E determined using (2.9) ($F_E = F_C$), it increases at higher F_A. This is because the pulling force generated by the support suspension system E increases with the lateral displacement of table C, and thus the pulling force lifts table C via the suspension system C. On the other hand, the same model of the 3D microstage was used to calculate the resonance frequency in another FEM analysis. As a result, the first resonance mode was found at about 16 KHz, which is likely to be higher than the resonance frequency of a PZT scanner. If the stiffness of the suspension systems is increased, the resonance frequency will further increase. Thus, the 3D microstage can move faster than a PZT scanner and achieve high speed scanning if it is used as a scanner of an SPM.

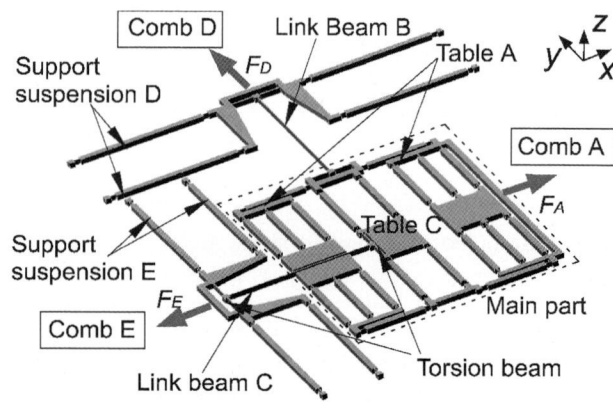

Fig. 2.10. Schematic of improved 3D microstage structure based on model 3 (Fig. 2.9). Supporting suspensions and deformable beams have been added to the main structure (within *dashed lines*) as shown in Fig. 2.8

Fig. 2.11. Analytical driving force
and displacements of table C in the
x-z-plane calculated using FEM for improved 3D microstage with supporting
suspensions

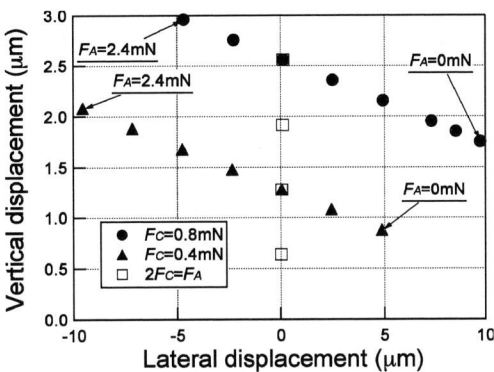

Table 2.1. Dimension and stiffness of deformable parts

	Type	L [μm]	w [μm]	K [N/m]
Suspension A	(a)	248	3	112
Suspension B	(a)	270	3	42
Suspension C	(b)	270	4	100 (x)
Suspension D	(a)	470	4	26
Suspension E	(a)	470	4	26
Beam B	(c)	485	4	
Beam C	(c)	495	4	
Torsion beam	(d)	170	4	–

Table 2.2 summarizes the interaction of the applied forces of F_A, F_D and F_E
and the displacements of table C and combs D and E. When applying the force of
F_A to F_D, and paying attention to the displacement of table C in a plane parallel
to the substrate, the maximum displacement in the direction perpendicular to the
applied force is 6 nm in the y-direction at $F_A = 500\,\mu\text{N}$. For the displacement of
combs D and E, the maximum displacement in the direction perpendicular to the
applied force is 3 nm, which is found in comb E at $F_D = 500\,\mu\text{N}$. From the above
findings, the introduction of support suspension systems and link beams suppresses
the interaction between combs and increases the displacement of table C more than
the air gap in the comb actuator.

Table 2.2. Displacements of table C and comb D and E

				Displacement					
Applied force [μN]			Table C [μm]			Comb [nm]			
F_A	F_D	F_E	x	y	z		x	y	z
500			−3.0	−0.006	0.25	Comb D	1	−1	0
	500		0.002	7.0	0.004	Comb E	3	−1	1
		500	6.1	0.002	1.1	Comb D	−1	1	0
1000		500	0.073	−0.003	1.6	Comb D	1	−2	0

2.3.3
Fabrication Using ICP-RIE

In the real z-stage shown in Fig. 2.6, FIB processing was used to fabricate inclined leaf springs. Although FIB is useful to fabricate (without masks) an inclined part at any angle against a substrate, it requires 5 to 10 minutes to mill and fabricate each inclined leaf spring in a 20-μm-thick device layer at an ion beam current of 5 to 10 nA and accelerating voltage of 30 kV. Therefore, using FIB to form the 8 to 16 inclined leaf springs needed for a 3D microstage is time consuming. Another disadvantage of FIB is that moving parts sometime fuse to the substrate, probably by redeposition of sputtered silicon.

Anisotropic wet etching is usually used to form sloped structures in single-crystal silicon. To form inclined leaf springs, however, a complex process using this etching from the backside would be required. If the angle of DRIE (deep reactive ion etching) can be controlled and then used to form inclined structures, the fabrication procedure would be simple and fabrication accuracy would be improved. In ICP-RIE, etching proceeds along the direction perpendicular to the wafer surface, but inclined milling was reportedly found at the periphery of the wafer because the gradient of the electrical field was not perpendicular to the wafer plane [18]. We therefore tried to form an inclined structure by using ICP-RIE as follows. First, an SOI wafer with a patterned SiO$_2$ mask was cut into separate plates measuring 8×8 mm^2 or 8×4 mm^2. Then, each diced SOI plate was set on another supporting wafer at an incline angle of 30° or 60°, as shown in Fig. 2.12. Finally, the wafer with the diced plates was loaded into an ICP chamber (STS Multiplex ICP System), and subjected to ICP-RIE using a standard recipe.

Figure 2.13 shows the angle of this inclined structure (i.e., structure angle, θ_S in Fig. 2.12) fabricated using inclined ICP-RIE. The abscissa shows the position of the formed inclined structures in each diced SOI plate, where the top-side edge of the plate is defined as zero and θ_S is plotted as a function of distance of the inclined structure from the top-side edge. For all plates, θ_S rapidly decreased as the position of the inclined structure approached the top-side edge and was the maximum near the center of the plate. When comparing the two 8-mm-wide plates set at different inclination angles ($\theta_P = 30°$ and 60°), θ_S of the 60° plate was lower than that of the 30° plate close to the bottom-side edge, although the maximum θ_S near the center

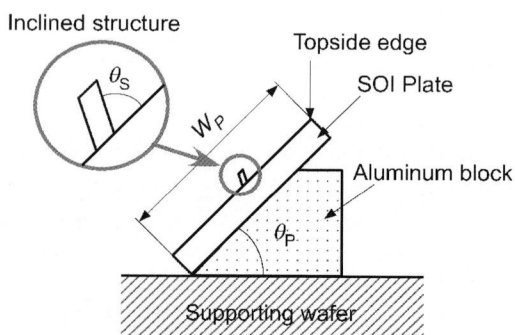

Fig. 2.12. Schematic of diced SOI plate on a supporting wafer. Supporting wafer with the diced SOI plate is set in an ICP-RIE chamber and then subjected to inclined ICP-RIE

Fig. 2.13. Structure angle θ_S on diced plate fabricated using inclined ICP-RIE. Diced plate was set on a supporting wafer at $\theta_P = 30°$ or $60°$. θ_S changed with distance of the fabricated inclined structure from the top-side edge of the inclined plate

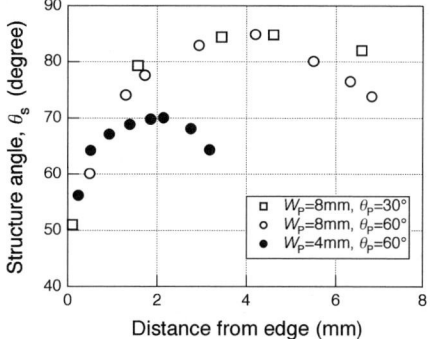

of the plate was similar for both plates. At the same θ_P of $60°$, θ_S of the 4-mm-wide plate was much lower than that of the 8-mm-wide plate at the center part of the plate.

In conventional ICP-RIE, plasma particles (ions and radicals) collide perpendicular to the silicon wafer, and thus vertical structures are formed. In the proposed method, on the diced plate set on a supporting wafer at a given θ_P, the direction of plasma flow was affected by both the diced SOI plate and supporting wafer. Therefore, inclined structures were formed on the diced plate. Because the direction of plasma was affected more by the supporting wafer that had a narrower width plate (i.e., the 4-mm-wide plate), this plate had a lower θ_S.

A larger vertical displacement can be obtained if the inclined leaf spring suspension system is formed close to the top-side edge. However, the angle of the inclined leaf spring should be the same for each suspension, when table C must provide pure translation without tilt as shown in Fig. 2.2. To apply the 3D microstage to an SPM as a scanner, the change in angle of table C must be as small as possible. Therefore, suspension components should be formed at the center of the plate where θ_S is relatively constant for approximately $500\,\mu m$ (distance between the suspension components). The vertical displacements generated by $\theta_S = 83°$ (for the 8-mm-wide plate) and $70°$ (for the 4-mm-wide plate) will be approximately $0.6\,\mu m$ and $1.7\,\mu m$, respectively, for an applied lateral displacement of $5\,\mu m$. Although either of these vertical displacements is adequate for an SPM, it is easier for a probe to approach table C when the vertical displacement is larger. Moreover, measurement of the force-distance curve sometimes requires large displacement in the z-direction. Thus, a 4-mm-wide SOI plate is better for fabricating a 3D microstage using inclined ICP-RIE.

For applying inclined ICP-RIE model 3 is superior to models 1 and 2. First, models 2 and 3 need only one time of the inclined ICP-RIE processing, whereas model 1 needs two times. Second, comparing models 2 and 3, the angle of each inclined leaf spring in model 3 is more uniform compared with that in model 2, because the distance between suspension C is narrower than between suspension A (Fig. 2.8), and because a difference in θ_S between two inclined structures tends to increase with increasing distance between the inclined structures (Fig. 2.13).

Figure 2.14 shows a schematic of the fabrication process of 3D microstages in a 20-μm-thick device layer on SOI wafers by using conventional and inclined ICP-RIE. There are a few things to note. First, a photoresist film is coated on an SOI wafer by spin coating and then patterned by contact exposure and development (Fig. 2.14a).

Second, the unmasked part of the device layer is removed by conventional ICP-RIE (Fig. 2.14b). Third, after the photoresist film is removed, the surface of the wafer is simply oxidized thermally at $1100\,°C$. The thickness of the oxidized film is approximately $0.3\,\mu m$ (Fig. 2.14c). Fourth, after the wafer is cut into separate plate of $4 \times 8\,mm^2$, part of the SiO_2 film on suspension system C is removed by using FIB at an accelerating voltage of $30\,kV$ and ion current of $5\,nA$ (Fig. 2.14d). Because only the SiO_2 film is removed, the processing time for a 3D microstage is approximately 20 minutes for patterning 8 leaf springs. Then, the diced wafer is fixed on an aluminum block and set on supporting wafer at $\theta_P = 60°$. Inclined ICP-RIE is then used to form inclined structures (Fig. 2.14e). Finally, to release the moving parts (e.g., tables, suspensions, mobile combs, etc.) of the device layer (Fig. 2.14f), the wafer is immersed in buffered HF for about 2 hours to remove the SiO_2 layer between the device layer and the substrate. Simultaneously, the oxidized film on the wafer is also removed.

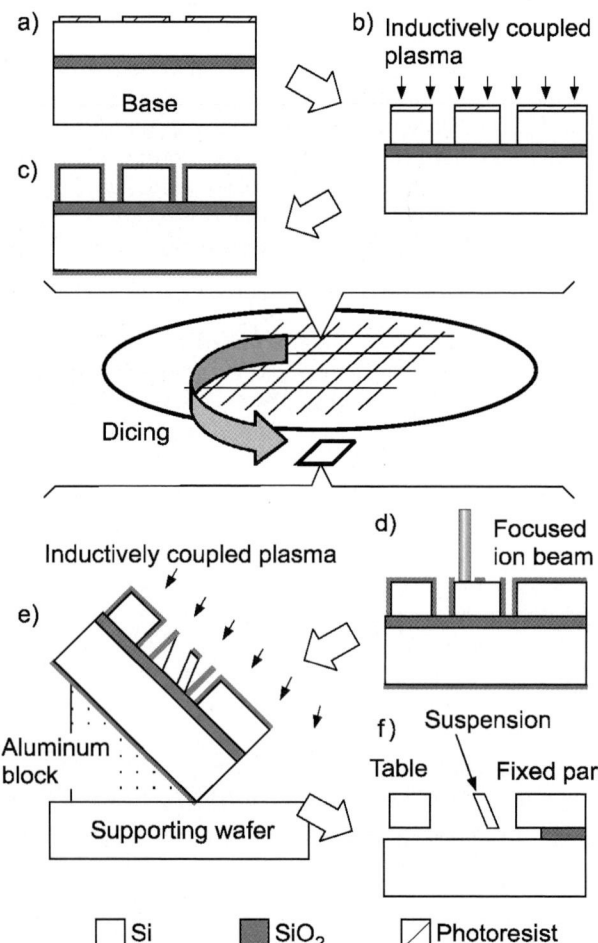

Fig. 2.14. Schematic of processes used to fabricate 3D microstages in a 20-μm-thick device layer on SOI wafers. (**a**) Photoresist patterning. (**b**) ICP-RIE to form vertical structures. (**c**) Oxidization to form SiO_2 film as protective mask. (**d**) FIB to remove SiO_2 film from diced plate. (**e**) ICP-RIE to form inclined leaf springs. (**f**) Wet etching in buffered HF to remove the SiO_2 layer

2.3.4
Evaluation of Motion of 3D Microstage

Figure 2.15a shows an SEM (scanning electron microscopy) image of the entire structure of the 3D microstage (model 3 with support suspension systems). Figure 2.15b,c shows close-up views of an inclined leaf spring in suspension system C, as well as the vertical leaf spring in suspension system A. The 3D microstage is fabricated on an SOI plate measuring 8 mm × 4 mm. The size of table C is 180 μm × 160 μm (Fig. 2.15a). The number of fingers in the comb actuator on support suspension systems D and E is 133 each. The number of fingers on table A is 471. Each suspension consists of a beam and a pair of leaf springs. Suspension system C has a total of 8 inclined leaf springs. Based on the SEM image (Fig. 2.15b), the thicknesses of the vertical and inclined leaf spring are 3.6 μm and 3 μm, respectively, and θ_S of the inclined leaf springs ranges from 68° to 69°.

Figure 2.16a–c shows the measured lateral and vertical displacements of table C as a function of the driving voltage applied to combs A, D, and E, respectively.

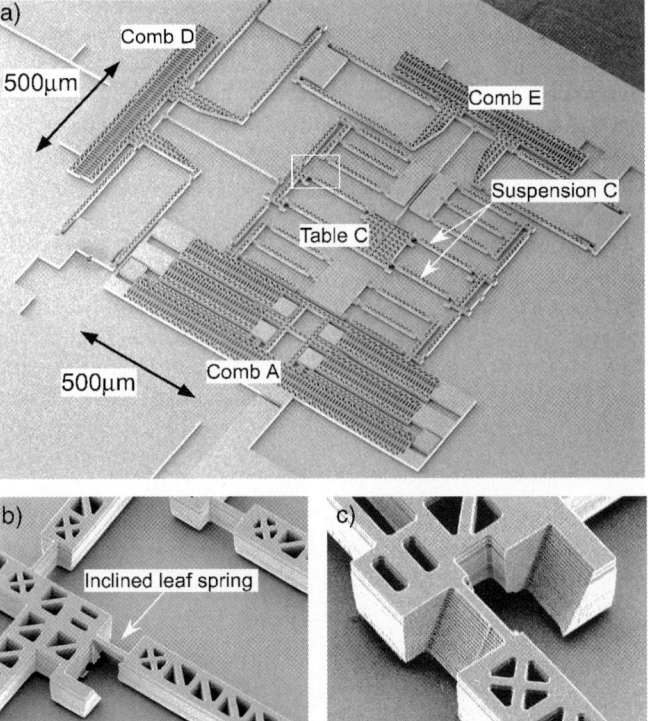

Fig. 2.15. Secondary electron images of fabricated 3D electrostatic stages on SOI wafer observed by using an SEM. (**a**) Entire structure. Thickness of the 3D electrostatic stage was 20 μm. (**b**) Close-up views of suspension systems A and C including inclined and vertical leaf springs, respectively. (**c**) Close-up view of inclined leaf spring in suspension systems C. θ_C of the inclined leaf spring was 69°

Figure 2.16d shows the displacements when the driving voltage was simultaneously applied to combs A and E, where the ratio of driving voltage to comb A to E was 9 to 10 and the driving voltage on comb E is plotted on the abscissa. The measurements were done by using a confocal scanning laser microscope. The maximum lateral displacement is 7.3 μm in the x-direction (Fig. 2.16a) and 6.0 μm in the y-direction (Fig. 2.16b). The maximum vertical displacement (z-direction) is 1.3 μm when a driving voltage of 220 V is applied only to comb E (Fig. 2.9c), which increases to 2.6 μm when a driving voltage of 198 V is added simultaneously to comb A (Fig. 2.16d).

Figure 2.16a–c also shows the lateral displacement perpendicular to the applied force. The change in displacement in the y-direction is less than the measurement error when the driving voltage is applied to comb A or E (Fig. 2.16a,c). When the driving voltage was applied to comb D, the displacement in both the x- and z-directions was negligible and only displacement in the y-direction was observed (Fig. 2.16b). Therefore, there is negligible crosstalk between the x- and y-directions and between the y- and z-directions. Although the lateral displacement was measured when the driving voltages were applied to combs A and E (Fig. 2.16d), the lateral displacement can be minimized by adjusting each driving voltage.

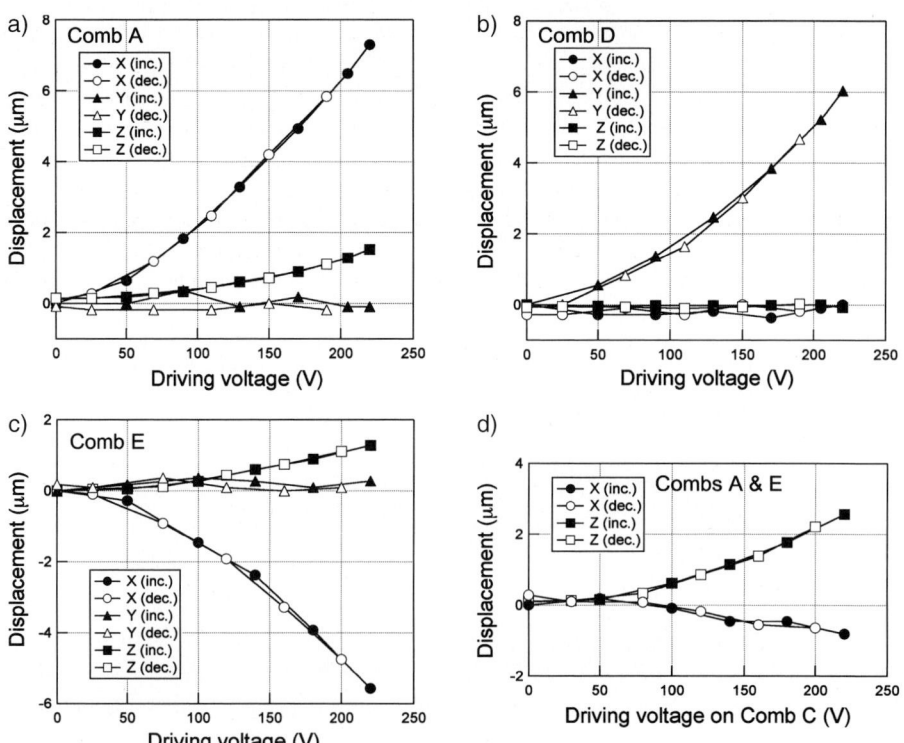

Fig. 2.16. Lateral and vertical displacements of table C as a function of driving voltage applied to (**a**) comb A, (**b**) comb B, (**c**) comb C, and (**d**) both combs A and C

Vertical displacement is generated when the driving voltage is applied to comb A, which was predicted by the FEM analysis and confirmed by the measurement of the fabricated 3D microstage. The crosstalk between the x- and z-directions might affect the measurement when a 3D microstage is used as a scanner. The hysteresis in the displacement, however, is negligible for the vertical and lateral directions and is less than the measurement error when the driving voltage is applied to any comb (Fig. 2.16a–d). Compared to a PZT scanner, the negligible hysteresis of the electrostatic actuators gives a considerable advantage to the 3D microstage for use as a scanner for SPM.

2.4
Applications to AFM

2.4.1
Operation by Using Commercial Controller

The 3D microstage is applied to an AFM and is operated as a scanner. Figure 2.17 shows the overall structure of an electrostatic microscanner, which consists of a 3D microstage, four silicon pads and electrical connection structures between the 3D microstage and the silicon pads, that is fabricated on an SOI plate measuring 8 mm × 4 mm. Three of the four silicon pads are electrically connected to the fixed electrodes of combs A, D and E by a one-to-one scheme as shown in Fig. 2.17. The rest of the pad is for the ground electrode and is electrically connected to the moving parts of the 3D microstage (e.g., tables, suspensions, mobile combs, etc.). The electrostatic microscanner is frangible and loses its function if tweezers or other apparatus touches the electrostatic microscanner. To protect the microstructures against rough handling and install the electrostatic microscanner more easily on an AFM, an adapter is combined with the electrostatic microscanner as shown in Fig. 2.18. The adapter consists of a ceramic plate, four stainless

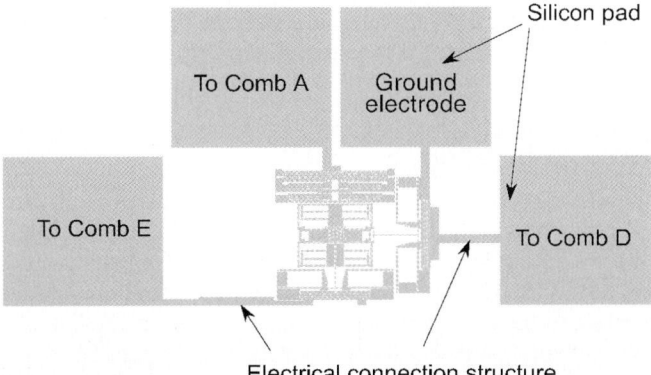

Fig. 2.17. Entire microstructure of electrostatic microscanner, consisting of 3D microstage, four silicon pads and electrical connection structures, as fabricated on an SOI plate measuring 8 × 4 mm^2

Fig. 2.18. Adapter or electrostatic microscanner. Stainless
steel sheets on the ceramic
plate make contact with silicon
pads of the electrostatic microscanner. Both adapter and
microscanner can be heated by
a resistive heater

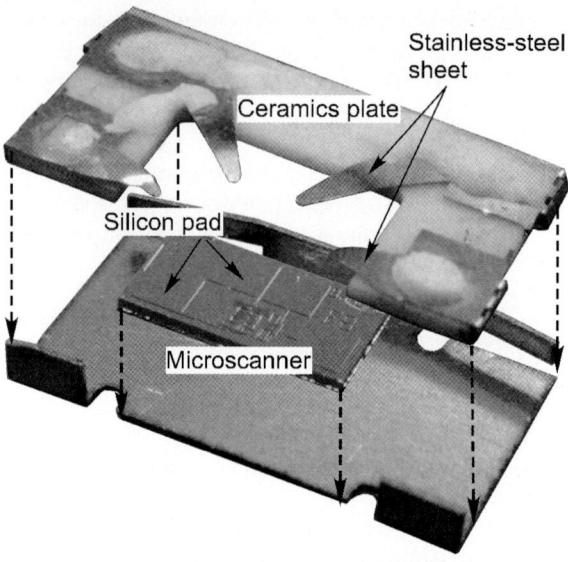

steel sheets and a metal base plate. The ceramic plate measures 16 mm × 8 mm
and has an opening of 8 mm × 4 mm. The four stainless steel sheets are bonded
to the ceramic plate. The tip of each stainless steel sheet makes contact with a silicon pad of the electrostatic microscanner. The ceramic plate with the stainless
steel sheets and the electrostatic microscanner are elastically held by a metal base
plate.

When using a commercial AFM to control the electrostatic microscanner, the
major differences between control voltages for a PZT scanner and an electrostatic
microscanner are as follows:

1. Driving voltages V_X, V_Y, and V_Z are individually applied to three sets of electrodes in a PZT scanner to generate motions along the x-, y- and z-directions,
   respectively. In an electrostatic microscanner, the driving voltages are simultaneously applied to two electrodes (comb A and C) to obtain displacement in the
   z-direction.

2. A PZT tube scanner requires bipolar voltage, whereas an electrostatic microscanner requires unipolar voltage.

To convert the driving voltage on PZT tube scanner, a calculating circuit is used.
Figure 2.19 shows a block diagram for converting the driving voltage. The driving
voltages to the PZT scanner are customized and then applied to the electrostatic
microscanner. In the calculating circuit, the driving voltage applied to comb A is
adjusted according to the driving voltage applied to comb E, and an offset is added
to each driving voltage. Table 2.3 shows the relationship between the control voltages V_X, V_Y, and V_Z supplied from the AFM controller and the driving voltages
applied to the electrostatic microscanner. The offset voltage V_{OFFSET} is set to obtain
maximum scanning area and factor α is adjusted to achieve pure vertical motion of
table C.

Fig. 2.19. Block diagram showing how the driving voltage on PZT was converted to 3D microstage

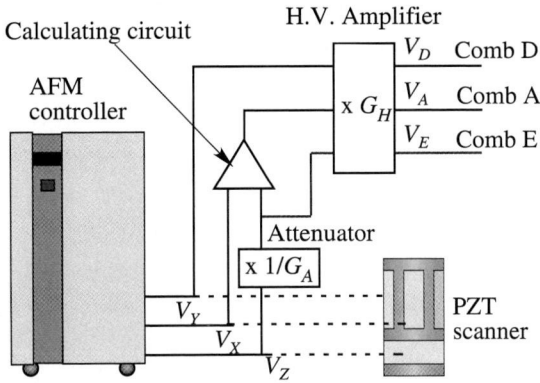

Table 2.3. Relationships between PZT scanner control voltages and comb actuator driving voltages ($V_O = 110$–120 V, $\alpha = 0.5$–0.9)

AFM controller		Microscanner		
Control voltage	Electrode of PZT	Driving voltage	Comb	Limits (V)
V_X	X	$V_O + G_H V_X + \alpha(G_H/G_A)V_Z$	A	
V_Y	Y	$V_O + G_H V_Y$	D	0–230
V_Z	Z	$V_O + (G_H/G_A)V_Z$	E	

2.4.2
Evaluation of Microscanner Using Grating Image

The electrostatic microscanner with an adapter was set on the AFM sample holder for evaluation. Figure 2.20 shows the appearance of the microscanner installed on a commercial AFM (SII NanoTechnology Inc., SPA-300HV), where the connection of the driving voltage is arranged as shown in Fig. 2.17 (no voltage is applied to the PZT scanner). To locate the cantilever probe above table C, an optical microscope was used. Then the probe was moved to table C using the approach mechanism prepared for the AFM in the same way as when using a PZT scanner. Raster scanning and topography measurements using the microscanner are also quite similar to the use of a PZT scanner.

Figure 2.21 shows AFM images of the part of table C that was measured by raster scanning using an electrostatic microscanner. Before placing the microscanner into the AFM system, part of table C in the 3D microstage was milled using FIB and a grating was formed. In this grating the distance between the adjacent grooves was 330 nm or 250 nm. The scanning areas from the grating for Fig. 2.21a–c were $7.6 \times 5.3 \, \mu m^2$, $2.9 \times 2.7 \, \mu m^2$ and $6.7 \times 5.7 \, \mu m^2$, respectively. For each AFM image, a 1D tilt correction was applied. A linear correction was not applied.

The scanning area shown in Fig. 2.21a,c corresponds relatively well to the maximum displacement range of the 3D microstage The grating is wider in the y-direction at the lower part of the figure because the comb actuators show a smaller displacement at a lower driving voltage (Fig. 2.16b). Based on the displacement measurement (by using a confocal laser microscope) when a driving voltage is applied to comb A

Fig. 2.20. External appearance
of electrostatic microscanner
set in a commercial AFM.
Driving voltage is applied to the
electrostatic microscanner via
an adapter

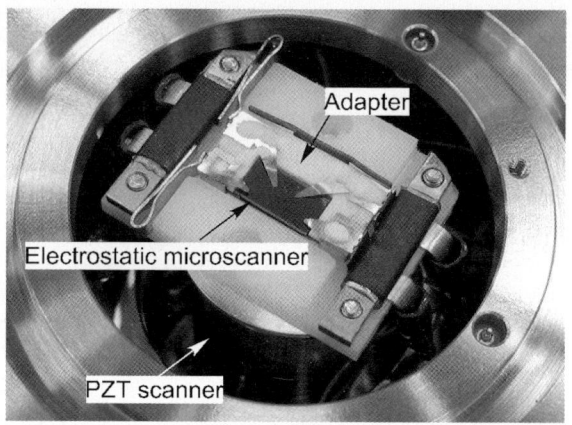

(Fig. 2.16a), table C moves upward with increasing driving voltage and with increasing displacement in the x-direction. Although interference between displacements in the x- and z-directions was expected to affect the measurement using AFM, such interference can be almost canceled by incorporating a 1D tilt correction. Figure 2.21b shows a close-up view of the grating measured by scanning a narrower area. The spacing between grooves in the x-direction is relatively uniform. For the spacing in the y-direction, the maximum spacing is 25% wider than minimum spacing.

For the 250-nm pitch grating, the AFM images were obtained at each forward and reverse scan for the main scanning direction of x and y (with a triangular voltage applied on comb A and D, respectively). The position of each peak surrounded by grooves was then read along each direction. The actual displacement was then calculated assuming that the distance between adjacent peaks was 250 nm. The driving voltage at each peak position was obtained from the waveform of the driving voltage recorded in a storage scope. Figure 2.22 shows the relationship of the actual displacement (D) as a function of the driving voltage (V_D). The relationships between D and V_D in the x- and y-directions are shown in Fig. 2.22a,b. In each figure, the triangles pointing up and down represent the displacement obtained while the driving voltage increased and decreased, respectively. The measured points are approximated by a line represented by:

$$D = D_0 + CV_D^\beta ,\tag{2.10}$$

where D_0, C and β are constant. In Fig. 2.22, β is shown in the inset box. When Fig. 2.22a,b are compared, the displacement in the x-direction is more linear than in the y-direction. The difference in linearity between the x- and y-directions is also found in the AFM images (Fig. 2.21).

Figure 2.23 shows the changes of the driving voltages on comb A and E in the 3D microstage during scanning in the x-direction. Signals in two reciprocating phases are shown in this figure. The V_E on comb E changes at the opposite phase against V_A on comb A. Because table C moves in the $x(+)$-direction with decreasing V_E on comb E, the displacement of table C is amplified in the $x(+)$-direction when V_A on comb A increases. This phase inversion is caused by crosstalk between x and z as shown in Fig. 2.16a. When F_A increases table C is lifted up by the work of support

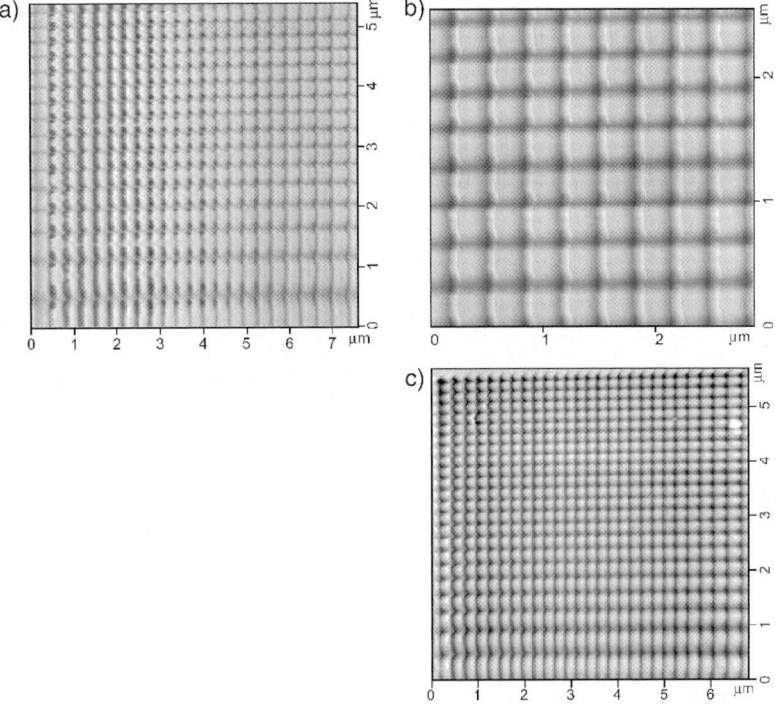

Fig. 2.21. AFM images of 330-nm and 250-nm gratings formed on table C (sample table) using FIB. Measurements were taken using the 3D microstage as the AFM scanner. Scanning areas were $7.6 \times 5.3\ \mu m^2$ (**a**), $2.9 \times 2.7\ \mu m^2$ (**b**) and $6.7 \times 5.7\ \mu m^2$ (**c**) from the grating

suspension system E. But the height of table C is kept constant by the feedback of AFM. Then the driving V_E on comb E decreases.

Figure 2.24 shows a comparison of the display displacement between the electrostatic microscanner and PZT scanner. In this figure, the display displacement is plotted against the actual displacement for each scanner. The display displacement was obtained from the AFM image of a 250-nm pitch grating assuming a linear scale for the entire displayed image. The position on the assumed linear scale was read for each peak surrounded by grooves. At the same time, the actual displacement was obtained only based on the distance between the adjacent peaks of 250 nm, i.e., irrespective of the assumed linear scale. The main scanning direction for the electrostatic microscanner was y (with a triangular voltage applied on comb D). The maximum scanning range of PZT scanner was $20\ \mu m$. The scanning frequency was the same for each scanner and was 0.2 Hz. The up-pointing triangles represent the displacement obtained while the driving voltage was increased. The down-pointing triangles represent the displacement obtained while decreasing the driving voltage. For the electrostatic microscanner the slope of the displayed displacement is higher at the lower stage displacement (the lower driving voltage), but no hysteresis is found between increasing and decreasing process of driving voltage. A maximum hysteresis of about 200 nm is found in the displacement for the PZT scanner.

Fig. 2.22. Relationship between table C (sample table) displacement and comb actuator driving voltage, as calculated from the grating image. Displacements in x- (**a**) and y-directions (**b**) were measured when the driving voltage was applied to comb A (**a**) and comb D (**b**). Displacement in the x-direction is more linear than in the y-direction

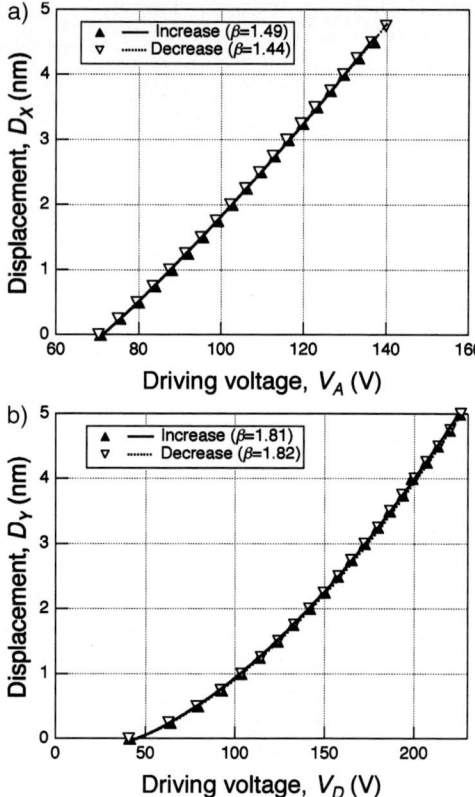

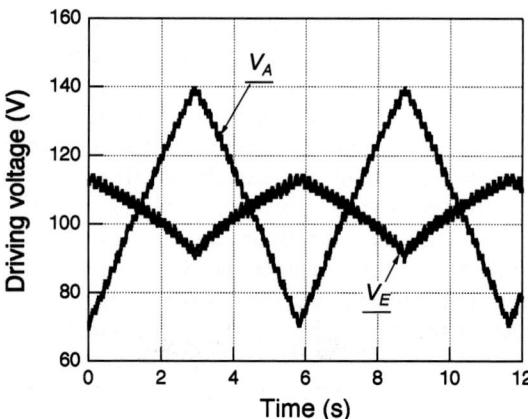

Fig. 2.23. Changes in driving voltages on comb A and E in the 3D microstage during scanning in the x-direction. Signals in two reciprocating phases are shown. V_E on comb E changes at the phase opposite to V_A on comb A

Fig. 2.24. Displayed displacement derived from AFM images as a function of stage displacement. Adjacent plot intervals are 250 nm along the abscissa and correspond to the grating pitch

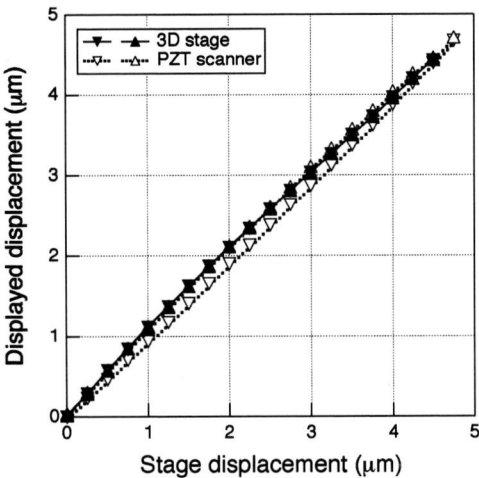

2.4.3
SPM Operation Using Microscanner

In the electrostatic microscanner, the air gap between the comb actuator and the clearance between the microstructures and the substrate is a couple of microns. If a speck of dust were trapped in this gap, short-circuiting would occur. Thus the electrostatic microscanner must be stored and operated in a clean environment. The electrostatic microscanner can be operated in air or in vacuum and can be possibly operated in an insulating liquid, but it cannot be operated in a conductive liquid such as water containing ions. The maximum operating voltage seemed to be limited by the dielectric breakdown strength of the SiO_2 layer between the microstructures and the substrate. When the thickness of the SiO_2 layer is 3 μm, the microscanner does not degrade at driving voltages under 300 V. It is possible to connect an output from the high voltage (HV) amplifier of an SPM controller to the electrostatic microscanner because the maximum driving voltage supplied by an SPM controller is usually around 200 V. Even in this case, however, short-circuit protection should be utilized.

A major issue when using the electrostatic microscanner is setting a sample on the sample table (table C). A suitable sample for the electrostatic microscanner is chemical vapor deposition (CVD) film or physical vapor deposition (PVD) film. Figure 2.25 shows an LFM image of gold film deposited on the sample table by sputtering. When sputtering a thick gold film on the entire structure of the microscanner, short circuits tend to occur between the microscanner and the substrate. If a thick film is required, parts other than the sample table should be covered with a mask. It is also possible to form a film on the stage by chemical adsorption in solution. For example, this approach can be used to form a self-assembled monolayer (SAM) on the sample table (or other parts of the electrostatic microscanner), although stiction should be avoided while drying the solution. If the sample object is smaller than the sample table, the object can be placed on the sample table using, for example, a micromanipulator. In this case, the method of fixing the object is a point to consider. If any problems in the fixing method can be overcome, the weight of the object is not

Fig. 2.25. LFM image of gold film
sputtered on the sample table (table C)
at a thickness of about 15 nm

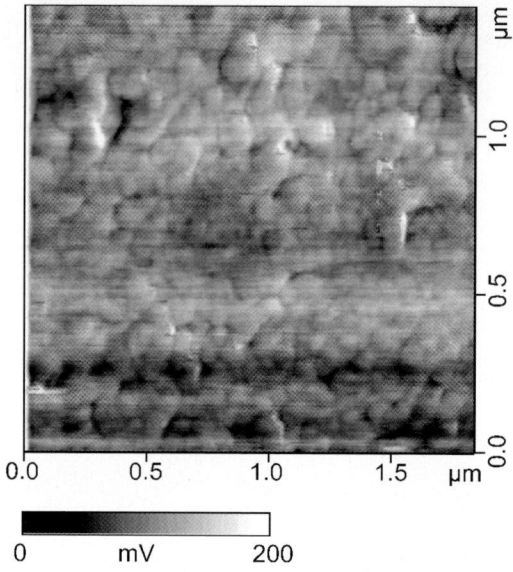

a critical point because the displacement caused by the object's gravity is constant
and is less than one nanometer even for a metal object of $100 \times 100 \times 100 \, \mu m^3$.

When applying the microscanner to a scanning device for AFM, contact mode
operation is likely to be more suitable than tapping mode. Tapping mode operation
was attempted using a 20-μm-thick 3D microstage, however it seemed to fail to detect
a reduction in the amplitude of the cantilever while the probe approached the sample
table. This is likely to have occurred because, when the probe touched the sample
table, the sample table vibrated with the cantilever and the amplitude of the cantilever
vibration did not clearly decrease. If the thickness of the 3D microstage were two
to four times thicker, tapping mode operation might be possible with the increased
mass of the sample table and/or an increase in the stiffness of the suspension systems
in the vertical direction. On the other hand, there is probably no crucial problem
using a microscanner for other operation modes such as STM, scanning near field
optical microscopy (SNOM) and magnetic force microscopy (MFM).

For a contact mode AFM, basic operations such as topography measurements
(Fig. 2.21), lateral force measurements (Fig. 2.25) and force-distance curve measurements (Fig. 2.26a) are possible. The friction force between the sample table
and the probe could be measured at a relatively high load of 400 nN as shown in
Fig. 2.27. In this figure, the friction force linearly increased with the applied load.
However, the AFM's operation failed at applied loads over 400 nN. This limitation
of the applied load is determined by the stiffness of the 3D microstage. Therefore,
the maximum operating load could be increased by fabricating the 3D microstage
in a thicker device layer on an SOI plate. In the force-distance curve measurements,
no hysteresis was found (Fig. 2.26a), which has a benefit for pull-off force measurements. Determining the pull-off force from the force-distance curve usually
requires evaluating the displacement of the cantilever from the output of the photosensor. A problem in utilizing the output of the photosensor is that it is impossible

Fig. 2.26. Comparison of force-distance curves measured by (**a**) electrostatic microscanner, and (**b**) PZT scanner

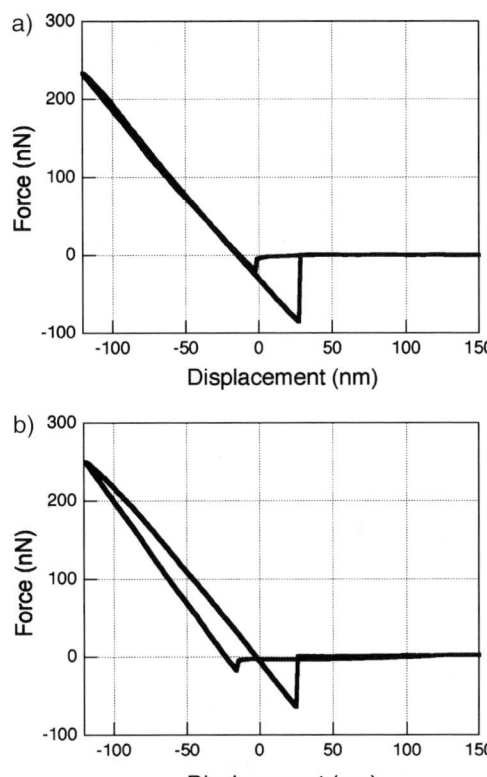

to know the displacement if the pull-off force is too high to evaluate the displacement due to the saturated output. Moreover, the output of the photosensor should be pre-calibrated. When using the electrostatic microscanner, the pull-off force can be evaluated from the vertical displacement of the sample table (i.e., the voltage applied to comb E). A high pull-off force can be measured and no calibration is required when the cantilever is changed. This can be accomplished because, while the electrostatic microscanner has no hysteresis affects, hysteresis errors are not negligible when using a PZT scanner (Fig. 2.26b).

The small size expands the applications of the electrostatic microscanner. Figure 2.28 shows AFM units specially designed for the electrostatic microscanner. In the unit shown in Fig. 2.28a, a commercially available self-detecting cantilever is incorporated with the microscanner, which measures about 80 mm × 40 mm × 30 mm. This unit can be placed on a sample stage on a conventional microscope or in a low-vacuum chamber after approaching the cantilever probe to the sample table. In Fig. 2.28b, the electrostatic microscanner is installed in a high-vacuum chamber as small as a coffee cup. Because of the small capacity of the chamber, the lower pressure can be easily obtained using a small pump. In this unit, a ceramic heater is set underneath the adapter containing the electrostatic microscanner. The heater heats the entire adapter as well as the electrostatic microscanner. Unlike the use of a PZT scanner, heat-insulating and heat-radiating designs are very simple.

Fig. 2.27. Relationship between applied load and friction force measured by electrostatic microscanner. Friction force is the average of five reciprocating motions in each raster scan. Scanning range was approximately 3.6 μm in primary scanning direction and scanning frequency was 1 Hz

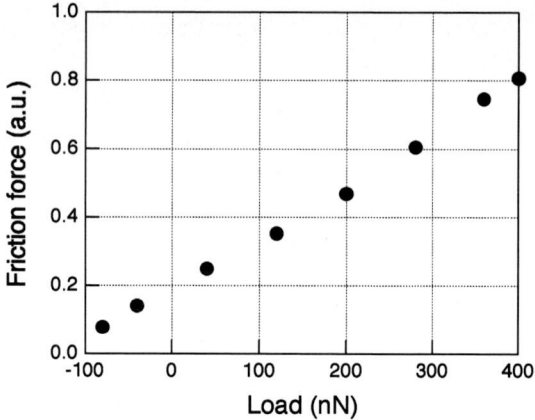

Fig. 2.28. External appearances of AFM units specially designed for the electrostatic microscanner. (**a**) An electrostatic microscanner combined with a self-detecting cantilever. (**b**) An electrostatic microscanner installed in a high-vacuum chamber

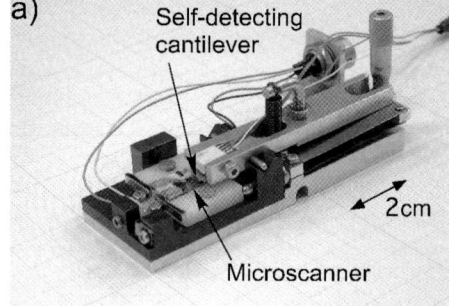

In summary, the electrostatic microscanner has the following advantages (1–3) and disadvantages (4,5):

1. **No hysteresis in motion** enables precise positioning and force measurement.
2. **Small size** enables various designs of the SPM apparatus (e.g., for high-vacuum, high-temperature applications).
3. **High resonance frequency** enables high-speed scanning.
4. **Small and fragile structure** restricts the size and fixing method of the samples.
5. **Low stiffness** is unsuitable for tapping mode operation of the AFM.

Another major feature is the flexible design of the electrostatic microscanner itself. By tuning the structural design, including the thickness of the microstructures, it is possible to suppress defects and enhance the other advantages. Electrostatic microscanner technology will see wide use in SPMs as well as in scientific applications.

References

1. Binning G, Rohrer H, Gerber C, Weibel E (1982) Phys Rev Lett 49:57
2. Uchihashi T, Sugawara Y, Tsukamoto T, Ohta M, Morita S, Suzuki M (1997) Phys Rev B 56:9834
3. Meyer G, Amer NM (1990) Appl Phys Lett 57:2089
4. Ando Y, Ino J (1998) Wear 216:115
5. Binnig G, Smith DPE (1986) Rev Sci Instrum 57:1688
6. Croft D, Shed G, Devasia S (2001) J Dyn Syst Meas Control Trans ASME 123:35
7. Akiyama Y (1986) Ultrasonic Motors/Actuators. Torikeppusu, Tokyo, p 19
8. Grétillat MA, Thiébaud P, de Rooij NF, Linder C (1994) IEEE Micro Electro Mechanical Systems. Oiso, p 97
9. Kim CJ, Pisano AP, Muller RS, Lim MG (1992) Sensor Actuat A Phys 33:221
10. Diem B, Rey P, Renard S, Bosson SV, Bono H, Michel F, Delaye MT, Delapierre G (1995) Sensor Actuat A Phys 46:8
11. Yee Y, Nam HJ, Lee SH, Bu JU, Lee JW (2001) Sens Actuator A Phys 89:166
12. Hirano T, Furuhata T, Gabriel KJ, Fujita H (1991) 7th Int Conf Solid-State Sensors and Actuators (Transducers '91). San Francisco, p 873
13. Lee D, Krishnamoorthy U, Yu K, Solgaard O (2003) 12th Int Conf Solid-State Sensors and Actuators (Transducers '03). Boston, p 873
14. Hah D, Huang STY, Tsai JC, Toshiyoshi H, Wu MC (2004) J Microelectromech S 13:279
15. Ando Y, Ikehara T, Matsumoto S (2002) Sens Actuator A Phys 97–98:579
16. Ando Y (2004) Sens Actuator A Phys 114:285
17. Ikehara T, Kato C, Suzuki Y, Fukuhara S, Watanabe T (1995) 13th Sensor Symp. Tokyo, p 149
18. Jansen H, Boer MD, Elwenspoek M (1996) 9th Int Workshop Micro Electro Mechanical Systems. San Diego, p 250

3 Low-Noise Methods for Optical Measurements of Cantilever Deflections

Tilman E. Schäffer

3.1
Introduction

The measurement of cantilever deflections is essential to all cantilever-based scanning probe and sensor methods. Several detection techniques have been devised for this purpose, including tunneling detection [1], laser-diode feedback detection [2], capacitive detection [3], piezo-resistive detection [4, 5], and optical interferometric detection [6–8]. Two years after the invention of the atomic force microscope (AFM) [1], a detector based on the deflection of an optical beam was first described [9, 10]. The optical beam deflection ("optical lever") method has since become the preferred detection method owing to its simplicity and versatility. It has a sensitivity and a signal-to-noise ratio comparable to those of optical interferometry [11, 12], and almost all of the current commercially available AFMs are based on it.

The optical beam deflection method spatially separates the detector from the cantilever–sample environment. The cantilever can therefore easily be submerged in transparent liquids [13]. Various cantilever sizes, shapes, and materials with a wide variety of different properties can be chosen to satisfy particular experimental requirements [14–23].

There is an ever-increasing need for high-resolution detection of cantilever deflections, for example with respect to the measurement of forces [24–26]. Force measurements are used, for example, for the characterization of sample stiffness [27], surface charges [28], and solvation forces [29]. The field of single-molecule force spectroscopy [30] evolved after it became apparent that the AFM can measure the interaction forces between individual ligand–receptor pairs [31–33] and that it can mechanically unfold biological macromolecules [34–36]. Such single-molecule force spectroscopy measurements are currently limited to a resolution of approximately 5–10 pN, enough to resolve intermolecular and intramolecular unfolding events, but not enough to resolve refolding events that occur at even smaller forces (less than 2 pN) [30].

One of the fundamentally limiting noise sources in cantilever-based force measurements is intrinsic thermal vibration noise of the cantilever [37]. Smaller cantilevers reduce this noise source so that smaller forces can be measured [17, 38]. With smaller forces (or with a larger spring constant of the cantilever), however, the deflections of the cantilever become smaller and detection noise becomes more significant. It is therefore important to have a good, low-noise detection system.

A low-noise detection system is also required in cases where thermal noise is not a limiting noise source but the actual subject of study [39–43].

This chapter discusses several methods for low-noise optical measurements of cantilever deflections. Section 3.2 outlines the physical principles of the optical beam deflection method. Section 3.3 discusses optical detection noise, particularly shot noise. Sections 3.4 and 3.5 show how a novel detector that is based on an array of photodiode segments can detect with an increased signal-to-noise ratio, an increased dynamic range and an increased linearity. Section 3.6 discusses how to optimize the sensitivity for the detection of higher-order normal cantilever vibration modes. Section 3.7 presents a calculation of the thermal vibration noise of a cantilever when measured with the optical beam deflection method. Finally, Sect. 3.8 shows how to use thermal noise to calibrate the spring constant of the cantilever.

3.2
The Optical Beam Deflection Method

The optical beam deflection method is based on an incident optical beam that is focused on and reflected from the cantilever. In the geometrical optics approximation, the incident beam is focused to a point on the cantilever (Fig. 3.1). The cantilever slope at this point determines the angle with which the beam is reflected. This reflection angle is detected with a position-sensitive photodetector. A two-segment (split) photodiode is usually used, generating a differential signal [9, 10]. The spatial separation between cantilever and detector allows for measurements in different media, including biological buffer solutions [13].

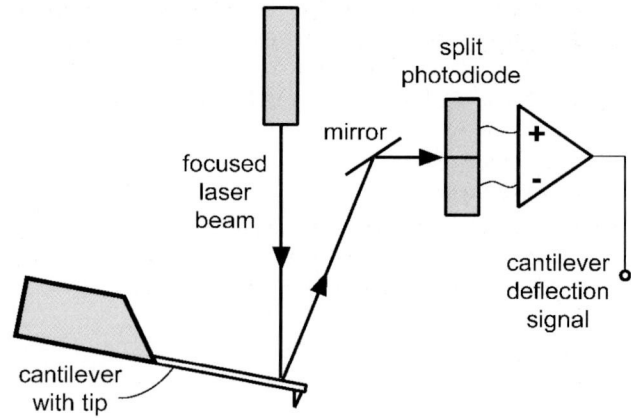

Fig. 3.1. Optical beam deflection ("optical lever") detection in an atomic force microscope, in the geometrical optics approximation. In this approximation, a laser beam is focused to a point on the cantilever. The reflected beam is projected onto a two-segment detector (split photodiode) that produces the cantilever deflection signal

3.2.1
Gaussian Optics

Typically, a solid-state laser diode with a collimating lens is used to provide the incident beam. The profile of this beam is approximately Gaussian in shape. Owing to the wave nature of light, this beam cannot be focused to a point, but only to

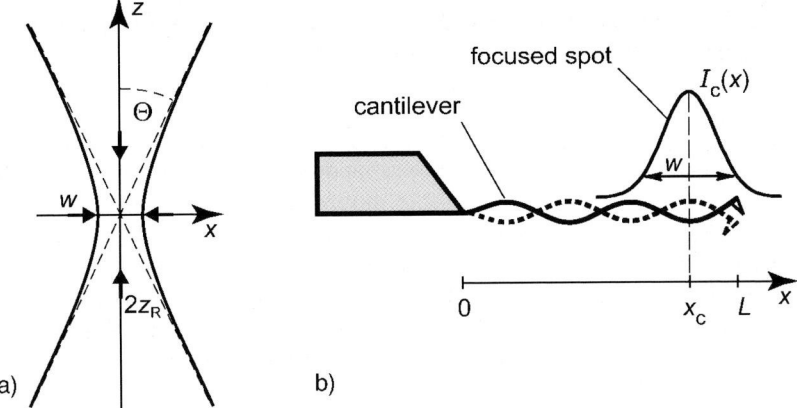

Fig. 3.2. (a) Profile of a focused Gaussian beam (cross section along the beam axis). The beam does not converge to a point in the focal plane ($z = 0$) but remains finite in size. The focused spot diameter, w, is defined as the beam diameter between the $1/e^2$-irradiance points (marked by the *two opposing horizontal arrows*). The Rayleigh range, $2z_R$, is the length in the z-direction over which the beam diameter remains below $\sqrt{2}w$ (marked by the *two opposing vertical arrows*). The larger the half-opening angle of the incident beam, Θ, the smaller the focused spot diameter and the smaller the Rayleigh range. **(b)** The focused optical spot on a vibrating cantilever. The focused spot has a Gaussian irradiance profile, $I_c(x)$, with a center position x_c and a diameter w between its $1/e^2$-irradiance points. The signal generated by the optical beam deflection method depends on the diameter and the position of the focused spot and on the functional shape of the cantilever below the spot

a spot of finite size (Fig. 3.2a) [44, 45]. In an ideal optical system, the focused spot is also Gaussian in shape. The relationship between the diameter of the focused spot (defined at the $1/e^2$-irradiance points), w, and the diameter of the collimated incident beam (also defined at the $1/e^2$-irradiance points), d, is given by

$$w = \frac{4\lambda f}{\pi n d}, \tag{3.1}$$

where λ is the wavelength (in vacuum) of the incident beam, f is the effective focal length of the focusing lens, and n is the index of refraction of the medium. The half-opening angle of the focused beam, Θ, is

$$\Theta = \arctan \frac{d}{2f}. \tag{3.2}$$

Another important property is the "depth of focus", which is a measure of the range in propagation direction of the focused beam in which the beam diameter does not spread significantly. The Rayleigh range, $2z_R$, serves as a measure for the depth of focus and is defined as the propagation length in which the beam diameter remains within $\sqrt{2}$ of its minimum value:

$$2z_R = \frac{\pi n w^2}{2\lambda}. \tag{3.3}$$

The Rayleigh range defines the size of the focused spot in the axial direction and depends quadratically on w. A small focused spot diameter therefore results in a small Rayleigh range as well. For example, a spot diameter of $w = 2\,\mu m$ results in a Rayleigh range of $2z_R \approx 12\,\mu m$ in aqueous solution (for $\lambda = 670\,nm$). It should be mentioned that when using a laser diode as the light source, the incident beam is not circular, but elliptical in shape (unless additional optical elements are used to circularize the beam). The consequence of this is that the focused spot is elliptical in shape too. This does, however, not usually cause a problem, as long as the long axis of the focused spot is aligned along the cantilever length. It is this spot diameter along the cantilever length that is denoted with "focused spot diameter" in the following.

3.2.2
Detection Sensitivity

Well within the Rayleigh range of the incident Gaussian beam ($|z| \ll z_R$), the one-dimensional irradiance distribution along x can be written as

$$I_c(x) = \sqrt{\frac{8}{\pi}}\frac{P_0}{w}\,e^{-2[2(x-x_c)/w]^2} \,, \tag{3.4}$$

where x_c is the position of the center of the focused spot on the x-axis, and P_0 is the total power of the incident beam (Fig. 3.2b). In (3.4), the positional variable in the direction along the cantilever width was integrated out. When using a split photodiode as a detector, the detection signal is the difference in incident power on each segment: $P_B - P_A$. The optical detection sensitivity for such a detector can be defined as the detection signal per unit tip deflection in direction of the z-scanner. For small deflections ($z \ll \lambda/(2\pi)$), it can be calculated using scalar diffraction theory as [46,47]

$$\sigma = \frac{4}{\lambda}\int\limits_0^L dx \int\limits_0^L dx'\,\sqrt{I_c(x)\,I_c(x')}\,\frac{h(x/L) - h(x'/L)}{x - x'} \,. \tag{3.5}$$

$h(q)$ is the functional shape of the cantilever, normalized so that at the cantilever base, $h(0) = 0$, and at the cantilever tip, $h(1) = 1$. Equation (3.5) takes account of the fact that the bent cantilever acts as a curved mirror for the finite-sized focused spot. A more general case that considers an arbitrary angle between the cantilever and the z-scanner, and an arbitrary angle between the cantilever and the axis of the incident beam was derived in [48]. In typical experiments, a large fraction of the incident light is collected at the detector, and optimizing the sensitivity usually results in an approximately optimized signal-to-noise ratio.

One common cantilever shape arises in force spectroscopy experiments, where a quasi-static force acts on the tip of the cantilever. In the case of a homogenous, rectangular cantilever, the resulting cantilever shape becomes [49]

$$h(q) = \frac{q^2(3 - q)}{2} \,. \tag{3.6}$$

Another set of common cantilever shapes arises in modes of operation where the cantilever is vibrated at its fundamental resonant frequency, for example, in noncontact [50–53] or tapping mode [54–57] operations. These shapes will be discussed in Sect. 3.6.

One result from (3.5) is that the size and shape of the focused spot must be tailored to the size and shape of the particular cantilever for optimum signal-to-noise ratio. An adjustable aperture in the incident beam path has been used to achieve this, thereby increasing the signal-to-noise ratio of a particular measurement by a factor of 3 [46].

3.3
Optical Detection Noise

3.3.1
Noise Sources

For the optical beam deflection method, the significant noise sources are connected with the generation and the detection of the optical beam: (1) time-correlated fluctuations/drift in the power of the beam ("intensity noise"), (2) time-correlated fluctuations/drift in the shape/direction of the beam ("pointing noise"), and (3) time-uncorrelated fluctuations in the photodetection process (shot noise). Other noise sources (e.g., dark current noise, electronic Johnson noise, backaction noise) are usually not limiting, but electronic $1/f$ noise can be a factor in low frequency measurements (such as contact mode or force spectroscopy). Intensity and pointing noise are caused by instabilities in the laser cavity, resulting in mode hopping and mode competition. Pointing noise can be transformed into intensity noise by the use of a single-mode optical fiber [40], and intensity noise can be minimized in the incident beam by a feedback system ("noise eater") or in the detection circuitry with the help of a fast divider. Shot noise arises from the fact that photons are emitted at random points in time (resulting in Poisson statistics). The current produced by the electrons in the photodetection process, where photons cause the release of electrons, therefore exhibits random statistical fluctuations (with a "white" spectral distribution). These fluctuations increase in amplitude with the square root of the beam power and are a fundamental limit to the detection sensitivity. They also impose a fundamental limit on the scan speed, since the faster the measurement, the larger the required bandwidth, and thus the higher the noise level in the measurement.

3.3.2
Shot Noise

Shot noise is quantum noise in the optical field. It is caused by the fact that optical energy is carried by discrete carriers (photons). The optical power in an ideal laser beam (coherent state) is shot-noise-limited. (We note that in the case of amplitude-squeezed light, noise levels below the shot noise level have been observed [58].) Assuming that the probability for the absorption of one photon at the detector in a certain time interval is constant and uncorrelated with previous counting events, we can use Poisson statistics to describe the fluctuations in the average detected optical power. The probability to detect m photons in a certain time interval is

$$p_m = e^{-n} \frac{n^m}{m!} \, , \tag{3.7}$$

where n is the expectation value for the total number of detected photons in that time interval. Using the property of Poisson distributions that the variance equals the expectation value, the relative shot noise (root-mean-square, rms, power fluctuations as a fraction of the total power at the detector) can be expressed as [47]

$$N_r = \sqrt{\frac{2hc\Delta f}{\lambda P_{det}}} = \sqrt{\frac{1}{n}} \, . \tag{3.8}$$

Here, h is Planck's constant, c the speed of light, λ the wavelength of the incident beam, P_{det} the optical power arriving at the detector, and Δf the detection bandwidth. A perfect detection responsivity is assumed. The relative shot noise equals the inverse square root of n. n can also be regarded as the mean number of photons counted within the measurement bandwidth. We note that, for typical measurement frequencies, shot noise is independent of frequency ("white noise").

3.4
The Array Detector

The conventional split photodiode (Fig. 3.1) works well for reflected optical beams that are of approximately Gaussian shape. Frequently, however, reflected beams are of less-than-ideal shape [59], especially those that arise in AFMs for operation in liquid and in AFMs for small cantilevers. Multiple optical interfaces and prototype cantilevers often cause the reflected beam to look scattered and spotted. In this case, centering the beam on the two-segment detector does not necessarily position the part of the beam with the highest intensity between the segments, a condition that is required for high detection sensitivity [11, 12, 46]. And clearly, if there are multiple intensity maxima, they cannot be centered all at the same time.

The detection sensitivity for almost any shape of the reflected beam can further be increased by the use of an array detector, in conjunction with a simple algorithm [60, 61]. Such an array detector consists of an array of N photodetector segments over which the beam is distributed (Fig. 3.3). The optical power P_i is incident on segment number i (Fig. 3.4, panel a). Upon a change in cantilever deflection, this optical power changes by a small amount ΔP_i, thereby constituting the signal for the respective segment (Fig. 3.4, panel b). This signal is first weighted by amplifying it with an individual gain factor, g_i (Fig. 3.3). Then all the weighted signals from all the segments are added, producing the cantilever deflection signal that is fed to the feedback or data acquisition system:

$$S = \sum_{i=1}^{N} g_i \Delta P_i \, . \tag{3.9}$$

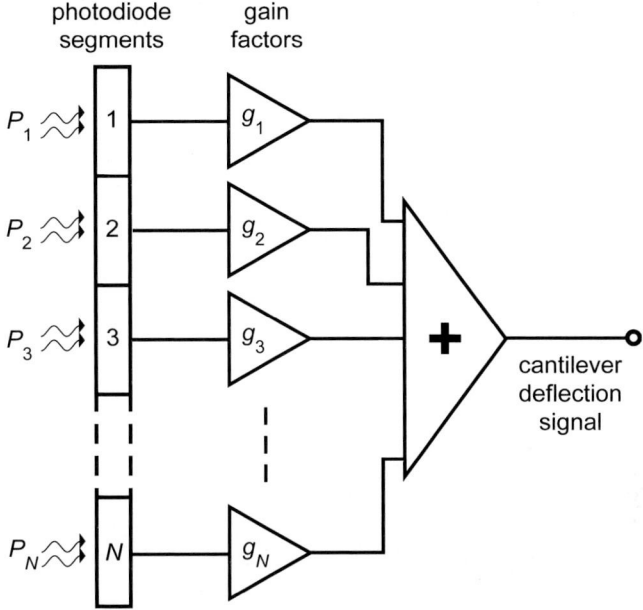

Fig. 3.3. The array detector. The reflected beam is spread across an array of N photodetector segments $(1, 2, \ldots, N)$. The signal from each segment is amplified by an individual gain factor $(g_1, g_2, \ldots, g_N)$ that can be set dynamically, weighting the signal from each segment. The weighted signals are then added to form the cantilever deflection signal. Such an array detector can be programmed to dynamically adapt to the experimental conditions in order to maximize the detection sensitivity

If we assume photonic shot noise as the fundamentally limiting noise source, the rms noise in the signal becomes [60]

$$N = \left(\frac{2hc}{\lambda} \Delta f \sum_{i=1}^{N} g_i^2 P_i \right)^{1/2} . \tag{3.10}$$

The remarkable property of such a detector is that a set of gain factors $(g_1, g_2, \ldots, g_N)$ can be found dynamically that optimizes the detection sensitivity for any particular experimental condition. Maximization of the signal-to-noise ratio is performed by simultaneously setting all partial derivatives of S/N to zero, thereby arriving at the maximizing conditions for the gain factors [60]:

$$g_i = \beta \frac{\Delta P_i}{P_i} , \tag{3.11}$$

where β is an arbitrary positive scalar that can be chosen so that $-1 \le g_i \le 1$ for all i (Fig. 3.4, panel c). The powerful aspect of this derivation is that no assumptions about the particular arrangement and shape of the detector segments are made. In particular, (3.11) works for both one- and two-dimensional arrays and is also

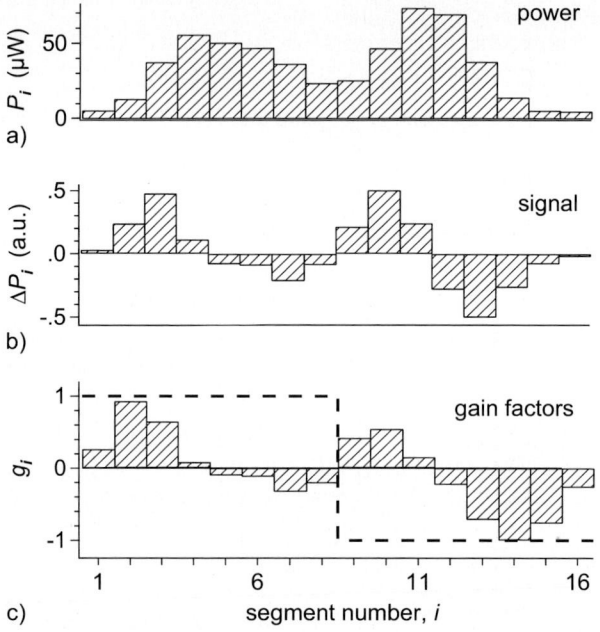

Fig. 3.4. Power, signal, and gain factors for a measurement of a particular 12-μm-long cantilever with an array detector. (**a**) Power distribution of the reflected optical beam on the array detector. The distribution exhibits a double peak and is significantly different from a Gaussian. (**b**) Change in optical power on each segment when the cantilever is deflected by a small amount. Some segments show a larger change in power than others, indicating that they are affected by the cantilever deflection more strongly. (**c**) Optimum gain factors, calculated from the measured powers (**a**) and signals (**b**). These gain factors are significantly different from those representing a conventional two-segment detector (*dashed line*) and optimize the measurement of signal-to-noise ratio in the case of shot noise. (Reprinted with permission from Schäffer et al. [60]. Copyright 2000 American Institute of Physics)

suitable for interferometric detection schemes. The key point is that (3.11) uses the knowledge of the set of P_is and ΔP_is to generate a larger signal-to-noise ratio, regardless of the particular details of the setup. Depending on the particular shape of the spot on the detector, we can obtain large improvements in the signal-to-noise ratio of measurements of cantilever deflection. The more distorted and "rough" the beam on the detector, the larger the improvement. In a particular case of a 12-μm-long cantilever, an increase in the signal-to-noise ratio by a factor of 5 was achieved (Fig. 3.5) [60]. The array detector can also be programmed to dynamically adapt to varying experimental conditions. Furthermore, previously necessary adjustment procedures such as mechanically centering the optical beam on the detector are obsolete for the array detector.

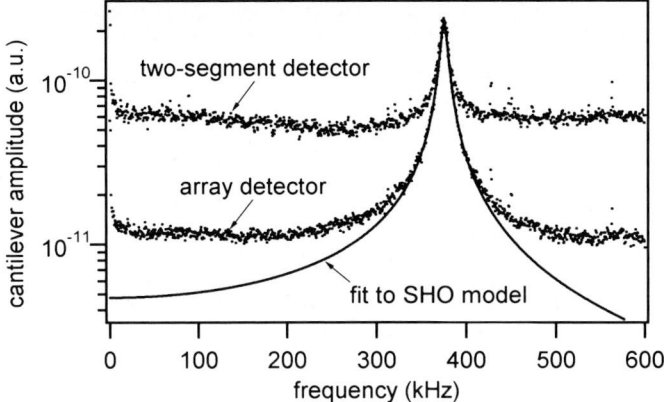

Fig. 3.5. Thermal noise amplitude density of a particular small cantilever ($L = 12\,\mu$m), measured with a two-segment detector and an array detector. Both detectors faithfully identify the thermal resonance peak, but the detection noise level (off-resonance) for the array detector is lower by a factor of 5. (Reprinted with permission from Schäffer et al. [60]. Copyright 2000 American Institute of Physics)

3.5
Dynamic Range and Linearity

3.5.1
The Two-Segment Detector

The important properties of detectors are not only sensitivity, but also linearity and dynamic range. High linearity and a large measurement range are especially important for force measurements [26]. When force–distance curves are recorded at increasing cantilever deflection, the reflected beam traverses an increasing distance on the detector plane (Fig. 3.6). This effect also comes about in theoretical calculations of the reflected beam profile at different cantilever deflections (Fig. 3.7). In the case of a conventional two-segment detector, the reflected beam eventually fully crosses the separation line between the two segments (illustrated symbolically by the dashed lines in Figs. 3.6, 3.7). This causes the measured cantilever deflection signal, z_{m}, to become nonlinear with respect to the actual cantilever deflection (Fig. 3.8, diamond-shaped markers). For a setup that is optimized for high sensitivity, it can be shown both experimentally and theoretically that the dynamic range (in which the measured deflection signal remains linear with the actual cantilever displacement) extends to an upper limit of only 115 nm (for less than 10% nonlinearity and for $\lambda = 670$ nm) [47]. It can also be shown that the relative width (ratio of upper detection limit, z_{max}, to lower detection limit, z_{min}) of the dynamic range depends only on the number of photons, n, counted within the measurement bandwidth [47]:

$$\frac{z_{\mathrm{max}}}{z_{\mathrm{min}}} = \sqrt{n}\,. \tag{3.12}$$

This result illustrates that there is a tradeoff between a high detection sensitivity (i.e., low z_{min}) and a large upper limit (i.e., large z_{max}) of the dynamic range: If the

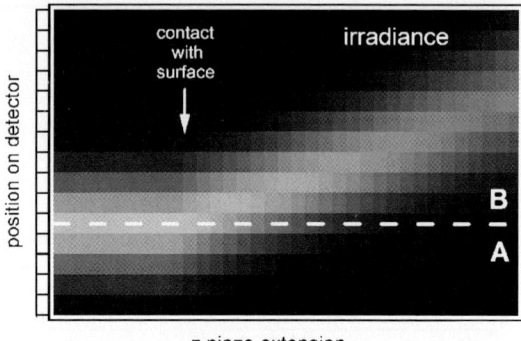

z-piezo extension

Fig. 3.6. One-dimensional irradiance distribution of the reflected optical beam (*red*) on the detector (*vertical direction*) as a function of z-piezo extension (*horizontal direction*). The translation of the beam on the detector is shown during a force curve. The cantilever initially is out of contact with the sample surface (*left-hand side*) and the reflected beam is centered on the line between the two segments (A and B). Upon contact with the surface, the cantilever deflects and the reflected beam moves away from the center (*upwards*). When the beam is completely shifted onto segment B at large cantilever deflections (*right-hand side*), the difference signal becomes saturated

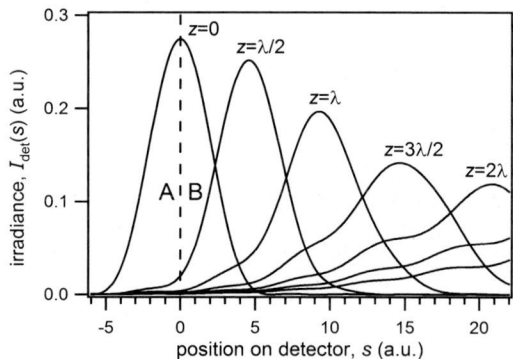

Fig. 3.7. Theoretical irradiance profiles of the reflected optical beam on the detector. The focused spot size and position were selected for optimized sensitivity. For an undeflected cantilever ($z = 0$), the beam on the detector is approximately Gaussian in shape and centered between segments A and B. For increasing cantilever deflections ($z = \lambda/2, \ldots, 2\lambda$), the beam moves onto segment B. Already at a cantilever deflection of $\lambda/2$, almost all of the optical power is shifted onto segment B. But the beam not only shifts in position, it also changes its shape: it increases in width and decreases in maximum irradiance. Furthermore, it becomes distorted. This is due to the fact that a deflected cantilever acts as a curved mirror, distorting the wave fronts of the reflected beam

sensitivity of the detection is increased (e.g., by varying the focused spot size or position), the upper limit of the dynamic range is automatically decreased, and vice versa (assuming constant power on the detector). For conventional AFM setups, the dynamic range covers 5–6 orders of magnitude (Fig. 3.9).

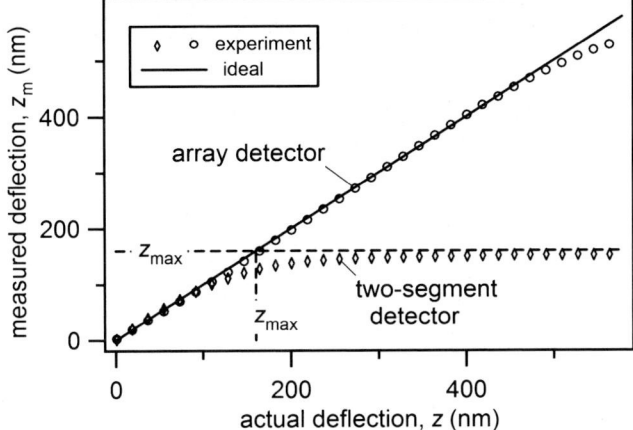

Fig. 3.8. Force curve of a 12-μm-long cantilever, recorded with the two-segment detector and with the array detector. For the two-segment detector, the measured cantilever deflection became 10% nonlinear at 115 nm and saturated at 182 nm. For the array detector, the measured deflection stayed within 10% nonlinearity up to 580 nm, even though the reflected beam became quite distorted. Adding more segments to the array detector would have allowed the dynamic measurement range to increase even more. (Reprinted with permission from Schäffer [47]. Copyright 2002 American Institute of Physics)

3.5.2
The Array Detector

An array detector significantly increases the linearity and the dynamic range of cantilever deflection measurements. The question is in which way to construct a signal from the set of array segments that best fulfills these requirements. Experimental measurements (Fig. 3.6) and theoretical calculations (Fig. 3.7) of the irradiance distribution of the reflected beam on the detector show that the beam not only shifts laterally, but also deforms at large cantilever deflections. It was shown, however, that the mean of the irradiance distribution, $\bar{s}(z)$, is a linear function of the actual cantilever deflection z, even for relatively large deflections [47]:

$$\bar{s}(z) = z \frac{2f}{P_{\text{det}}} \int_0^L dx I_c(x) h'(x/L) \ . \tag{3.13}$$

This result is not obvious a priori, since the reflected beam deforms at large deflections. The quantity $\bar{s}(z)$ can be used as a simple measure for the cantilever deflection. Much larger cantilever deflections can thereby be measured than what is possible with the conventional two-segment detector. Practically, the array detector is set up in a way that it continually records the irradiance distribution on the detector during a measurement. The mean of the distribution is then calculated and used as the cantilever deflection signal. In the case of a small cantilever of 12-μm length, it was shown that the upper detection limit is larger by a factor of 5 for an array

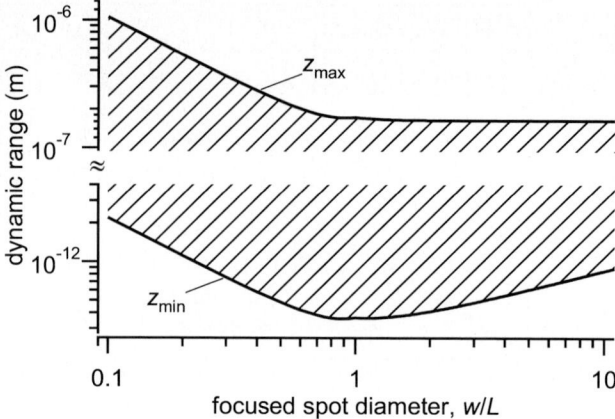

Fig. 3.9. Calculated dynamic measurement range of the conventional two-segment detector as a function of the focused spot diameter. The dynamic range is the range between the lower (z_{min}) and the upper (z_{max}) detection limits (*highlighted area*). It becomes apparent that a high detection sensitivity (small z_{min}) comes at the expense of a small upper detection limit (small z_{max}). (Reprinted with permission from Schäffer et al. [47]. Copyright 2002 American Institute of Physics)

detector (with 16 segments) than for the two-segment detector (Fig. 3.8) [47]. And this factor could easily be further increased by simply increasing the number of detector segments.

3.6
Detection of Higher-Order Cantilever Vibration Modes

Static deflection measurements are based on a deflection of the cantilever due to (quasi-) static applied forces. "Dynamic" measurement modes, on the other hand, usually utilize the first normal vibration mode of the cantilever, and have a number of advantages over static modes (e.g., insensitivity to low-frequency noise, reduction of lateral forces, prevention of instabilities). Examples are the tapping mode [54–57], noncontact modes [50–53], and dynamic force spectroscopy [62–64]. It has been shown that also higher-order normal vibration modes can be used for imaging applications [65–67], revealing additional structural and functional information about the sample. In atomic force acoustic microscopy, for example, ultrasonic sample surface vibrations in the megahertz range couple to the AFM cantilever and can be measured using higher-order modes [68–71]. Also, a quantitative understanding of tapping mode AFM was shown to be possible only if higher-order modes are considered [72]. The time course of tip–sample forces can then faithfully be extracted [73]. Furthermore, the simultaneous excitation of multiple higher-order normal modes was proposed to be essential in spatially resolved force spectroscopy [74]. And even in the conventional tapping mode in liquid, higher-order modes underlie the complex frequency spectrum of the cantilever [75]. One important question is how the vibration amplitudes of higher-order vibration modes can be accurately measured with high detection sensitivity. When using the optical beam deflection method, the detection sensitivities strongly depend on the shape, size, and position of the focused spot on the cantilever [46]. For example, it was shown that there can be (almost) pole-zero cancellations that lead to a significantly reduced detection sensitivity for higher-order modes [76]. A detailed theoretical investigation was performed with the purpose of calculating and optimizing these detection sensitivities [48,77]. Some of these issues will be outlined in the following.

3.6.1
Normal Vibration Modes

Flexural (transverse) vibrations of an undamped rectangular AFM cantilever can be described by the homogenous partial differential equation [78]

$$\frac{\partial^2 u(x,t)}{\partial t^2} + \frac{Eb^2}{12\rho} \frac{\partial^4 u(x,t)}{\partial x^4} = 0 \,, \tag{3.14}$$

where $u(x,t)$ is the displacement of the beam at position x and time t. The beam is assumed to have a uniform density ρ, a Young's modulus E, a length L, and a thickness b. As a bounded linear system, the eigenvalues of this system are discrete. The general solution of (3.14) can be obtained via separation of variables and be written as a linear expansion in flexural normal modes. Each mode behaves as an independent harmonic oscillator with sinusoidal time dependence. The spatial eigenfunctions (mode shapes) are

$$h_n(q) = \frac{(-1)^n}{2} \left[\cos \kappa_n q - \cosh \kappa_n q - \frac{\cos \kappa_n + \cosh \kappa_n}{\sin \kappa_n + \sinh \kappa_n} \left(\sin \kappa_n q - \sinh \kappa_n q \right) \right], \tag{3.15}$$

where n denotes the mode number, $q = x/L$, and the cantilever is assumed to be clamped at its base and free at its tip [corresponding to boundary conditions $h_n(0) = 0$; $h_n'(0) = 0$; $h_n''(1) = 0$; $h_n'''(1) = 0$]. The characteristic equation is $\cos \kappa_n \cosh \kappa_n + 1 = 0$, resulting in spatial eigenvalues

$$\kappa_n \cong 1.875, 4.694, \pi \left(n - \frac{1}{2} \right) \,, \quad \text{for } n = 1, 2, \geq 3 \,. \tag{3.16}$$

The mode shapes are orthogonal in the interval $[0; 1]$ and were normalized so that $h_n(1) = 1$ for all n. The functional shapes of the first five normal modes are display in Fig. 3.10. The temporal eigenvalues (resonant frequencies) can be obtained from the spatial eigenvalues via the dispersion relation

$$f_n = \frac{1}{2\pi} \frac{\kappa_n^2}{\sqrt{12}} \sqrt{\frac{E}{\rho} \frac{b}{L^2}} \,. \tag{3.17}$$

Fig. 3.10. Mode shapes for the first five flexural normal vibration modes of a clamped-free cantilever. The *crosshatched areas* outline the focused spots that optimize the detection sensitivity for each mode. For mode number three and higher, larger detection sensitivities are obtained for a spot close to the cantilever base (globally optimized spot, *left-hand side*) than for a spot close to the cantilever tip (tip-optimized spot, *right-hand side*). (Reprinted with permission from Schäffer and Fuchs [48]. Copyright 2005 American Institute of Physics)

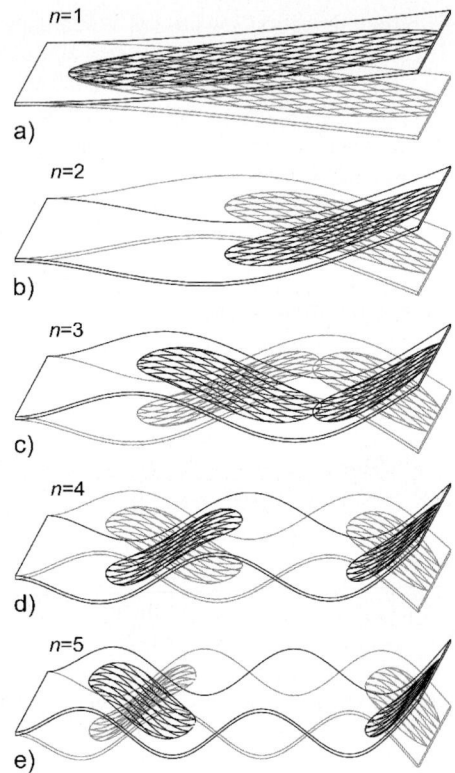

3.6.2
Optimization of the Detection Sensitivity

The question now is how the focused optical spot needs to be sized and positioned on the cantilever in order to obtain maximum detection sensitivity for a given normal mode. Using (3.5) and (3.15), we can calculate the optical detection sensitivity for the first five normal modes as a function of the focused spot diameter and position and display it as grayscale plots (Fig. 3.11a–e) [48]. Contour lines on which the detection sensitivity is zero are plotted as white lines. If the spot parameters (spot position, spot diameter) are chosen to lie on these lines, the motion of the respective mode is invisible to the detection. On the other extreme, the point in each grayscale plot where the detection sensitivity exhibits a global maximum is marked with a diagonal, red cross. Interesting is that the optimized sensitivities for the second and higher normal modes are about 2 times larger than that for the first mode. This suggests that higher-order normal modes can be advantageous for imaging and spectroscopic applications. In the case that multiple normal modes need to be detected simultaneously, it is important to also consider the position in the grayscale plot for each mode where the sensitivity is locally optimized and closest to the cantilever tip ("tip-optimized"). These positions are marked with a vertical, blue cross. For the first two normal modes, the globally optimized and the

Fig. 3.11. (**a**)–(**e**) Optical detection sensitivity as a function of focused spot diameter and position as a combined grayscale/contour plot, for the first five normal modes. Larger values of the brightness correspond to larger absolute values of the sensitivity. The globally optimized and the tip-optimized sensitivities are marked with *diagonal* (*red*) and *vertical* (*blue*) crosses, respectively. A constant scaling factor $\sigma_0 = \sqrt{8\pi}\,P_0/\lambda$ is used. (**f**) Focused spot diameters and positions for the optimized sensitivities. It becomes apparent that there is no spot diameter and position that optimizes the sensitivities of multiple normal modes simultaneously. (Reprinted with permission from Schäffer and Fuchs [48]. Copyright 2005 American Institute of Physics)

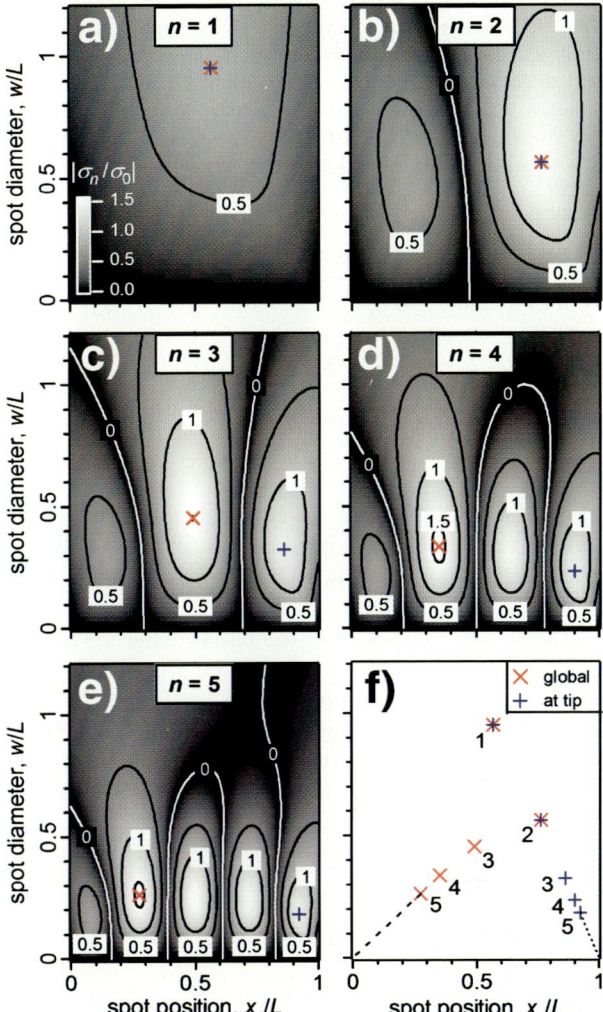

tip-optimized sensitivities coincide. The corresponding focused spots are outlined as the crosshatched areas on the cantilevers in Fig. 3.10a and b. For the third and higher normal modes, however, detection sensitivities that are approximately 14% larger are obtained for a spot position that is closer to the cantilever base than to the tip. The corresponding globally optimized and tip-optimized spots are outlined as crosshatched areas on the left-hand side and on the right-hand side, respectively, of each cantilever in Fig. 3.11c–e.

The spot parameters for the globally optimized and for the tip-optimized spots are shown in Fig. 3.11f and in Table 3.1. It becomes apparent that it is not possible to detect two or more modes simultaneously with optimum sensitivity for each mode, since the optimizing focused spot parameters vary significantly between the modes. If multiple normal modes need to be detected simultaneously, spot parameters need to

Table 3.1. Focused spot parameters for an optimized detection of cantilever deflections. For the static cantilever deflection and for the first five normal (transverse) vibration modes, the focused spot diameter, w, the focused spot position, x_c, and the resulting optimized value for the detection sensitivity, σ_n, are listed. Parameters for mode numbers larger than five can be calculated from the given formulas in the last row [48]. A scaling factor $\sigma_0 = \sqrt{8\pi} P_0/\lambda$ is used

Mode number n	Globally optimized			Tip-optimized		
	w/L	x_c/L	$|\sigma_n/\sigma_0|$	w/L	x_c/L	$|\sigma_n/\sigma_0|$
Static	Identical to tip-optimized			0.944	0.583	0.778
1	Identical to tip-optimized			0.952	0.569	0.783
2	Identical to tip-optimized			0.563	0.763	1.421
3	0.453	0.490	1.501	0.325	0.862	1.368
4	0.335	0.353	1.560	0.233	0.901	1.371
≥ 5	$1.17/$ $(n-1/2)$	$1.24/$ $(n-1/2)$	1.557	$0.814/$ $(n-1/2)$	$1 - 0.345/$ $(n-1/2)$	1.370

be chosen that are as far away as possible from all zero-contour lines of all considered modes. For the purpose of a unifying description of the optical detection sensitivity, a universal sensitivity function was constructed. This universal sensitivity function describes, for all modes, the optical detection sensitivity for a given spot diameter and spot position [48]. On the basis of the universal sensitivity function, several optimization strategies were devised for the simultaneous detection of multiple normal modes. One result can be summarized in the following way: (1) for small spot diameters, the strategy is to position the spot on the cantilever so that the distance from the spot center to the tip is approximately 0.424 times the spot diameter; (2) for large spot diameters, the strategy is to position the spot center exactly at the tip of the cantilever [48].

3.7
Calculation of Thermal Vibration Noise

Thermal noise of the cantilever is one of the most significant noise sources in an AFM. Thermal noise, however, can also be used for measurement purposes. One of the most prominent ways to measure the spring constant of a cantilever, for example, is based on thermal noise analysis [17, 39, 79, 80]. Also, tip–sample interaction potentials can be extracted from thermal cantilever noise [40–43]. Thermal noise can be reduced [81, 82] or amplified [83] by parametric feedback control.

3.7.1
Focused Optical Spot of Infinitesimal Size

The thermal displacement noise of an AFM cantilever was calculated by Butt and Jaschke [79] by performing a normal mode decomposition. They showed that the optical beam deflection method introduces additional, apparent thermal noise in the detection. This is due to the fact that this method detects the slope of the cantilever and not its displacement; therefore, nonzero thermal noise can appear in the measurement even if the actual tip displacement is zero. These calculations, however, are based on the approximation that the optical spot is focused to a point at the tip of the cantilever. It has already been shown in this chapter that this approximation is often not valid, in particular when higher-order normal modes are considered or when the spot diameter covers a significant fraction of the cantilever length. In these cases, the finite size of the focused spot must be taken into account.

3.7.2
Focused Optical Spot of Finite Size

In order to calculate the total apparent thermal noise of the cantilever, the noise contributions from all normal modes need to be added. The normal mode shapes for a clamped-free cantilever are given by (3.15). What is needed now is a description of the actual thermal noise for each mode. In thermodynamic equilibrium, the equipartition theorem states that each normal mode has the potential energy $\frac{1}{2} k_B T$ in temporal average, where k_B is the Boltzmann constant and T is the absolute temperature. The potential energy as a function of mode amplitude can be obtained by integrating the square of the curvature of the mode shape along the cantilever length. Thereby, the actual thermal amplitude of each normal mode can be calculated. In a second step, each mode is scaled by its respective optical detection sensitivity (3.5). The thermal displacement noise (rms) of the nth normal mode that occurs in the measurement ("measured displacement noise") can then be expressed as [77]

$$\sqrt{\langle z_n^{*2}(t)\rangle} = \left|\frac{\sigma_n}{\sigma_{FC}}\right| \frac{\sqrt{12}}{\kappa_n^2} \sqrt{\frac{k_B T}{k}} . \tag{3.18}$$

Here, k is the static spring constant of the cantilever, κ_n is the eigenvalue of the nth mode (3.16), σ_n is the optical detection sensitivity for the nth mode ((3.5) in conjunction with (3.15)), and σ_{FC} is the optical detection sensitivity for a statically deflected cantilever ((3.5) in conjunction with (3.6)). If the contributions from the first N modes are considered, the combined measured thermal displacement noise becomes

$$\sqrt{\langle Z_N^{*2}(t)\rangle} = \left(\sum_{n=1}^{N} \langle z_n^{*2}(t)\rangle\right)^{\frac{1}{2}} . \tag{3.19}$$

In Fig. 3.12a, the measured thermal noise for each normal mode of a clamped-free cantilever, $\sqrt{\langle z_n^{*2}\rangle}$, is plotted for different spot diameters as a function of mode number n. The spot positions were chosen so that the $1/e^2$-irradiance point of the spot coincides with the tip (or as $L/2$ if $w/L > 1$). It can be seen that $\sqrt{\langle z_n^{*2}\rangle}$ is largest for $n = 1$ and approaches zero for large n. The larger the spot diameter, the faster $\sqrt{\langle z_n^{*2}\rangle}$ converges to zero. Thermal noise contributions from higher-order modes are therefore suppressed for large spot diameters. The actual thermal noise is shown as the dotted line. The combined measured thermal noise is plotted in Fig. 3.12b. This combined measured thermal noise asymptotically approaches a constant that is the smaller the larger the focused spot size. It is interesting to note that the measured thermal noise for the first mode can even be slightly smaller than the actual thermal noise.

Fig. 3.12. Calculated thermal displacement noise (root mean square) that appears in atomic force microscopy measurements using the optical beam deflection method for a clamped-free cantilever. (**a**) Noise for each normal mode. (**b**) Combined noise for the first n normal modes. The noise strongly depends on the diameter of the focused optical spot. The traces for $w/L \to 0$ (infinitely small spot diameter) are identical to the results obtained by Butt and Jaschke [79]. (Reprinted with permission from Schäffer [77]. Copyright 2005 Institute of Physics Publishing)

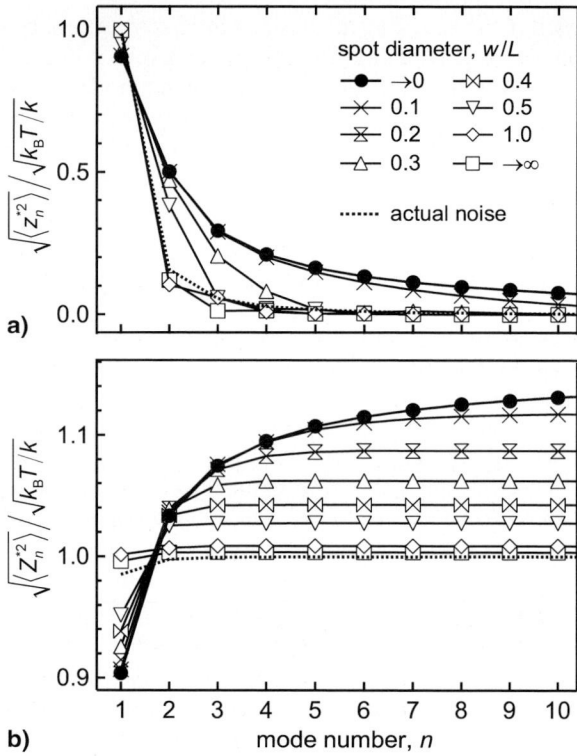

In some AFM imaging applications, the cantilever appears in a clamped-pinned configuration. This occurs when the cantilever tip is in contact with a stiff sample. The boundary conditions at the tip become $h_n(1) = 0$ and $h''_n(1) = 0$. The actual thermal displacement noise is therefore zero for any mode. But with the optical beam deflection method, the measured thermal displacement noise is nonzero. Similar to the case of the clamped-free cantilever, $\sqrt{\langle z_n^{*2} \rangle}$ converges to zero for large n, and the convergence is faster for larger spot diameters. A significant difference from the clamped-free case, however, is the larger influence of the spot diameter: whereas in the clamped-free case the combined measured thermal noise is similar in value for all spot diameters, for the clamped-pinned cantilever there is approximately 6–8 times less measured thermal noise for the largest spot diameter than for the smallest diameter. Also, the absolute levels of the measured thermal noise are significantly smaller. The details are treated in [77].

The results of this section can be used to outline strategies for choosing experimental conditions for reduced thermal noise in contact mode and force spectroscopy experiments. Three different experimental conditions corresponding to different magnitudes of the sample stiffness, k_s, can be considered: (1) soft sample ($k_s/k \lesssim 1$), (2) stiff sample ($k_s/k \gtrsim 1000$), and (3) intermediately stiff sample ($1 \lesssim k_s/k \lesssim 1000$). It can be shown that, in general, a soft cantilever and a large spot diameter are the most beneficial experimental parameters for reducing thermal cantilever noise in the measurement [77].

3.8
Thermal Spring Constant Calibration

Calibration of the spring constant of the AFM cantilever is an important task. An accurate value of the spring constant is required for almost all imaging and spectroscopy modes, including contact mode, noncontact mode, tapping mode, and force modulation. The determination of the spring constant is a challenging metrology problem because of the small dimensions of the cantilever. Several methods have been proposed and applied to date. For example, in the "added mass method" [84], a small object of known mass is added to the tip of the cantilever. The spring constant can then be derived from the shift in the resonant frequency of the cantilever. In the "thermal noise method" [17, 39], Brownian motion of the cantilever is used in conjunction with the equipartition theorem to infer the spring constant. In the "plane view method" [85–87], the planar cantilever dimensions are used together with a measurement of the resonant frequency and the Q factor to calculate the spring constant via a hydrodynamic theory [88]. Despite this wealth of methods [89], no standard has been established yet. Each method has its own advantages and disadvantages, and there can be significant spread in the results from different methods.

Of all the methods for spring constant calibration, the thermal noise method is one of the simplest and is the most popular one. In practice, calibration is performed in three steps: (1) measurement of the amplitude spectrum of thermal deflection noise; (2) calibration of the spectrum in units of meters per hertz to the half power; and (3) determination of the spring constant from a fit to this spectrum. As the spectrum contains the thermal contributions from all vibration modes of the cantilever (Sect. 3.7), a simple harmonic oscillator model is usually fit to the first vibration mode only. The fit reduces the influence of electronic and ambient noise. From this fit, the spring constant can be calculated using the equipartition theorem. One difficulty in this process is the determination of the optical detection sensitivity. Typically, a force curve is acquired for this purpose, but also alternative methods exist [90]. For the widespread optical beam deflection method, it was shown in Sect. 3.2.2 that the sensitivity depends on the size and position of the focused optical spot, and on the particular shape of the cantilever. This shape is different for the first normal vibration mode ((3.15) for $n = 1$) and for the statically deflected cantilever (3.6). By using the theoretical description of the detection sensitivity (3.5), a correction factor, $\chi = \sigma_{FC}/\sigma_1$, was derived that corrects for the effect of those different shapes (Fig. 3.13) [91, 92]. This factor requires knowledge of the size and position of the focused optical spot. It is possible, however, to have the size of the focused spot serve as a yardstick to measure the position [91]. With use of this correction factor, thermally measured spring constants that previously systematically deviated by 25–50% from those measured with other methods were brought into well within 10% agreement [91] (the spring constant depends quadratically on χ). This improvement is the more significant the smaller the cantilever. It is therefore expected that this correction factor will play an important role in the future when AFM technology based on small cantilevers will become more widespread.

Fig. 3.13. Correction factor for the detection sensitivity of thermal spring constant calibration, for a series of different spot diameters as a function of spot position along the cantilever. For small spot diameters positioned either at the base ($x_c/L = 0$) or at the tip ($x_c/L = 1$), the correction factor has the largest deviation from unity. By applying this correction factor, thermally measured spring constants were brought into better than 10% agreement with other methods. (Reprinted with permission from Proksch et al. [91]. Copyright 2004 Institute of Physics Publishing)

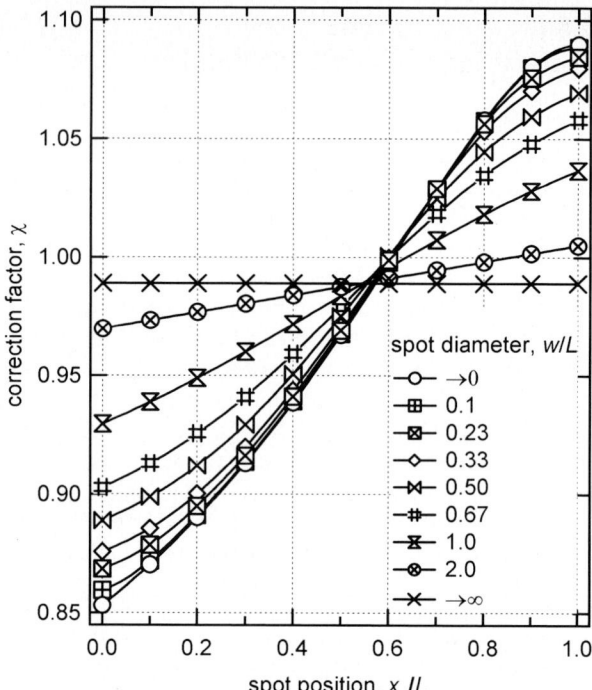

Acknowledgements. I thank Roger Proksch, Jason Cleveland, Harald Fuchs, and Boris Anczykowski for fruitful discussions, Asylum Research for AFM support, and the Gemeinnützige Hertie-Stiftung in the Stifterverband für die Deutsche Wissenschaft for financial support.

References

1. Binnig G, Quate CF, Gerber C (1986) Atomic force microscope. Phys Rev Lett 56:930–933
2. Sarid D, Iams D, Weissenberger V, Bell LS (1988) Compact scanning-force microscope using a laser diode. Opt Lett 13:1057–1059
3. Neubauer G, Cohen SR, McClelland GM, Horne D, Mate CM (1990) Force microscopy with a bidirectional capacitance sensor. Rev Sci Instrum 61:2296–2308
4. Tortonese M, Yamada H, Barret RC, Quate CF (1991) The proceedings of Transducers 1991. IEEE, Pennington, pp 448–451
5. Minne SC, Manalis SR, Quate CF (1995) Parallel atomic force microscopy using cantilevers with integrated piezoresistive sensors and integrated piezoelectric actuators. Appl Phys Lett 67:3918–3920
6. Rugar D, Mamin HJ, Erlandsson R, Stern JE, Terris BD (1988) Force microscope using a fiber-optic displacement sensor. Rev Sci Instrum 59:2337–2340
7. Rugar D, Mamin HJ, Guethner P (1989) Improved fiber-optic interferometer for atomic force microscopy. Appl Phys Lett 55:2588–2590
8. Schönenberger C, Alvarado SF (1989) A differential interferometer for force microscopy. Rev Sci Instrum 60:3131–3134
9. Meyer G, Amer NM (1988) Novel optical approach to atomic force microscopy. Appl Phys Lett 53:1045–1047

10. Alexander S, Hellemans L, Marti O, Schneir J, Elings V, Hansma PK, Longmire M, Gurley J (1989) An atomic-resolution atomic-force microscope implemented using an optical lever. J Appl Phys 65:164–167
11. Putman CAJ, De Grooth BG, Van Hulst NF, Greve J (1992) A detailed analysis of the optical beam deflection technique for use in atomic force microscopy. J Appl Phys 72:6–12
12. Gustafsson MGL, Clarke J (1994) Scanning force microscope springs optimized for optical-beam deflection and with tips made by controlled fracture. J Appl Phys 76:172–181
13. Drake B, Prater CB, Weisenhorn AL, Gould SA, Albrecht TR, Quate CF, Cannell DS, Hansma HG, Hansma PK (1989) Imaging crystals, polymers, and processes in water with the atomic force microscope. Science 243:1586–1589
14. Albrecht TR, Akamine S, Carver TE, Quate CF (1990) Microfabrication of cantilever styli for the atomic force microscope. J Vac Sci Technol A 8:3386–3396
15. Wolter O, Bayer T, Greschner J (1991) Micromachined silicon sensors for scanning force microscopy. J Vac Sci Technol B 9:1353–1357
16. Pechmann R, Kohler JM, Fritzsche W, Schaper A, Jovin TM (1994) The Novolever – a new cantilever for scanning force microscopy microfabricated from polymeric materials. Rev Sci Instrum 65:3702–3706
17. Walters DA, Cleveland JP, Thomson NH, Hansma PK, Wendman MA, Gurley G, Elings V (1996) Short cantilevers for atomic force microscopy. Rev Sci Instrum 67:3583–3590
18. Schäffer TE, Viani M, Walters DA, Drake B, Runge EK, Cleveland JP, Wendman MA, Hansma PK (1997) An atomic force microscope for small cantilevers. Proc SPIE 3009: 48–52
19. Kulisch W, Malave A, Lippold G, Scholz W, Mihalcea C, Oesterschulze E (1997) Fabrication of integrated diamond cantilevers with tips for SPM applications. Diamond Relat Mater 6: 906–911
20. Berger R, Delamarche E, Lang HP, Gerber C, Gimzewski JK, Meyer E, Güntherodt HJ (1997) Surface stress in the self-assembly of alkanethiols on gold. Science 276:2021–2024
21. Chand A, Viani MB, Schaffer TE, Hansma PK (2000) Microfabricated small metal cantilevers with silicon tip for atomic force microscopy. J Microelectromech Syst 9:112–116
22. Despont M, Brugger J, Drechsler U, Durig U, Haberle W, Lutwyche M, Rothuizen H, Stutz R, Widmer R, Binnig G, Rohrer H, Vettiger P (2000) VLSI-NEMS chip for parallel AFM data storage. Sens Actuators A 80:100–107
23. Oesterschulze E, Abelmann L, Bos Avd, Kassing R, Lawrence N, Wittstock G, Ziegler C (2006) In: Bushan B, Fuchs H (eds) Applied scanning probe methods, vol II. Springer, Berlin Heidelberg New York, pp 165–203
24. Weisenhorn AL, Hansma PK, Albrecht TR, Quate CF (1989) Forces in atomic force microscopy in air and water. Appl Phys Lett 54:2651–2653
25. Rugar D, Stipe BC, Mamin HJ, Yannoni CS, Stowe TD, Yasumura KY, Kenny TW (2001) Adventures in attonewton force detection. Appl Phys A 72:S3–S10
26. Butt H-J, Cappella B, Kappl M (2005) Force measurements with the atomic force microscope: Technique, interpretation and applications. Surf Sci Rep 59:1–152
27. Tao NJ, Lindsay SM, Lees S (1992) Measuring the microelastic properties of biological material. Biophys J 63:1165–1169
28. Ducker WA, Senden TJ, Pashley RM (1991) Direct measurement of colloidal forces using and atomic force microscope. Nature 353:239–241
29. Butt H-J (1991) Measuring electrostatic, van der Waals, and hydration forces in electrolyte solutions with an atomic force microscope. Biophys J 60:1438–1444
30. Rief M, Grubmüller H (2002) Force spectroscopy of single biomolecules. Chem Phys Chem 3:255–261
31. Florin EL, Moy VT, Gaub HE (1994) Adhesion forces between individual ligand-receptor pairs. Science 264:415–417

32. Hinterdorfer P, Baumgartner W, Gruber HJ, Schilcher K, Schindler H (1996) Detection and localization of individual antibody-antigen recognition events by atomic force microscopy. Proc Natl Acad Sci USA 93:3477–3481

33. Dammer U, Hegner M, Anselmetti D, Wagner P, Dreier M, Huber W, Güntherodt HJ (1996) Specific antigen/antibody interactions measured by force microscopy. Biophys J 70: 2437–2441

34. Lee GU, Chrisey LA, Colton RJ (1994) Direct measurement of the forces between complementary strands of DNA. Science 266:771–773

35. Rief M, Oesterhelt F, Heymann B, Gaub HE (1997) Single molecule force spectroscopy on polysaccharides by atomic force microscopy. Science 275:1295–1297

36. Rief M, Gautel M, Oesterhelt F, Fernandez JM, Gaub HE (1997) Reversible unfolding of individual titin immunoglobulin domains by AFM. Science 276:1109–1112

37. Gittes F, Schmidt CF (1998) Thermal noise limitations on micromechanical experiments. Eur Biophys J 27:75–81

38. Viani MB, Schaffer TE, Chand A, Rief M, Gaub HE, Hansma PK (1999) Small cantilevers for force spectroscopy of single molecules. J Appl Phys 86:2258–2262

39. Hutter JL, Bechhoefer J (1993) Calibration of atomic-force microscope tips. Rev Sci Instrum 64:1868–1873

40. Cleveland JP, Schäffer TE, Hansma PK (1995) Probing oscillatory hydration potentials using thermal-mechanical noise in an atomic-force microscope. Phys Rev B 52:R8692–8695

41. Roters A, Gelbert M, Schimmel M, Ruhe J, Johannsmann D (1997) Static and dynamic profiles of tethered polymer layers probed by analyzing the noise of an atomic force microscope. Phys Rev E 56:3256–3264

42. Heinz WF, Antonik MD, Hoh JH (2000) Reconstructing local interaction potentials from perturbations to the thermally driven motion of an atomic force microscope cantilever. J Phys Chem B 104:622–626

43. Benmouna F, Johannsmann D (2004) Viscoelasticity of gelatin surfaces probed by AFM noise analysis. Langmuir 20:188–193

44. Born M, Wolf E (1980) Principles of optics. Pergamon, Oxford

45. Goodman JW (1968) Introduction to Fourier optics. McGraw-Hill, San Francisco

46. Schäffer TE, Hansma PK (1998) Characterization and optimization of the detection sensitivity of an atomic force microscope for small cantilevers. J Appl Phys 84:4661–4666

47. Schäffer TE (2002) Force spectroscopy with a large dynamic range using small cantilevers and an array detector. J Appl Phys 91:4739–4746

48. Schäffer TE, Fuchs H (2005) Optimized detection of normal vibration modes of atomic force microscope cantilevers with the optical beam deflection method. J Appl Phys 97:083524

49. Sarid D (1994) Scanning force microscopy: with applications to electric, magnetic, and atomic forces. Oxford University Press, New York

50. Martin Y, Wickramasinge HK (1987) Magnetic imaging by "force microscopy" with 1000 Å resolution. Appl Phys Lett 50:1455–1457

51. Albrecht TR, Grütter P, Horne D, Rugar D (1991) Frequency modulation detection using high-Q cantilevers for enhanced force microscope sensitivity. J Appl Phys 69:668–673

52. Giessibl FJ (1995) Atomic resolution of the silicon (111)-(7 × 7) surface by atomic force microscopy. Science 267:68–71

53. Sugawara Y, Ohta M, Ueyama H, Morita S (1995) Defect motion on an InP(110) surface observed with noncontact atomic force microscopy. Science 270:1646–1648

54. Zhong Q, Inniss D, Kjoller K, Elings VB (1993) Fractured polymer/silica fiber surface studied by tapping mode atomic force microscopy. Surf Sci 290:L688–692

55. Hansma PK, Cleveland JP, Radmacher M, Walters DA, Hillner PE, Bezanilla M, Fritz M, Vie D, Hansma HG, Prater CB, Massie J, Fukunaga L, Gurley J, Elings V (1994) Tapping mode atomic force microscopy in liquids. Appl Phys Lett 64:1738–1740

56. Putman CAJ, Werf KOV, Grooth BGD, Hulst NFV, Greve J (1994) Tapping mode atomic force microscopy in liquid. Appl Phys Lett 64:2454–2456
57. Lantz MA, O'Shea SJ, Welland ME (1994) Force microscopy imaging in liquids using ac techniques. Appl Phys Lett 65:409–411
58. Walls D (1983) Squeezed states of light. Nature 306:141
59. Pierce M, Stuart J, Pungor A, Dryden P, Hlady V (1994) Adhesion force measurements using an atomic force microscope upgraded with a linear position sensitive detector. Langmuir 10:3217–3221
60. Schäffer TE, Richter M, Viani MB (2000) Array detector for the atomic force microscope. Appl Phys Lett 76:3644–3646
61. Schäffer TE, Hansma PK (2002) High sensitivity deflection sensing device. US Patent 6,455,838
62. Anczykowski B, Krüger D, Fuchs H (1996) Cantilever dynamics in quasinoncontact force microscopy – spectroscopic aspects. Phys Rev B 53:15485
63. Liu YZ, Leuba SH, Lindsay SM (1999) Relationship between stiffness and force in single molecule pulling experiments. Langmuir 15:8547–8548
64. Dürig U (1999) Relations between interaction forces and frequency shift in large-amplitude dynamic force microscopy. Appl Phys Lett 75:433–435
65. Minne SC, Manalis SR, Atalar A, Quate CF (1996) Contact imaging in the atomic force microscope using a higher order flexural mode combined with a new sensor. Appl Phys Lett 68:1427–1429
66. Stark RW, Drobek T, Heckl WM (1999) Tapping-mode atomic force microscopy and phase-imaging in higher eigenmodes. Appl Phys Lett 74:3296–3298
67. Hillenbrand R, Stark M, Guckenberger R (2000) Higher-harmonics generation in tapping-mode atomic-force microscopy: Insights into the tip–sample interaction. Appl Phys Lett 76:3478–3480
68. Rabe U, Arnold W (1994) Acoustic microscopy by atomic force microscopy. Appl Phys Lett 64:1493
69. Yamanaka K, Ogiso H, Kolosov O (1994) Ultrasonic force microscopy for nanometer resolution subsurface imaging. Appl Phys Lett 64:178
70. Rabe U, Janser K, Arnold W (1996) Vibrations of free and surface-coupled atomic force microscope cantilevers: theory and experiment. Rev Sci Instrum 67:3281–3293
71. Rabe U, Turner J, Arnold W (1998) Analysis of the high-frequency response of atomic force microscope cantilevers. Appl Phys A 66:S277–S282
72. Stark RW, Heckl WM (2000) Fourier-transformed force microscopy: tapping-mode atomic force microscopy beyond the Hookian approximation. Surf Sci 457:219–228
73. Stark M, Stark RW, Heckl WM, Guckenberger R (2002) Inverting dynamic force microscopy: from signals to time-resolved interaction forces. Proc Natl Acad Sci USA 99:8473–8478
74. Rodríguez TR, García R (2004) Compositional mapping of surfaces in atomic force microscopy. Appl Phys Lett 84:449–451
75. Schäffer TE, Cleveland JP, Ohnesorge F, Walters DA, Hansma PK (1996) Studies of vibrating atomic force microscope cantilevers in liquid. J Appl Phys 80:3622–3627
76. Stark RW (2004) Optical lever detection in higher eigenmode dynamic atomic force microscopy. Rev Sci Instrum 75:5053–5055
77. Schäffer TE (2005) Calculation of thermal noise in an atomic force microscope with a finite optical spot size. Nanotechnology 16:664–670
78. Timoshenko S, Young DH, Weaver W (1974) Vibration problems in engineering. Wiley, New York
79. Butt HJ, Jaschke M (1995) Calculation of thermal noise in atomic force microscopy. Nanotechnology 6:1–7

80. Drobek T, Stark RW, Heckl WM (2001) Determination of shear stiffness based on thermal noise analysis in atomic force microscopy: passive overtone microscopy. Phys Rev B 64:045401
81. Rugar D, Grütter P (1991) Mechanical parametric amplification and thermomechanical noise squeezing. Phys Rev Lett 67:699–702
82. Liang S, Medich D, Czajkowsky DM, Sheng S, Yuan J-Y, Shao Z (2000) Thermal noise reduction of mechanical oscillators by actively controlled external dissipative forces. Ultramicroscopy 84:119–125
83. Muralidharan G, Mehta A, Cherian S, Thundat T (2001) Analysis of amplification of thermal vibrations of a microcantilever. J Appl Phys 89:4587–4591
84. Cleveland JP, Manne S, Bocek D, Hansma PK (1993) A nondestructive method for determining the spring constant of cantilevers for scanning force microscopy. Rev Sci Instrum 64:403–405
85. Sader JE, Larson I, Mulvaney P, White LR (1995) Method for the calibration of atomic force microscope cantilevers. Rev Sci Instrum 66:3789–3798
86. Sader JE (1995) Parallel beam approximation for V-shaped atomic force microscope cantilevers. Rev Sci Instrum 66:4583–4587
87. Sader JE, Chon JWM, Mulvaney P (1999) Calibration of rectangular atomic force microscope cantilevers. Rev Sci Instrum 70:3967–3969
88. Sader JE (1998) Frequency response of cantilever beams immersed in viscous fluids with applications to the atomic force microscope. J Appl Phys 84:64–76
89. Sader JE (2002) In: Hubbard A (ed) Encyclopedia of surface and colloidal science. Dekker, New York, pp 846–856
90. D'Costa NP, Hoh JH (1995) Calibration of optical lever sensitivity for atomic force microscopy. Rev Sci Instrum 66:5096–5097
91. Proksch R, Schäffer TE, Cleveland JP, Callahan RC, Viani MB (2004) Finite optical spot size and position corrections in thermal spring constant calibration. Nanotechnology 15:1344–1350
92. Cook SM, Schäffer TE, Chynoweth KM, Wigton M, Simmonds RW, Lang KM (2006) Practical implementation of dynamic methods for measuring atomic force microscope cantilever spring constants. Nanotechnology 17:2135–2145

4 Q-controlled Dynamic Force Microscopy in Air and Liquids

Hendrik Hölscher · Daniel Ebeling · Udo D. Schwarz

4.1
Introduction

Today, atomic force microscopy (AFM) [1] has been established as an important imaging tool for nanoscale applications. For soft sample imaging, it has been proven to be advantageous to avoid lateral forces between tip and sample by the application of dynamic force microscopy (DFM), where the cantilever is vibrated near the sample surface. In environments like air and liquids, the cantilever is typically oscillated with a *fixed excitation frequency* [2–5]. An often-applied imaging technique is the use of the actual value of the oscillation amplitude as a measure of the tip–sample distance, which is kept constant during imaging by a feedback. This mode, which is referred to as amplitude modulation mode (also known as tapping mode [3]), is the workhorse among the various different scanning probe microscopy techniques applied in ambient conditions and liquids.

During the last few years, however, it has been found that the active modification of the cantilever damping by the controlled increase or decrease of the apparent or "effective" Q-factor of the system exhibits very interesting properties, which will be further explored in this chapter. Such a setup was pioneered by Anczykowski et al. and has been named Q-control. Its features have been be used in different ways. For example, the Q-factor can be increased to lower the maximum forces acting between tip and sample in air [6]. In similar experiments carried out in liquids, the active decrease of the damping was shown to enhance image quality [7–10]. In contrast, the possibility to actively reduce the quality factor allows the scan speed to be increased in DFM experiments performed in air [11, 12]. Other applications include the use in shear force microscopy [13, 14], ultrasonic AFM [15, 16], or Q-controlled dynamic force spectroscopy [17]. Here, we focus on three especially important aspects of Q-control: the reduction of tip–sample forces in air, the enhanced image quality in liquids, and the possible increase of the scan speed.

Among the different fields of application, the ability of Q-control to provide enhanced image quality attracted most interest and has been discussed in a multitude of experimental [7, 9, 10, 18–21] and theoretical [22–26] studies. Our detailed theoretical analysis of Q-controlled DFM (QC-DFM) is based on analytical as well as numerical methods. In Sect. 4.2, we first review the theoretical background of Q-control with respect to conventional amplitude modulation (tapping mode) DFM. This comparison shows how Q-control allows the cantilever damping to be increased or decreased and how the peak in the resonance curves can be shifted. Subsequently, analytical formulas describing QC-DFM imaging are developed that explicitly include tip–sample interactions. This analysis is then used in Sects. 4.2.4 and 4.2.5, where we explain how Q-control helps to control tip–sample forces in ambient conditions and liquids, respectively. In Sect. 4.3 we show some examples of Q-control obtained under ambient conditions and in liquids. A summary is given in Sect. 4.4.

4.2
Theory of Q-controlled Dynamic Force Microscopy

In this section, we will develop the theoretical framework of QC-DFM. First, we introduce the equation of motion of a dynamic force microscope with Q-control. Then, we obtain analytical formulas describing amplitude and phase as a function of driving frequency. Finally, we include tip–sample model forces for air and liquids into our analysis. All analytical results are compared with numerical simulations based on the equation of motion.

4.2.1
Equation of Motion of a Dynamic Force Microscope with Q-control

A sketch of the experimental setup of an atomic force microscope driven in amplitude modulation mode with Q-control is shown in Fig. 4.1. The deflection of the cantilever is typically measured with the laser beam deflection method as indicated, but other displacement sensors, such as interferometric sensors, can be applied as well. During operation in conventional tapping mode, the cantilever is driven at a fixed frequency by a constant-amplitude signal originating from an external function generator, while

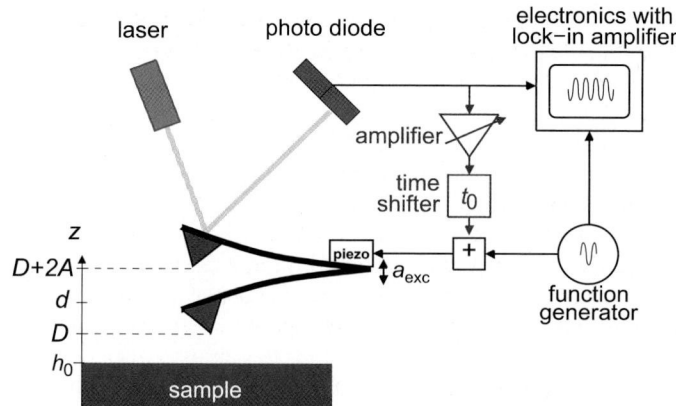

Fig. 4.1. The experimental setup of a dynamic force microscope with Q-control (not to scale). In contrast to a conventional dynamic force microscope driven in the so-called tapping mode, this experimental setup exhibits an additional feedback loop consisting of a time ("phase") shifter and an amplifier. The tip oscillates between the nearest tip–sample position D and $D + 2A$. The equilibrium position of the tip is denoted as d. For the analysis we assume the sample surface position at h_0

the resulting oscillation amplitude and/or the phase shift are detected by a lock-in amplifier. The function generator supplies not only the signal for the dither piezo; its signal serves simultaneously as a reference for the lock-in amplifier.

A dynamic force microscope with Q-control has an additional feedback circuit consisting of an amplifier and a time shifter, which is also often referred to as "phase" shifter (see later for details). The signal of the displacement sensor is first fed into the amplifier and subsequently used to excite the dither piezo driving the cantilever in addition to the fixed-frequency, constant-amplitude signal from the function generator. Before the addition, however, the properly amplified displacement signal can be delayed in time by a preset but variable time delay t_0 in the time ("phase") shifter.

Naturally, the two driving mechanisms will be reflected in the corresponding equation of motion for the cantilever, which is given by [17, 23]

$$m\ddot{z}(t) + \frac{2\pi f_0 m}{Q_0}\dot{z}(t) + c_z[z(t) - d] + \underbrace{g\,c_z\,z(t - t_0)}_{Q\text{-control}} =$$

$$\underbrace{a_d\,c_z\,\cos(2\pi f_d t)}_{\text{external driving force}} + \underbrace{F_{ts}[z(t), \dot{z}(t)]}_{\text{tip–sample force}} . \quad (4.1)$$

Here, $z(t)$ is the position of the tip at the time t; c_z, m, Q_0, and $f_0 = \sqrt{(c_z/m)}/(2\pi)$ are the spring constant, the effective mass, the quality factor, and the eigenfrequency of the cantilever, respectively. The equilibrium position of the tip is denoted as d. The active feedback of the system is described by the retarded amplification of the displacement signal, i.e., the tip position z is measured at the retarded time $t - t_0$ and amplified by a *gain factor* g. The first term on the right-hand side of the equation represents the external driving force of the cantilever. It is modulated with the constant excitation amplitude a_d at a fixed frequency f_d. The (nonlinear) tip–sample interaction force F_{ts} is introduced by the second term.

The key feature of the Q-control is the positive feedback, which can be used to increase or decrease the effective Q-factor of the system. The basic idea of the feedback loop is to reduce the damping force acting on the cantilever by the surrounding medium. Mathematically, the damping force $(2\pi f_0 m/Q_0)\dot{z}$ has to be compensated by the active feedback term $g c_z z(t - t_0)$ [6].

The time ("phase") shifter in the electronic setup controls the time delay between the sensor signal and the piezo excitation. This results in a phase shift between cantilever oscillation and excitation. We describe this feature explicitly by a time shift t_0. Nonetheless, a consideration of the time shift by a phase difference θ_0 is also possible, giving equivalent results. Therefore, we use "time shift" and "phase shift" as synonyms throughout this chapter and notice that both parameters are scaled by $\theta_0 = 2\pi f_d t_0$.

In the following, we solve the equation of motion by an analytical approach and consider only the steady-state solution given by the ansatz

$$z(t \gg 0) = d + A\,\cos(2\pi f_d t + \phi) , \quad (4.2)$$

where ϕ is the phase difference between the excitation and the oscillation of the cantilever. For comparison, we also performed computer simulations, calculating the numerical solutions of the equation of motion with a fourth-order Runge-Kutta method [27]. These computations were carried out until a stable steady-state oscillation of the cantilever was reached for each set of parameters.

4.2.2
Active Modification of the Q-factor

In order to establish the basic principles of a dynamic force microscope with Q-control, we start with the analysis of a cantilever oscillating far away from the sample surface. In this case, there are no tip–sample forces (i.e., $F_{ts} = 0$). Further simplification arises from restricting ourselves to the steady-state solutions for $t \gg 0$, where the cantilever oscillates sinusoidally and with constant amplitude. With these assumptions, it is straightforward to introduce the ansatz (4.2) into the equation of motion (4.1). As a result, we obtain the following two equations for the amplitude and phase vs. driving frequency curves [23]:

$$A = \frac{a_{\mathrm{d}}}{\sqrt{\left(1 - \frac{f_{\mathrm{d}}^2}{f_0^2} + g\cos(2\pi f_{\mathrm{d}} t_0)\right)^2 + \left(\frac{1}{Q_0}\frac{f_{\mathrm{d}}}{f_0} - g\sin(2\pi f_{\mathrm{d}} t_0)\right)^2}}, \quad (4.3a)$$

$$\tan\phi = \frac{\frac{1}{Q_0}\frac{f_{\mathrm{d}}}{f_0} - g\sin(2\pi f_{\mathrm{d}} t_0)}{1 - \frac{f_{\mathrm{d}}^2}{f_0^2} + g\cos(2\pi f_{\mathrm{d}} t_0)}. \quad (4.3b)$$

If the Q-control feedback is switched off ($g = 0$), both equations reduce to the well-known resonance curves of an externally driven harmonic oscillator. With Q-control, however, the gain factor is nonzero and the resonance curves are modified by the terms including g and t_0.

If the cantilever is driven at or near the resonance frequency, as is mostly the case, we can use the approximation $f_{\mathrm{d}} \approx f_0$ to define an effective Q-factor,

$$Q_{\mathrm{eff}}(g, t_0) = \frac{1}{1/Q_0 - g\sin(2\pi f_{\mathrm{d}} t_0)}, \quad (4.4)$$

which depends only on the gain factor g and the time ("phase") shift t_0. From this definition, it is obvious that the gain factor must be limited to $0 \leq g < 1/Q_0$. For larger values ($g \geq 1/Q_0$), the oscillation becomes unstable [23, 24], and the solution of the equation of motion cannot be described by the ansatz (4.2) anymore.

In order to get an overview of the oscillation properties of the system, it is convenient to first analyze the four special situations discussed in the next two subsections.

4.2.2.1
Increase and Decrease of the Q-factor

If the time delay is set to $t_0 = 1/4 f_0$ (i.e., $\theta_0 = 90°$), $3/4 f_0$ (270°), ..., the trigonometric functions in (4.3a) and (4.3b) simplify to $\sin(2\pi f_{\mathrm{d}} t_0) = \pm 1$ and $\cos(2\pi f_{\mathrm{d}} t_0) = 0$. This results in an effective Q-factor of

$$Q_{\mathrm{eff}} = \frac{1}{1/Q_0 \mp g}, \quad (4.5)$$

where the minus sign corresponds to 90° and the plus sign to 270°. The resulting amplitude and phase vs. driving frequency curves are plotted in Fig. 4.2, demonstrating nicely the name-giving feature of Q-control. With $g = 0.8/Q_0$ and a phase shift of 90°, the resonance peak of the amplitude curve is significantly enhanced and the effective Q-factor is increased to $Q_{\mathrm{eff}} = 1500$. For 270°, the Q-factor is decreased to $Q_{\mathrm{eff}} = 167$ and the amplitude curve is broadened. Corresponding behavior can be observed for the phase curves, which change their steepness with the increase or decrease of the effective Q-factor.

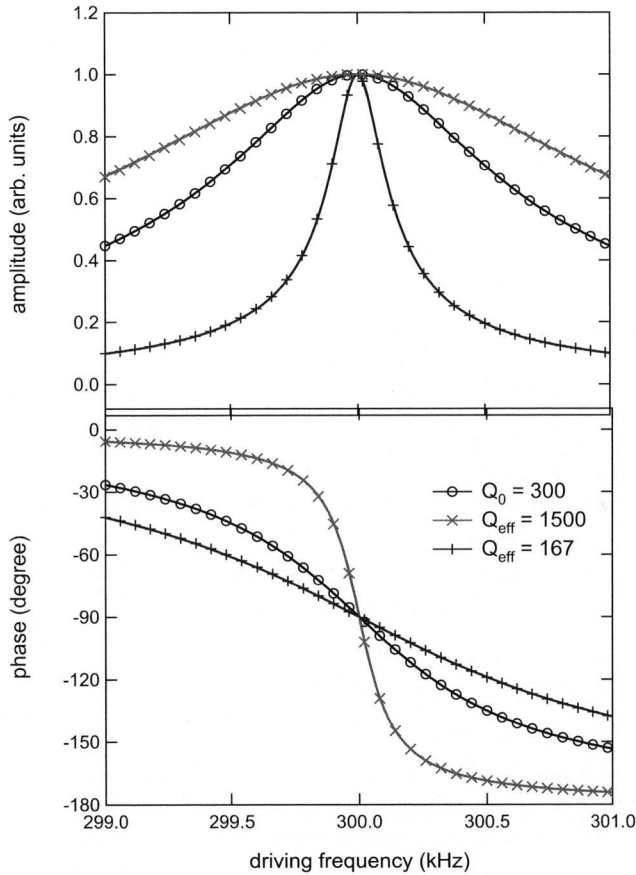

Fig. 4.2. Amplitude and phase vs. frequency curves with and without Q-control. For this example, the eigenfrequency and the natural Q-factor of the cantilever in air were assumed to be $f_0 = 300\,\mathrm{kHz}$ and $Q_0 = 300$, respectively. Using additionally a gain factor of $g = 0.8/Q_0$ and setting the time ("phase") shift to 90 or 270°, we can increase or decrease the effective Q-factor of the system to 1500 or to 167, respectively. The *solid lines* represent the analytical solutions (4.3a) and (4.3b), while the *symbols* reflect the numerical solution of the equation of motion (4.1)

4.2.2.2
Shifting the Resonance Curves

Another situation occurs for $t_0 = 1/2 f_0$ (180°), $1/f_0$ (360°), ... In such a case, the effective Q-factor of the system remains unchanged, but the resonant frequency of the system shifts to

$$f_{max} = f_0 \sqrt{1 \mp g} \approx f_0 \mp 0.5 f_0 g, \tag{4.6}$$

where the minus sign corresponds to 180° and the plus sign to 360°. Figure 4.3 displays resonance curves for these values. By setting the time ("phase") shift t_0 to 180 or 360°, the resonance curves are shifted along the frequency axis by ∓ 0.41 kHz for the specific values chosen in this example.

At the end of this section, let us note for completeness that the graphs displayed in Figs. 4.2 and 4.3 naturally reflect only the four most noticeable particular situations of the behavior of the resonance curves in comparison with the conventional tapping mode ($Q_0 = 300$). By choosing intermediate values for t_0, the amplitude and phase curves can be modified to any value within the limits discussed before.

4.2.3
Including Tip–Sample Interactions

During surface imaging, the vibrating cantilever is brought closely to the sample in order to monitor changes in the oscillation behavior induced by the tip–sample interaction. However, the mathematical form of realistic tip–sample forces is highly nonlinear for nearly all practical cases. This fact complicates the analytical solution of the equation of motion (4.1) even for simplified model forces. However, for the analysis of DFM experiments, we need to focus on steady-state solutions of the equation of motion with sinusoidal cantilever oscillation. Therefore, it is advantageous to expand the tip–sample force into a Fourier series:

$$
\begin{aligned}
F_{ts}[z(t), \dot{z}(t)] \approx & f_d \int_0^{1/f_d} F_{ts}[z(t), \dot{z}(t)] dt \\
& + 2 f_d \int_0^{1/f_d} F_{ts}[z(t), \dot{z}(t)] \cos(2\pi f_d t + \phi) dt \times \cos(2\pi f_d t + \phi) \\
& + 2 f_d \int_0^{1/f_d} F_{ts}[z(t), \dot{z}(t)] \sin(2\pi f_d t + \phi) dt \times \sin(2\pi f_d t + \phi) \\
& + \dots ,
\end{aligned}
\tag{4.7}
$$

where $z(t)$ is given by (4.2). With the assumption that the tip touches the surface only slightly during an individual oscillation cycle, it is sufficient to consider only the first harmonics of the system.

The first term in the Fourier series reflects the average tip–sample force over one full oscillation cycle, which shifts the equilibrium point of the oscillation from d to d_0. Actual values of this offset, however, are typically quite small (some picometers). For

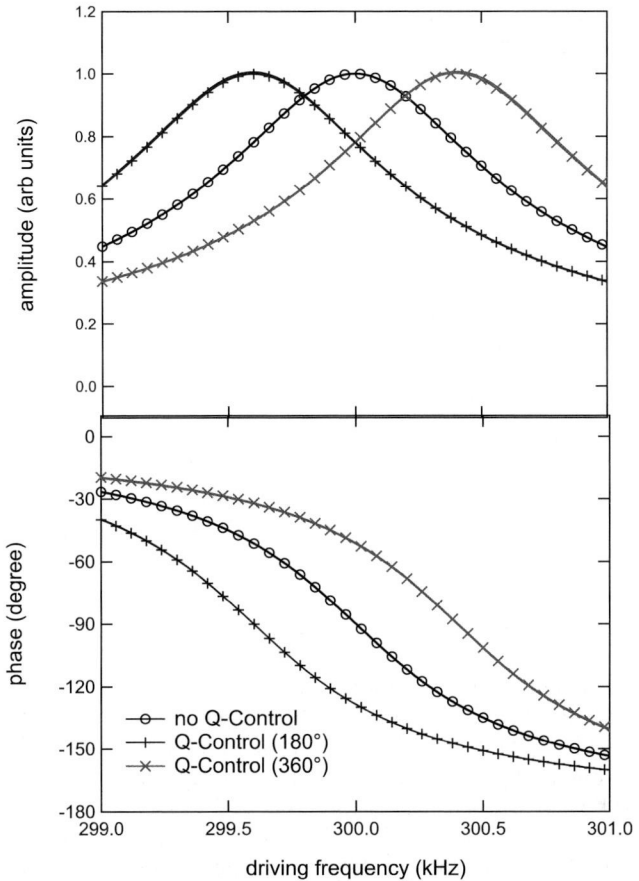

Fig. 4.3. Resonance curves for the same parameters as in Fig. 4.2 ($f_0 = 300\,\text{kHz}$, $Q_0 = 300$, $g = 0.8/Q_0$), but with time ("phase") shifts of 180 and 360°, respectively. The *solid lines* and *symbols* reflect again the analytical and numerical results. The amplitude and the phase curves do not change their shape compared with shapes for the solution without *Q*-control, but they are shifted along the frequency axis

typical amplitudes used in DFM in air (some nanometers), the average tip–sample force is in the range of some picometer. Since this is well beyond the resolution limit of a DFM experiment in air, we neglect this effect in the following and assume $d \approx d_0$. The tip–sample distance is therefore given by $D = d - A$.

For the further analysis, we now insert the first harmonics of the Fourier series (4.7) into the equation of motion (4.1), obtaining two coupled equations:

$$\frac{f_0^2 - f_\text{d}^2}{f_0^2} + g\cos(2\pi f_\text{d} t_0) = I_\text{odd}(d, A) + \frac{a_\text{d}}{A}\cos\phi\,, \tag{4.8a}$$

$$-\frac{1}{Q_0}\frac{f_\text{d}}{f_0} + g\sin(2\pi f_\text{d} t_0) = I_\text{even}(d, A) + \frac{a_\text{d}}{A}\sin\phi\,, \tag{4.8b}$$

where we defined the following integrals

$$
\begin{aligned}
I_{\text{odd}}(d, A) &= \frac{2f_{\text{d}}}{c_z A} \int_0^{1/f_{\text{d}}} F_{\text{ts}}[z(t), \dot{z}(t)] \cos(2\pi f_{\text{d}} t + \phi) \mathrm{d}t \\
&= \frac{1}{\pi c_z A} \int_0^{2\pi} F_{\text{ts}}[z(\tau), \dot{z}(\tau)] \cos(\tau) \mathrm{d}\tau \; ,
\end{aligned}
\tag{4.9a}
$$

$$
\begin{aligned}
I_{\text{even}}(d, A) &= \frac{2f_{\text{d}}}{c_z A} \int_0^{1/f_{\text{d}}} F_{\text{ts}}[z(t), \dot{z}(t)] \sin(2\pi f_{\text{d}} t + \phi) \mathrm{d}t \\
&= \frac{1}{\pi c_z A} \int_0^{2\pi} F_{\text{ts}}[z(\tau), \dot{z}(\tau)] \sin(\tau) \mathrm{d}\tau \; .
\end{aligned}
\tag{4.9b}
$$

Both integrals are functions of the actual oscillation amplitude A and cantilever sample distance d. The labeling "odd" for the integral containing the cosine and "even" for integral containing the sine was chosen in accordance with Dürig [28]. It is important to note that I_{even} is strongly connected to the dissipative tip–sample forces and vanishes for purely conservative tip–sample forces (see the discussions in [28–30]).

In order to ease the analysis of (4.8a) and (4.8b), we make three simplifying assumptions: (1) we assume an ideal time delay corresponding to 90° for an optimal increase of the Q-factor; (2) we restrict ourselves to the situation that the driving frequency is set *exactly* to the eigenfrequency of the cantilever ($f_{\text{d}} = f_0$); (3) the tip–sample force F_{ts} is postulated to be purely conservative, which causes the integral I_{even} to vanish. As a consequence, we get a handy relationship between the free oscillation amplitude A_0, the actual amplitude A, and the equilibrium tip position d:

$$
A_0 = A \sqrt{1 + [Q_{\text{eff}} I_{\text{odd}}(d, A)]^2} \; .
\tag{4.10}
$$

For the derivation of this formula, we used the approximation that the maximal value of the free oscillation amplitude at resonance is given by $A_0 \approx a_{\text{d}} Q_{\text{eff}}$. We will use mainly this equation together with numerical simulations to analyze the features of Q-control in ambient conditions and liquids in the next two sections.

4.2.4
Prevention of Instabilities by Q-control in Air

For the analysis of Q-control under ambient conditions, we choose a simplified model describing the tip–sample interactions. As with other authors [31–35], we assume that a sphere with radius R interacts with a flat surface and experiences long-range attractive forces described by a van der Waals term and short-range repulsive forces upon contact.

The long-range van der Waals forces are described by

$$
F_{\text{vdW}}(z) = -\frac{A_{\text{H}} R}{6z^2} \; ,
\tag{4.11}
$$

where A_{H} is the Hamaker constant. If the tip comes very close to the sample surface, the repulsive forces between tip and sample become significant. For simplicity, we assume that the geometrical shape of tip and sample does not change until contact has been established at $z = h_0$ and that afterwards the tip–sample forces are given by the well-known Hertz theory. In this approach, an offset $F_{\mathrm{vdW}}(h_0)$ is added considering the adhesion force between tip and sample surface; therefore, this model is often also referred to as the *Hertz-plus-offset model* [36]. The resulting overall force law is given by

$$F_{\mathrm{air}}(z) = \begin{cases} F_{\mathrm{vdW}}(z) & \text{for } z \geq h_0 \,, \\ \dfrac{4}{3}E^*\sqrt{R}(h_0 - z)^{3/2} + F_{\mathrm{vdW}}(h_0) & \text{for } z < h_0 \,. \end{cases} \quad (4.12)$$

The effective modulus E^*

$$\frac{1}{E^*} = \frac{\left(1 - \mu_{\mathrm{t}}^2\right)}{E_{\mathrm{t}}} + \frac{\left(1 - \mu_{\mathrm{s}}^2\right)}{E_{\mathrm{s}}} \quad (4.13)$$

depends on the elastic moduli $E_{\mathrm{t,s}}$ and the Poisson ratios $\mu_{\mathrm{t,s}}$ of tip and sample, respectively.

Figure 4.4 displays the assumed tip–sample model force. For the plot, the following parameters were used, which are typical for DFM measurements in air: $A_{\mathrm{H}} = 0.2\,\mathrm{aJ}$, $R = 10\,\mathrm{nm}$, $h_0 = 0.3\,\mathrm{nm}$, $\mu_{\mathrm{t}} = \mu_{\mathrm{s}} = 0.3$, $E_{\mathrm{t}} = 130\,\mathrm{GPa}$, and $E_{\mathrm{s}} = 1\,\mathrm{GPa}$. The eigenfrequency, the natural Q-factor, and the spring constant of the cantilever were chosen to be $f_0 = 300\,\mathrm{kHz}$, $Q_0 = 300$, and $c_z = 40\,\mathrm{N/m}$, respectively.

Introducing now the tip–sample force F_{air} into (4.10) and solving the resulting equation, we can study amplitude vs. distance curves for different effective Q-factors, as has been done in Fig. 4.5 for a natural Q-factor of 300 and an increased Q-factor of 1500 (dashed lines). The resulting curves differ significantly for the two different Q-values. Stable and unstable branches manifest, which can unambiguously be identified by a comparison with numerical results (symbols).

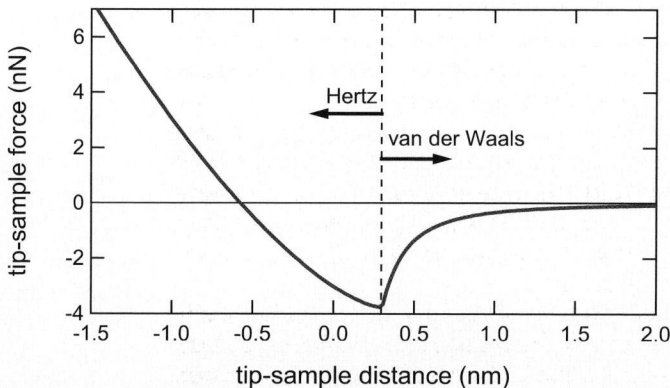

Fig. 4.4. Tip–sample model force for air (4.12) using typical parameters for ambient conditions. The *dashed line* at h_0 marks the borderline between the van der Waals (*right*) and the Hertz (*left*) forces

Fig. 4.5. Amplitude vs. distance curves for tapping-mode dynamic force microscopy (DFM) (*top*) and Q-controlled DFM (*bottom*) for $A_0 = 10$ nm, $f_0 = 300$ kHz, and a tip–sample interaction force as given in Fig. 4.4. The *dashed lines* represent the analytical result, while the *symbols* are obtained from the numerical solution of the equation of motion (4.1). The overall amplitudes decrease during an approach towards the sample surface in both cases, but the frequently discussed instability occurs only in the conventional tapping mode

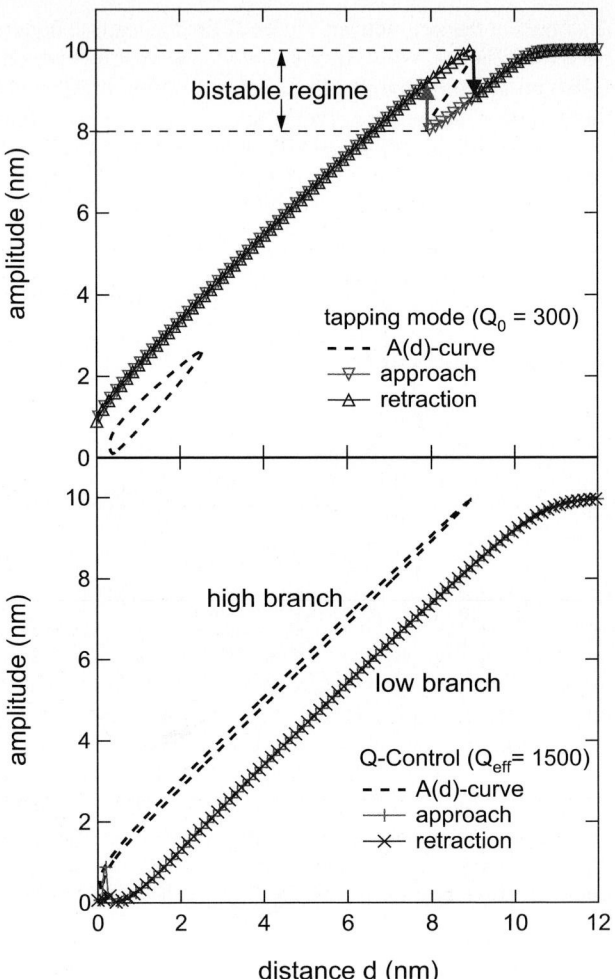

Most noticeably, the tapping mode curve exhibits jumps at unstable positions, which occur at different locations for approach and retraction. The resulting bistable regime then causes a hysteresis between approach and retraction, which was well examined in numerous experimental and theoretical studies (e.g., [33, 34, 37–40]). The Q-control curve, however, departs significantly from the qualitative behavior of the tapping mode curve. It is found that the oscillating cantilever stays always on the lower branch of the amplitude vs. distance curve and does not jump to the higher branch. Consequently, there is no hysteresis between approach and retraction.

This considerable difference between the amplitude vs. distance curves has a remarkable consequence for the strengths of the tip–sample interaction. As shown by various authors [33, 34, 39, 40], the instability in conventional amplitude modulation DFM divides the tip–sample interaction into two regimes. Before the instability occurs, the tip tracks exclusively the attractive part of the tip–sample force. After jumping to the higher branch, however, the tip senses also the repulsive part of the tip–sample interaction. Since this instability is not present for the increased Q-factor of $Q_{\mathrm{eff}} = 1500$, one might assume that the tip remains under these circumstances only within the attractive part of the tip–sample interaction. This is indeed the case, as we will show in the following.

The different interaction regimes realized by conventional DFM and QC-DFM, respectively, are visualized in Fig. 4.6, which summarizes our theoretical analysis. The oscillation amplitude is plotted as a function of the nearest tip–sample distance

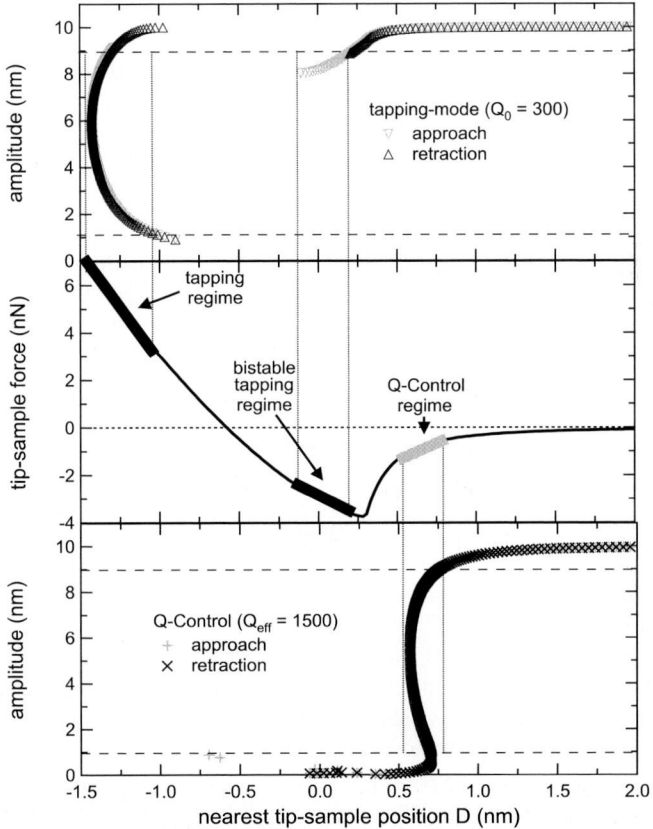

Fig. 4.6. A comparison between the maximum tip–sample forces (tip–sample forces acting at the point of closest tip–sample approach/nearest tip–sample position D) experienced by tapping-mode DFM and Q-controlled DFM assuming the same parameters as in Fig. 4.5. The *top* and *bottom graphs* show the nearest tip–sample position D vs. the actual oscillation amplitude A for tapping mode and Q-control, respectively. The *middle graph* reveals the force regimes sensed by the tip in the two imaging modes. The maximal tip–sample forces in tapping mode are on the repulsive (tapping regime) as well as the attractive (bistable tapping regime) part of the tip–sample force curve, but in both cases are beyond the contact point at $h_0 = 0.3$ nm. In contrast, the Q-control regime is not only entirely located in the attractive part of the force curve, but also always well within the noncontact regime ($D > 0.3$ nm). For any condition, the nearest tip–sample position is larger with active Q-control compared with conventional tapping mode imaging

for tapping mode DFM (top) and QC-DFM (bottom). These curves can easily be generated from the numerical simulations. In addition, the middle graph depicts the corresponding tip–sample force (cf. Fig. 4.4). The origin of the nearest tip–sample position D is defined by this force curve. Since the amplitude curves and the tip–sample force curve are all plotted as a function of the nearest tip–sample position, it is possible to identify the resulting maximum tip–sample interaction force for a given oscillation amplitude.

A closer look at the $A(D)$ curves helps now to identify the different interaction regimes in conventional DFM and QC-DFM. First, we analyze the tapping mode curve at the top. During the approach of the vibrating cantilever towards the sample surface, this curve shows a discontinuity for the nearest tip–sample position D (point of closest approach during an individual oscillation) between 0 and −1 nm. This gap corresponds to the bistability and the resulting jumps in the amplitude vs. distance curve. After the jump from the attractive to the repulsive regime has occurred, the amplitude decreases continuously. The nearest tip–sample position, however, does not reduce accordingly, remaining roughly between −0.8 and −1.5 nm. In addition, it is interesting to note that the minimum distance D is reached around 6-nm amplitude; for smaller amplitudes, D even increases again (smaller negative values)!

For practical applicability, it is reasonable to assume that the setpoint of the amplitude used for imaging has been set to a value between 90% (= 9 nm) and 10% (= 1 nm) of the free oscillation amplitude. With this condition, we can identify the accessible imaging regimes indicated by the horizontal (dashed) lines and the corresponding vertical (dotted) lines. In tapping mode, two imaging regimes are realized: the *tapping regime* (left) and the *bistable tapping regime* (middle). The first one can be accessed by any amplitude setpoint between 9 and 1 nm and results in maximum tip–sample forces well within the repulsive regime. The second one, belonging to the bistable imaging state, is only accessible during approach. The corresponding amplitude setpoint is between 9 and 8 nm. Please note that even though the maximum tip–sample forces are attractive, the tip already touches the sample in this regime, since $D < h_0 = 0.3$ nm. Imaging in this regime is possible with the limitation that the oscillating cantilever might jump into the repulsive regime [35,39].

Quite different behavior is observed for the Q-controlled curve, which is continuous and shows no jump. Interestingly, the nearest tip–sample position reduces not further than 0.5 nm, although the amplitude decreases from 10 to 1 nm at the same time. Using the same graphical analysis as for the tapping mode (shown by the dashed and dotted lines), we get only one accessible imaging regime (right) where the nearest tip–sample distance is well before the bistable tapping mode regime. This *Q-control regime* is entirely well inside the attractive van der Waals part of the tip–sample force curve, thus representing pure noncontact imaging conditions. Consequently, Q-control suppresses too close an approach of the tip towards the sample surface, leading to low maximum tip–sample interaction forces.

4.2.5
Reduction of Tip–Sample Indentation and Force by Q-control in Liquids

In the previous section, we suggested that the main benefit of Q-control in air is its ability to effectively minimize tip–sample forces by restricting them to certain parts of the attractive regime. However, since attractive surface forces are significantly reduced in liquids, if present at all, the influence of the Q-factor in a liquid environment is carefully reinvestigated in this section.

The main difference between measurements in liquids as opposed to measurements in air other than a much reduced Q-factor due to the enhanced damping caused by the fluid is the fact that surface forces in liquids are generally considerably reduced [41–43]. Therefore, we neglect the attractive part of the former tip–sample interaction force and describe the interaction force solely by the well-known Hertzian model force

$$
F_{\text{liquid}}(z) = \begin{cases} 0 & \text{for} \quad z \geq h_0 , \\ \dfrac{4}{3} E^* \sqrt{R}(h_0 - z)^{3/2} & \text{for} \quad z < h_0 , \end{cases}
\tag{4.14}
$$

which reflects a good first approximation to realistic force laws and can be expected to allow a reasonable simulation of the actual dynamic behavior of the cantilever. Thereby, the tip is assumed to feature a spherical apex with radius R, while h_0 represents the position of the undeformed sample surface. The effective modulus E^* is again given by (4.13).

For the further analysis, we used the following parameters: $R = 10$ nm, $f_0 = 20$ kHz, $c_z = 1$ N/m, $\mu_t = \mu_s = 0.3$, and $E_t = 130$ GPa. As a model sample, we assumed a soft island ($E_{\text{island}} = 1$ GPa) of 2-nm height, which is supported by a considerably harder substrate ($E_{\text{substrate}} = 20$ GPa). The geometry is plotted as an inset in Fig. 4.7b. The height of the island was chosen to coincide with the expected diameter of adsorbed DNA, which frequently served as a "benchmark sample" for the analysis of QC-DFM experiments in the past [10,19]. To represent typical values for Q in liquids, we further assumed a natural Q-factor of $Q_0 = 5$. The function generator frequency is set to $f_d = f_0$, which is a common choice in actual DFM

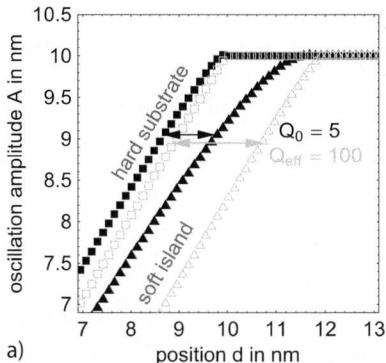
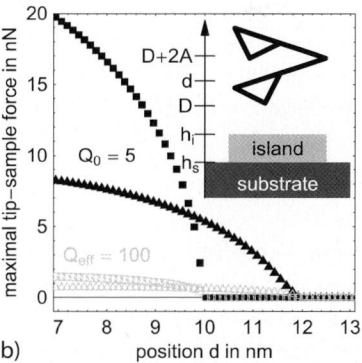

a)   b)

Fig. 4.7. (a) Amplitude vs. distance curves for tapping mode DFM (*filled symbols*) and Q-controlled DFM (*open symbols*). The *triangles* and *squares* reflect the amplitudes on the soft island and the hard substrate, respectively. The two *arrows* indicate the measured apparent height differences in the two different cases. (b) A The corresponding maximum tip–sample forces. They are found to be considerably smaller with active Q-control compared with operation without Q-enhancement. The *inset* shows the assumed geometry. A soft island ($h_i = 2$ nm) is adsorbed on a considerably harder substrate ($h_s = 0$ nm)

experiments and ensures defined imaging conditions. As shown in Sect. 4.2.2, the activation of Q-control with gain factor $g = 0.95/Q_0$ and time delay $t_0 = 1/(4f_0)$ (which reflects the optimal phase shift of 90°) then increases the effective Q-factor of the system to $Q_{\text{eff}} = 1/(1/Q_0 - g) = 100$.

In Fig. 4.7, we present the numerical solution of the equation of motion using a fourth-order Runge-Kutta method [27] with a free oscillation amplitude of $A_0 = 10$ nm. The amplitude vs. distance curves are plotted in Fig. 4.7a for the hard substrate and the soft island with and without active Q-control. The overall shape of the curves is quite similar: first, the oscillation amplitude is constant before it decreases almost linearly as the oscillating tip approaches the sample surface. The final slope of the curves, however, is nearly independent of the actual Q-factor. Please note that in contrast to corresponding spectroscopy curves in air (see the previous section), there are no unstable branches due to the absence of attractive surface forces. As a consequence, there is no hysteresis between approach and retraction.

The further analysis of these curves allows us to directly determine the apparent step height that would be measured between the substrate and the island during an actual measurement. For a setpoint amplitude of $A = 9$ nm, for example, an apparent height of approximately 1.8 nm is obtained with increased Q-factor (lower arrow), which is considerably larger than the corresponding value for conventional tapping mode (upper arrow, height approximately 1.1 nm).

Figure 4.7b reflects the corresponding maximum tip–sample force F_{max}, i.e., the tip–sample force at the lower turning point D as a function of the actual cantilever position d. Values up to 9 nN are revealed on the soft island in conventional tapping mode for $d \geq 7$ nm, which is about 2 orders of magnitude larger than the piconewton regime necessary to achieve molecular resolution on two-dimensional protein arrays [44]. The forces even increase to 20 nN on the hard substrate. Note, however, that F_{max} is significantly reduced by activating Q-control. On the soft island, for example, interaction forces drop to below 700 pN, which is over 10 times less than the corresponding values in conventional tapping mode.

For a better understanding of the numerical findings, we present an additional analytical analysis. It is again based on (4.10), which gives a relationship between all important parameters. Owing to our description of the tip–sample force by the Hertz model, we are able to derive a very accurate approximation of the integral function I_{odd} (4.9):

$$I_{\text{odd}}(d, A) = -\frac{E^*\sqrt{R}(d - A - h_0)^2}{\sqrt{2}c_z A^{3/2}} . \tag{4.15}$$

The approximation (4.15) is valid as long as the oscillation amplitude is considerably larger than the indentation of the tip into the sample surface ($A \gg h_0 - D$) [45]. Introducing this result into (4.10), we obtain a simple expression for the nearest tip–sample position as a function of the oscillation amplitude:

$$D(A) = h_0 - \sqrt{\frac{c_z}{Q_{\text{eff}}E^*}\sqrt{\frac{2A\left(A_0^2 - A^2\right)}{R}}} . \tag{4.16}$$

Fig. 4.8. Amplitude A vs. nearest position D curves for tapping-mode DFM (*filled symbols*) and Q-controlled DFM (*open symbols*). The *lines* reflect the analytical approximation (4.16), while the numerical results for the different cases are indicated by the same symbols as in Fig. 4.7

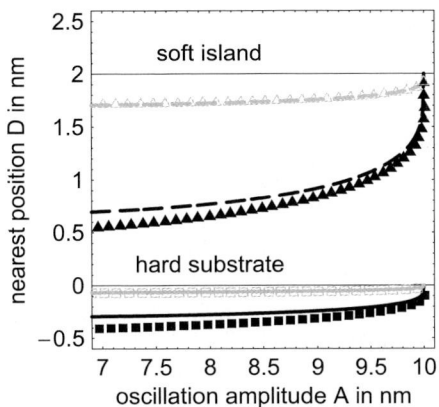

This equation reveals how the tip–sample indentation depends on the various experimental parameters. Obviously, the softer the sample, the larger the indentation. On the other hand, it also becomes evident that not only higher Q-factors help reduce excessive sample deformation, as we already know from the numerical results, but also softer cantilevers and tips with larger apex radii. Large tip radii, however, are not desirable since they negatively affect the lateral resolution. Similarly, a reduction of the tip–sample indentation can also be obtained by decreasing the free oscillation amplitude A_0. Very small values for A_0, however, might negatively affect the signal-to-noise ratio.

The accuracy of (4.16) was cross-checked by comparison with numerical results and turned out to be very reasonable (Fig. 4.8). In all cases, the nearest tip–sample position D reduces nonlinearly as the oscillating tip approaches the sample: after an initial sharp decrease, it somehow levels out to almost constant values. It is also striking that while the absolute indentation on the soft island reaches values as high as 1.5 nm for a natural Q-factor of 5, drastically lower deformation is found for the increased Q-factor of 100 (less than 0.3 nm). This strong dependence of the indentation on the Q-factor can be also observed for the substrate.

4.3
Experimental Applications of Q-control

In this section, we present some illustrative examples of QC-DFM applications in air and liquids. First, we review some results obtained in ambient conditions before discussing results obtained in liquids. The experimental data shown here were obtained with a commercial atomic force microscope (NanoScope IIIa with MultiMode head, Veeco Instruments). An open liquid cell for acoustic excitation of the cantilever was used for imaging in liquids. The Q-factor was controlled by Q-control electronics (nanoAnalytics), which were added externally to the existing Nanoscope IIIa electronics.

4.3.1
Examples for Q-control Applications in Ambient Conditions

Increasing and Decreasing the Q-factor

The active increasing and decreasing of the Q-factor theoretically examined in Sect. 4.2.2 is demonstrated in Fig. 4.9. A silicon cantilever with a nominal spring constant of $40\,\mathrm{N/m}$ was used to measure resonance curves in ambient conditions. The recorded data points of the amplitude and phase vs. frequency curves were fitted using (4.3a) and (4.3b), respectively. The fits delivered the corresponding Q-factors for conventional tapping mode and Q-controlled operation. The values obtained are summarized in Fig. 4.9.

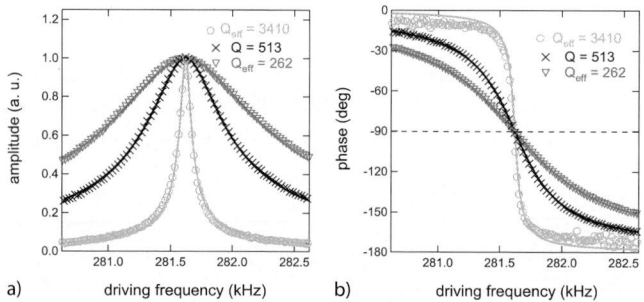

a)   b)

Fig. 4.9. Experimental resonance curves obtained with and without active Q-control under ambient conditions. The *solid lines* are fits to the experimental data plotted as *symbols*. The Q-factor in the conventional tapping mode was 513, while the effective Q-factor could be increased to 3410 or decreased to 262 using the Q-control electronics

Increase of the Scan Velocity

The active damping of the cantilever by Q-control was introduced by Sulchek et al. [11, 12]. As already shown by Albrecht et al. [46], the maximum accessible scan velocity in amplitude modulation AFM is strongly correlated with the cantilever damping. The response of the amplitude A to a sudden change in the topography such as, e.g., a step, can be described by a time constant $\tau = Q/(\pi f_0)$ [46], i.e., the larger the Q-factor, the longer it takes until the cantilever reaches again a steady-state oscillation. Consequently, it is advantageous to decrease the effective Q-factor for a fast response.

This effect is demonstrated in Fig. 4.10. A typical scan speed in air was about $1\,\mathrm{Hz}$ per scan line for the example presented, a $4\,\mu\mathrm{m}$ scan on a calibration test grid. Reducing the Q-factor from 416 to 151, we could increase the scan speed to $4\,\mathrm{Hz}$, while an increase of the factor to 2457 resulted in disturbed scan lines at 1-Hz scan speed. Scan speeds had to be decreased to $0.3\,\mathrm{Hz}$ to reestablish stable scanning. In general, however, it has to be considered that the maximal scan velocity is also influenced by other factors such as the bandwidth of the feedback electronics controlling the actual cantilever sample distance; a detailed discussion of this issue can be found in [12].

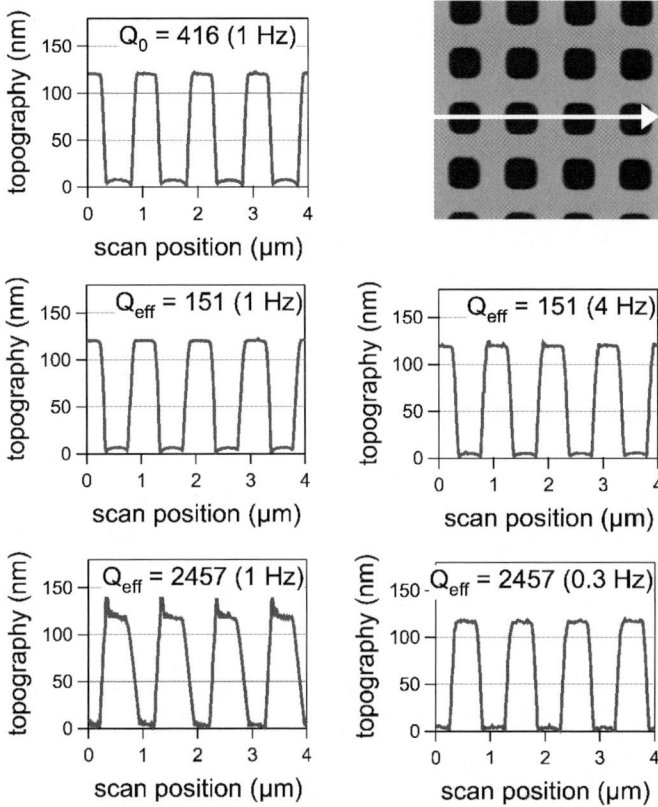

Fig. 4.10. Measurements performed on a test grid demonstrating the influence of the *Q*-factor on the possible scan speed in tapping mode. The *upper-left plot* shows a typical scan line obtained with a scan speed of 1 Hz (equivalent to 4 μm/s) and a natural *Q*-factor of 416 in ambient conditions. The *upper-right image* displays the topography of the corresponding sample (a test grid). The *plots in the middle row* show scan lines for a decreased effective *Q*-factor of 151. The scan speed could be increased up to 4 Hz. The opposite effect is observed for an increased *Q*-factor of 2457 (*bottom row*). Here, the topography is severely disturbed, making a decrease of the scan speed to 0.3 Hz necessary

Controlling the Tip–Sample Forces

An example for the reduction and control of tip–sample forces has been already shown in the chapter of Schirmeisen et al. in the first book of this series [47]. In their Fig. 1.19, they presented a surface scan of an ultrathin organic film acquired in tapping mode under ambient conditions. First, the inner square of the sample was scanned without the *Q*-enhancement, and then a wider surface area was scanned with applied *Q*-control. The high quality factor provides a larger parameter space for operating the atomic force microscope in the net-attractive regime, allowing good resolution of the delicate organic surface structure. Without *Q*-control, surface structures are deformed or even destroyed owing to the strong repulsive tip–sample interactions.

This effect also allowed imaging of DNA structures without predominantly depressing the soft material during imaging. With active Q-control feedback, Pignataro et al. [21] were able to obtain images from DNA molecules featuring diameters close to the theoretically expected value.

Examples for Q-control Applications in Liquid Environments

As reported by different authors, employing the Q-control detection scheme turned out to be helpful also in liquids environments [9, 10, 18–20]. Even though cantilevers are heavily damped in fluids, their Q-factors can nevertheless be increased to values comparable to those in ambient conditions. Here, we present applications of Q-control in water.

Resonance Curves

Figure 4.11 illustrates the increase of the effective Q-factor in water. Both resonance curves and all further data sets presented were measured with silicon cantilevers (Tap-Multi75, BudgetSensors) featuring nominal spring constants of $c_z = 3\,\text{N/m}$ and eigenfrequencies of $f_0 = 75\,\text{kHz}$. The symbols in Fig. 4.11 represent the measurements with (circles) and without (squares) active Q-control. The solid lines are fits to these data points, representing the Lorentzian shape of the resonance peak of an externally driven harmonic oscillator. Fitting the experimental data points, we obtained Q-factors of $Q_{\text{eff}} = 203$ with and $Q = 34$ without Q-control.

Fig. 4.11. Typical resonance curves obtained with and without active Q-control in water. The *solid lines* are fits to the experimental data plotted as *symbols*, which result in Q-factors of $Q_{\text{eff}} = 202.7 \pm 2.2$ with active Q-control and $Q = 33.6 \pm 1.2$ without Q-control, respectively

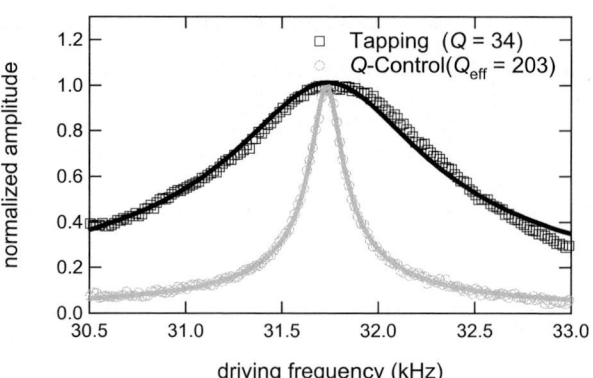

Dipalmitoylphosphatidycholine in Water

As a final example, we present a comparison of dynamic force microscopy images obtained with and without Q-Control on a soft sample. As a biologically relevant benchmark sample, we choose bilayers of dipalmitoylphosphatidycholine (DPPC), which is frequently used as a model system for biological membranes [48]. In order to allow a meaningful comparison between the results obtained with and without active Q-control, the images displayed were recorded at identical sample positions with Q-control on and off. All images were measured with setpoints and gains optimized for the respective imaging mode at the resonant frequency of the freely oscillating cantilever.

Figure 4.12 shows a $7 \times 7\,\mu m^2$ area of the DPPC sample. The data displayed in Fig. 4.12a were recorded applying conventional tapping mode with a Q-factor of 39, while data shown in Fig. 4.12b were taken with active Q-control and an increased Q-factor of $Q_{eff} = 209$ (see [10] for more experimental details). The two scan lines displayed in at the bottom of Fig. 4.12 reveal a significant change in the apparent height difference between the individual terraces if measured with and without active Q-control. In conventional tapping mode, the apparent bilayer height is (3.3 ± 0.4) nm, which is significantly less than the value of (5.2 ± 0.4) nm measured with active Q-control.

This result is in agreement with the theory presented in Sect. 4.2.5. Since the application of Q-control actually reduces the tip–sample interaction forces, it is ensured that the measured film thickness is systematically higher with Q-control. This result is supported by a study of Hartig et al. [49]. They systematically measured the the apparent layer height of Langmuir–Blodgett films as a function of the applied load and reported that the correct film thickness could only be measured in contact mode by minimizing the tip–sample force, while measurements in tapping mode without Q-control always led to decreased film thicknesses.

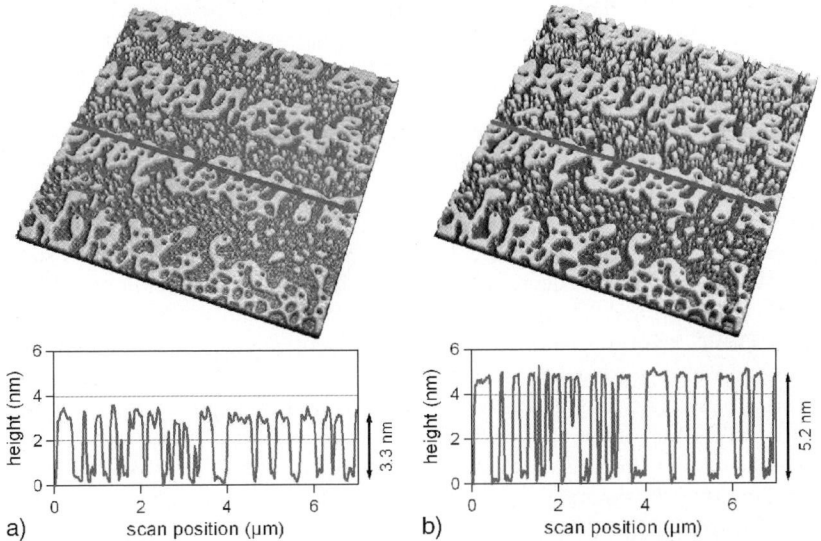

Fig. 4.12. Dipalmitoylphosphatidycholine bilayers on mica imaged in pure water (**a**) without and (**b**) with active Q-control. The scan lines plotted at the *bottom* were obtained at the positions marked by the *arrows* in the images at the *top*. The measured apparent bilayer height in (**a**) is (3.3 ± 0.4) nm and grows to (5.2 ± 0.4) nm in (**b**) when Q-control is switched on

4.4
Summary

In this chapter, we reviewed the theoretical background of QC-DFM in comparison with conventional amplitude modulation DFM ("tapping mode"). It was demonstrated how the positive feedback of Q-control can be used to modify the resonance curves of the oscillating cantilever by varying the phase shift of the positive feedback. In this way, the effective Q-factor of the system can either be increased or decreased and the peak of the resonance can be shifted. The accuracy of the derived formulas could be demonstrated by comparison with numerical simulations.

In an additional step, we included a model for tip–sample forces in air into the theory and calculated the stable and unstable oscillation regimes of the cantilever with respect to the effective Q-factor. With this analysis, we were able to show that an increased Q-factor suppresses the occurrence of the bistable imaging regime usually observed in conventional tapping mode in air. Therefore, Q-control prevents the oscillating cantilever from jumping into the repulsive imaging regime. As a consequence, the maximum tip–sample forces are restricted to the attractive noncontact regime, and a penetration of the tip into the sample surface can be successfully avoided. The comparison in Fig. 4.6 also shows that the nearest tip–sample distance with Q-control is larger even if compared with that for the attractive but bistable tapping mode regime.

In a next step, an analysis of Q-control in liquids was performed. Using a tip–sample model force that neglects attractive surface forces, we derived an explicit analytical formula for the nearest tip–sample position, and this was compared with numerical simulations. It could be shown that increasing the effective Q-factor limits the elastic deformation of the sample surface by reducing the maximum tip–sample forces. As a result, absolute heights of soft matter deposited on hard substrates are reflected more accurately if measured with high effective Q-factors.

Finally, we gave some experimental examples for Q-control applications in air and liquids and compared them with results of conventional ("tapping mode") amplitude modulation AFM imaging. Theoretically examined features like the increase and decrease of the Q-factor could be demonstrated by experimental resonance curves. We showed how a decreased Q-factor helps to increase the possible scan speed on a test grid. As a last example, we presented an analysis of the topography of DPPC bilayers obtained with and without Q-control using individually optimized parameters. We observed a significant height difference in the two operational modes. The heights measured with Q-control were reproducibly higher compared with those measured with conventional amplitude modulation AFM. In agreement with our theoretical analysis in Sect. 4.2.5, this effect could be attributed to the reduction of tip–sample forces by Q-control.

Acknowledgements. The authors would like to thank Boris Anczykowski (nanoAnalytics) for continuous support and many helpful discussions. This project was supported financially by the German Federal Ministry of Education and Research (BMBF) (grant no. 03N8704).

References

1. Binnig G, Quate CF, Gerber Ch (1986) Atomic force microscopy. Phys Rev Lett 56:930–933
2. Martin Y, Williams CC, Wickramasinghe HK (1987) Atomic force microscope–force mapping and profiling on a sub 100-Å scale. J Appl Phys 61:4723–4729
3. Zhong QD, Inniss D, Kjoller K, Elings VB (1993) Fractured polymer/silica fiber surface studied by tapping mode atomic force microscopy. Surf Sci Lett 290:L688–L692
4. Hansma PK, Cleveland JP, Radmacher M, Walters DA, Hillner PE, Bezanilla M, Fritz M, Vie D, Hansma HG, Prater CB, Massie J, Fukunaga L, Gurley L, Elings VB (1994) Tapping mode atomic force microscopy in liquids. Appl Phys Lett 64:1738–1740
5. Putman CAJ, Vanderwerf KO, Degrooth BG, Vanhulst NF, Greve J (1994) Tapping mode atomic force microscopy in liquid. Appl Phys Lett 64:2454–2456
6. Anczykowski B, Cleveland JP, Krüger D, Elings VB, Fuchs H (1998) Analysis of the interaction mechanisms in dynamic mode SFM by means of experimental data and computer simulation. Appl Phys A 66:S885–S889
7. Tamayo J, Humphris ADL, Miles MJ (2000) Piconewton regime dynamic force microscopy in liquid. Appl Phys Lett 77:582–584
8. Humphris ADL, Tamayo J, Miles MJ (2000) Active quality factor control in liquids for force spectroscopy. Langmuir 16:7891–7894
9. Grant A and McDonnell L (2003) A non-contact mode scanning force microscope optimised to image biological samples in liquid. Ultramicroscopy 97:177–184
10. Ebeling D, Hölscher H, Fuchs H, Anczykowski B, Schwarz UD (2006) Imaging of biomaterials in liquids: a comparison between conventional and Q-controlled amplitude modulation ("tapping mode") atomic force microscopy. Nanotechnology 17:S221–S226
11. Sulchek T, Hsieh R, Adams JD, Yaralioglu GG, Minne SC, Quate CF, Cleveland JP, Atalar A, Adderton DM (2000) High-speed tapping mode imaging with active Q control for atomic force microscopy. Appl Phys Lett 76:1473–1475
12. Sulchek T, Yaralioglu GG, Quate CF, Minne SC (2002) Characterization and optimization of scan speed for tapping-mode atomic force microscopy. Rev Sci Inst 73:2928–2936
13. Antognozzi M, Szczelkun MD, Humphris ADL, Miles MJ (2003) Increasing shear force microscopy scanning rate using active quality-factor control. Appl Phys Lett 82:2761–2763
14. Lei FH, Nicolas J-L, Troyon M, Sockalingum GD, Rubin S, Manfait M (2003) Shear force detection by using bimorph cantilever with enhanced Q-factor. J Appl Phys 93:2236–2243
15. Yamanaka K, Maruyama Y, Tsuji T, Nakamoto K (2001) Resonance frequency and Q factor mapping by ultrasonic atomic force microscopy. Appl Phys Lett 78:1939–1941
16. Fukuda K, Irihama H, Tsuji T, Nakamoto K, Yamanaka K (2003) Sharpening contact resonance spectra in UAFM using Q-control. Surf Sci 532–535:1145
17. Hölscher H (2002) Q-controlled dynamic force spectroscopy. Surf Sci 515:526
18. Gao S, Chi LF, Lenert S, Anczykowski B, Niemeyer C, Adler M, Fuchs H (2001) High-quality mapping of DNA-protein complexes by dynamic scanning force microscopy. Chem Phys Chem 6:384–388
19. Humphris ADL, Round AN, Miles MJ (2001) Enhanced imaging of DNA via active quality factor control. Surf Sci 491(3):468–472
20. Tamayo J, Humphris ADL, Owen RJ, Miles MJ (2001) High-Q dynamic force microscopy in liquid and its application to living cells. Biophys J 81:526–537
21. Pignataro B, Chi LF, Gao S, Anczykowski B, Niemeyer C, Adler M, Fuchs H (2002) Dynamic force microscopy study of self-assembeld dna-protein nanostructures. Appl Phys A 74:447
22. Jäggi RD, Franco-Obregon A, Studerus P, Ensslin K (2001) Detailed analysis of forces influencing lateral resolution for Q-control and tapping mode. Appl Phys Lett 79:135–137

23. Rodriguez TR, Garcia R (2003) Theory of Q-control in atomic force microscopy. Appl Phys Lett 82:4821–4823
24. Kokavecz J, Horváth ZL, Melcher A (2004) Dynamical properties of the Q-controlled atomic force microscope. Appl Phys Lett 85:3232–3234
25. Tamayo J (2005) Study of the noise of micromechanical oscillators under quality factor enhancement via driving force control. J Appl Phys 97:044903
26. Hölscher H, Ebeling D, Schwarz UD (2006) Theory of Q-controlled dynamic force microscopy in air. J Appl Phys 99:084311
27. Press WH, Tekolsky SA, Vetterling WT, Flannery BP (1992) Numerical Recipes in C. Cambridge University Press
28. Dürig U (2000) Interaction sensing in dynamic force microscopy. N J of Phys 2:5.1–5.12
29. Hölscher H, Gotsmann B, Allers W, Schwarz UD, Fuchs H, Wiesendanger R (2001) Measurement of conservative and dissipative tip-sample interaction forces with a dynamic force microscope using the frequency modulation technique. Phys Rev B 64:075402
30. Sader JE, Uchihashi T, Farrell A, Higgins MJ, Nakayama Y, Jarvis SP (2005) Quantitative force measurements using frequency modulation atomic force microscopy – theoretical foundations. Nanotechnology 16:S94–S101
31. Aimé JP, Boisgard R, Nony L, Couturier G (1999) Nonlinear dynamic behavior of an oscillating tip-microlever system and contrast at the atomic scale. Phys Rev Lett 82:3388–3391
32. García R, San Paulo A (1999) Attractive and repulsive tip-sample interaction regimes in tapping-mode atomic force microscopy. Phys Rev B 60:4961–4967
33. Lee SI, Howell SW, Raman A, Reifenberger R (2002) Nonlinear dynamics of microcantilevers in tapping mode atomic force microscopy: a comparison between theory and experiment. Phys Rev B 66:115409
34. Zitzler L, Herminghaus S, Mugele F (2002) Capillary forces in tapping mode atomic force microscopy. Phys Rev B 66:155436
35. Stark RW, Schitter G, and Stemmer A (2003) Tuning the interaction forces in tapping mode atomic force microscopy. Phys Rev B 68:085401
36. Schwarz UD (2003) A generalized analytical model for the elastic deformation of an adhesive contact between a sphere and a flat surface. J Coll Interf Sci 261:99–106
37. Anczkowski B, Krüger D, Fuchs H (1996) Cantilever dynamics in quasinoncontact force microscopy: Spectroscopic aspects. Phys Rev B 53:15485–15488
38. García R, San Paulo A (2000) Dynamics of a vibrating tip near or in intermittent contact with a surface. Phys Rev B 61(20):R13381–R13384
39. San Paulo A, García R (2000) High-resolution imaging of antibodies by tapping-mode atomic force microscopy: Attractive and repulsive tip-sample interaction regimes. Biophys J 78:1599–1605
40. San P A, García R (2002) Unifying theory of tapping-mode atomic-force microscopy. Phys Rev B 66:041406
41. Weisenhorn AL, Maivald P, Butt HJ, Hansma PK (1992) Measuring adhesion, attraction, and repulsion between surfaces in liquids with an atomic-force microscope. Phys Rev B 45:11226–11232
42. Israelachvili J (1992) Intermolecular and Surface Forces. Academic Press, London
43. Ohnesorge F, Binnig G (1993) True atomic-resolution by atomic force microscopy through repulsive and attractive forces. Science 260:1451–1456
44. Engel A, Müller DJ (2000) Observing single biomolecules at work with the atomic force microscope Nature Struct Biol 7:715–718
45. Bielefeldt H, Giessibl FJ (1999) A simplified but intuitive analytical model for intermittent-contact-mode microscopy based on hertzian mechanics. Surf Sci Lett 440:L863–L867

46. Albrecht TR, Grütter P, Horne D, Rugar D (1991) Frequency modulation detection using high-Q cantilevers for enhanced force microscope sensitivity. J Appl Phys 69:668–673
47. Schirmeisen A, Anczykowski B, Fuchs H (2004) Dynamic force microscopy. In Bhushan B, Fuchs H, and Hosaka S, editors, Applied Scanning Probe Methods, pages 3–39, Springer
48. Sackmann E (1996) Supported membranes: scientific and practical applications. Science 271:43–48
49. Hartig M, Chi LF, Liu XD, Fuchs H (1998) Dependence of the measured monolayer height on applied forces in scanning force microscopy. Thin Solid Films 327–329:262–267

5 High-Frequency Dynamic Force Microscopy

Hideki Kawakatsu

5.1
Introduction

One objective in the instrumentation of atomic force microscopy (AFM) [1] is to measure and control the force acting between the tip apex and the sample surface with as high a resolution as possible in terms of force as well as the volume concerned. Lateral resolution has been pursued by using a sharp tip, and vertical resolution has been pursued by using a small amplitude of drive of the cantilever, choosing parameters that allow probing for a wide range of tip–sample distances, hence enabling different contribution of the short and long range forces to be sensed. Low-noise detection schemes have been studied to give sufficient signal-to-noise margin for a given set of imaging parameters which mainly comes from the mechanical characteristics of the cantilever. This chapter explains some of the recent trends in dynamic force microscopy where stiffer cantilevers are used at higher frequencies and with lower amplitude of drive to improve the local probing capability of the microscope.

5.2
Instrumental

5.2.1
Cantilever

In dynamic force microscopy (DFM) [2], an oscillator, such as a cantilever, is oscillated to measure the force gradient around the tip apex. Since the local force gradient acts as an additional spring, a change in the force gradient results in a change in the oscillating frequency of a self-excited cantilever. The detection technique employed in DFM is a differential measurement technique, where the component of the force gradient in the direction of the position modulation is detected. In order to acquire a local force gradient with as little averaging by distance as possible, it is preferable to use a smaller amplitude of drive of the cantilever. The common DFM cantilever, whose spring constant is typically in the 40-N/m range, is not stiff enough to overcome the surface force gradient, which may result in snap-in of the tip to the surface, or the necessity to employ a certain level of amplitude of drive to maintain stable self-excitation of an oscillator placed in a nonlinear force gradient.

Detailed explanation and simulation can be found in the literature [3]. Since the force gradient can be up to around 100 N/m, a stiffer cantilever in the 100–1000 N/m range is a more favourable choice for probing of force with a small amplitude, at once allowing a wider range of working points or centres of oscillation to be chosen. However, a cantilever with a high spring constant gives higher demands on the required signal-to-noise margin of the detection scheme.

The minimum detectable force gradient of a free-oscillation cantilever is given by

$$\delta F'_{min} = (2kk_{B}TB/\omega_{o}QA)^{1/2} ,$$ (5.1)

where k is the spring constant, k_{B} is the Boltzmann constant, T is temperature, B is the measurement bandwidth, Q is the quality factor and ω_{0} the natural frequency of the oscillator, and A is the mean-square amplitude of the cantilever oscillation [4].

Increases of Q, $1/k$ and ω_{0} contribute to improvement of sensitivity. Since the choice of a stiffer cantilever acts against improving force sensitivity, we are left with the choice of increasing ω_{0} and/or Q of the cantilever. A smaller cantilever acts in favour of increasing ω_{0} for a given value of the spring constant k. Decreasing the amplitude of drive or A goes against improving sensitivity and frequency noise, but the merit of having the tip apex within the distance of short-range forces throughout the cycle of the oscillation is more important [3]. Various attempts have been made so far in fabricating small cantilevers with high natural frequency [5–18]. In the fabrication of small cantilevers, various issues, such as alignment accuracy and base overhang, which were not important or were overlooked in the case of 100-μm-sized cantilevers become important and non-negligible. Some of the important issues are:

1. The support should be well defined in comparison to the dimensions to improve the Q factor.
2. Overhang of the cantilever base should be zero or made small enough compared with the length of the cantilever to avoid shading the top surface of the cantilever for optical detection.
3. The shoulders of the cantilever base need to be made narrower to avoid shoulder-to-sample contact.
4. Alignment error of the tip to the cantilever.
5. The length of the cantilever should be accurate.

Video rate biomolecular imaging has been accomplished by Ando [19] using a silicon nitride cantilever supported on a ridge to avoid the shoulders touching the sample. Yang et al. [18] implemented a fabrication method that ensures high stability of cantilever length. Figure 5.1 shows a cantilever fabricated by silicon-to-silicon bonding. The method eliminates unwanted overhang and wide shoulders of the silicon base which become nonnegligible as the structure becomes smaller. Saya et al. [15] fabricated a series of small cantilevers based on anisotropic etching of silicon by KOH. Figure 5.2 shows a self-assembling cantilever structure where a bellows structure is shut by the meniscus force upon drying of water, causing the cantilever to protrude a designated length over the edge. Figure 5.3 shows a cantilever with a natural frequency of 10.5 MHz fabricated by conventional techniques [20].

Fig. 5.1. A 10-μm-long cantilever fabricated by silicon-to-silicon direct bonding

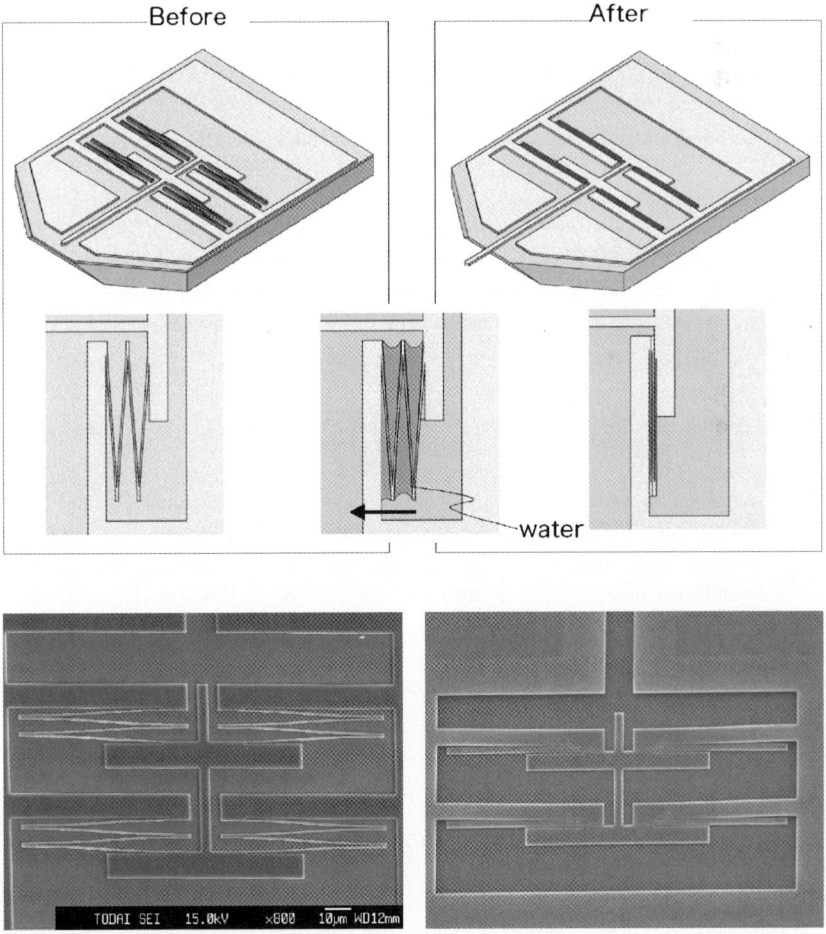

Fig. 5.2. A self-assembling cantilever structure. The cantilever protrudes and sticks to the base by the meniscus force of water upon drying

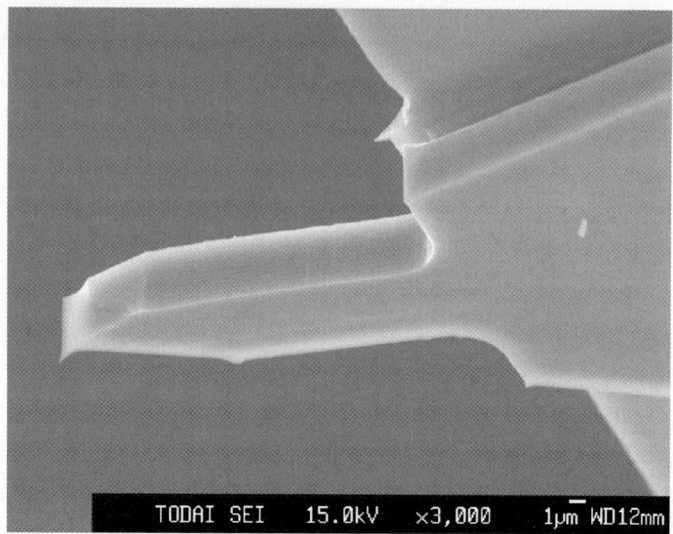

Fig. 5.3. A cantilever with 10-μm length fabricated by conventional techniques using a silicon on insulator wafer

Cantilevers for frequency modulation detection should in principle have a high Q factor for improved force sensitivity. The relation of annealing, hydrogen termination and crystal orientation can be found in the literature. Annealing commercial cantilevers in a vacuum at around 500 °C for 5 h has been known to give a Q factor as high as 5×10^6.

5.2.2
Detection

One important factor in detection is the noise floor. It can be defined as the level of noise in terms of displacement for a given bandwidth, and is plotted for a frequency range of interest. Lowering the noise floor is crucial for improving image quality, imaging speed and decreasing the amplitude of drive of the cantilever. Detection schemes used in AFM are:

1. Tunnelling detection [1,21]
2. Homodyne laser interferometery
3. Heterodyne interferometry [22]
4. Optical fibre type homodyne laser interferometery [23]
5. Optical lever [24]
6. Dual optical lever [25]
7. Heterodyne Dopper interferometry [26]
8. Piezo and piezo-resistive cantilever
9. Knife-edge detection
10. Electrostatic displacement sensing
11. Fabry–Perot interferometry
12. Optical fibre type Fabry–Perot interferometry [8]

13. Refraction grating
14. Tuning fork [27, 28]

Among the many methods listed above, the common methods are the optical fibre type homodyne laser interferometry and the optical lever. The methods applicable to small cantilevers are those that allow focusing down to the diffraction limit, namely heterodyne laser interferometry, optical fibre type homodyne laser interferometry with a focusing lens, heterodyne laser Doppler interferometry, and optical fibre type Fabry–Perot interferometry.

Fukuma et al. [29] succeeded in obtaining atomic resolution of mica in water by decreasing the noise floor of a conventional optical lever system below $20\,\text{fm/Hz}^{1/2}$. This was accomplished by (1) increasing the power of the laser diode to a level just below the noise increases due to mode hopping, and (2) superposition of high frequency to the laser diode drive current to lower the coherence length of the laser, and hence lower noise due to parasitic interference in the optical path. Hoogenboom et al. [8] implemented a Fabry–Perot interferometer by attaching a concave lens at the end of a cleaved optical fibre. With the use of a gold-coated cantilever for improved reflectivity, a noise level also in the range of $10\,\text{fm/Hz}^{1/2}$ was attained. The optics was successfully applied to atomic-resolution imaging in water.

Heterodyne laser Doppler interferometery depicted in Fig. 5.4 can also be used to detect motion of the atomic force microscope cantilever. As shown in Fig. 5.5, its noise floor in terms of displacement shows a $1/f$ decrease with frequency because

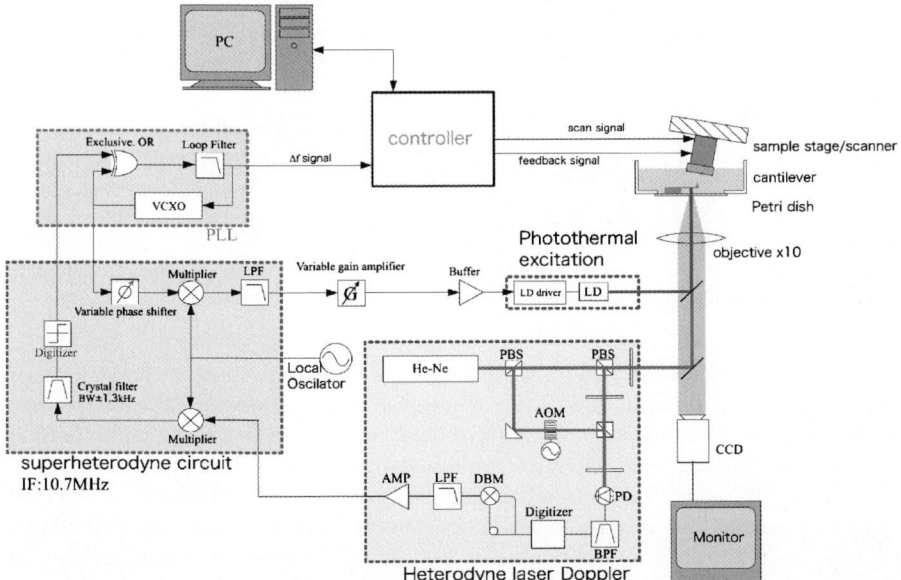

Fig. 5.4. Heterodyne laser Doppler interferometer with a photothermal excitation laser diode. It can excite and measure vibrations up to around 200 MHz. The figure is that applied to a liquid atomic force microscope (AFM) with an inverted optical microscope. The cantilever is attached to the bottom of a Petri dish

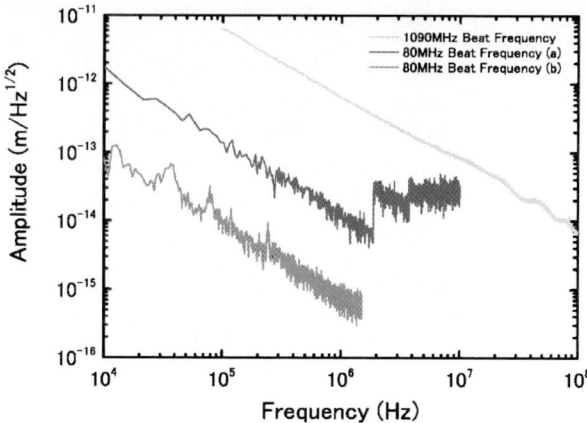

Fig. 5.5. Noise floor of a heterodyne laser Doppler interferometer

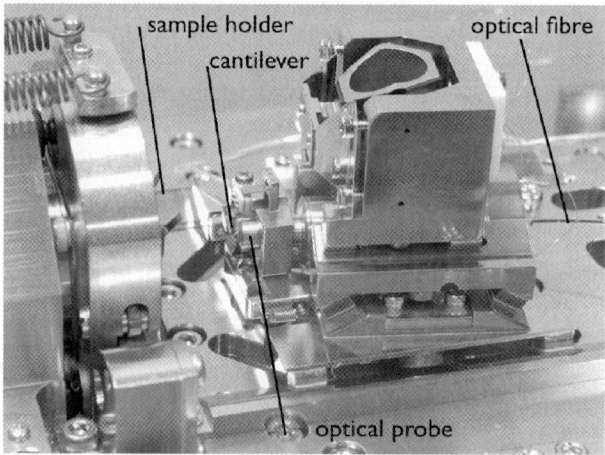

Fig. 5.6. An AFM head with a heterodyne laser Doppler optical probe. It can accommodate cantilevers with natural frequency up to 200 MHz

the velocity signal increases with frequency for a given amplitude. A noise level of $0.5\,\mathrm{fm/Hz^{1/2}}$ is attained around 2 MHz. Figure 5.6 shows an atomic force microscope with a heterodyne laser Doppler interferometer and a laser diode for photothermal excitation of the cantilever. The merit of these two methods is that the laser spot can be focused close to the diffraction limit, enabling the use of small cantilevers on the order of microns. In the case of the optical lever, decreasing the size of the laser spot acts against sensitivity, though it is true that a decrease of sensitivity can, in some cases, be compensated by the use of a smaller cantilever which acts to enhance sensitivity of detection owing to the leverage between the cantilever length and optical lever. Figure 5.7 shows sequential detection of cantilever arrays by scanning the heterodyne laser Doppler laser spot with a galvanoscanner. Though the cantilever width was on the order of 100 nm, clear peaks could be detected.

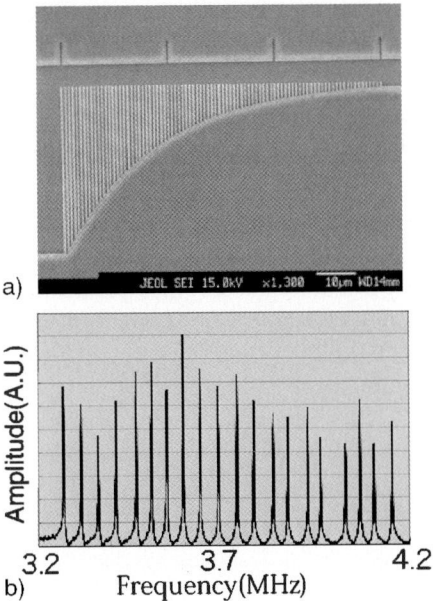

Fig. 5.7. Sequential excitation and measurement of a cantilever array with a frequency gradient. (**a**) The nanopiano, with 88 cantilevers, and (**b**) frequency spectrum of each cantilever. Each cantilever had a mass resolution better than 1 fg in vacuum

5.2.3
Excitation

Excitation of the cantilever by attaching a piezoelement at its base becomes difficult as the frequency exceeds 10 MHz. This is because the frequency that can be excited with the piezoelement is limited by the thickness of the element, as well as by the increasing loss and instability of clamping the cantilever for high-frequency drive. Photothermal excitation, first introduced to AFM by Umeda et al. [30], is an effective method that allows excitation even above 100 MHz, as well as in water. The merit of photothermal excitation in water is no spurious signal due to direct

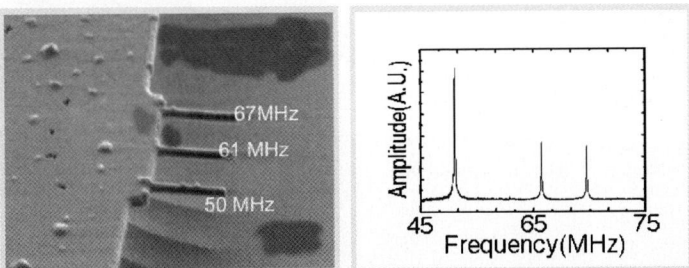

Fig. 5.8. *Left*: Cantilevers fabricated by stencil deposition; he cantilever width was 300 nm. *Right*: Natural frequencies of the cantilevers measured by photothermal excitation and heterodyne laser Doppler interferometry

excitation. Figure 5.8 shows small cantilevers measuring 300 nm in width that were excited sequentially by photothermal excitation [17]. Excitation and measurement of small cantilevers were confirmed up to 200 MHz. Ando et al. [31] adopted optical excitation for alternating current and direct current control of the cantilever motion for faster biosample imaging.

5.2.4
Circuitry

Superheterodyne circuitry is a powerful tool in implementing AFM that can accommodate cantilevers with natural frequencies above 1 MHz. The vibration signal of the atomic force microscope cantilever, whatever its frequency, can be shifted to the neighbourhood of a fixed intermediate frequency. The method is similar to that used in FM radios. Measurement of amplitude, phase and frequency shift can be carried out with the intermediate frequency converted signal, which eliminates the necessity to adopt filters and circuit designs per operating frequency. By using two or more local oscillators, we can monitor multiple vibrational modes simultaneously and use them for imaging. For example, deflection and torsion can be monitored for vertical and lateral force microscopy.

For fast amplitude detection, various schemes have been introduced to detect change in amplitude per cycle [31]. Extremely fast biosample imaging has been realized by such an approach. For frequency shift detection, a phase-locked-loop circuit is commonly used. However, for applications requiring faster FM detection, circuitry based on quadrature multiplication depicted in Fig. 5.9 is effective [32]. Deviation of 1 Hz with modulation of 10 kHz could be detected.

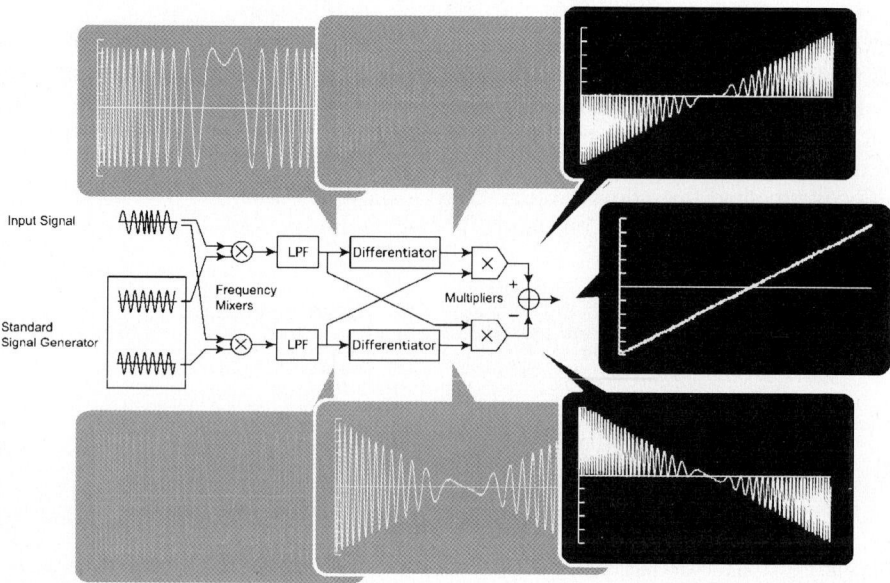

Fig. 5.9. Circuit for fast measurement of frequency shift by symmetric quadrature detection

5.3
Experimental

The low-noise detection schemes with sensitivity in the megahertz to the 100-MHz region allows the use of a small amplitude of drive with a relatively small and stiff cantilever, or with the higher modes of a commercial cantilever.

5.3.1
Low-Amplitude Operation

A stiff cantilever on the order of 100–1000 N/m enables probing of the sample surface with a small amplitude of drive below 0.5 nm, while maintaining stable oscillation. A commercial dynamic mode cantilever was set to self-oscillate at the second mode of deflection at 1.6 MHz. This was done to benefit from the higher spring constant and the low noise level of the Doppler interferometer at elevated frequencies. A heterodyne laser Doppler interferometer and a superheterodyne circuit was used for self-excitation and tip–sample distance control. Figure 5.10 shows a series of images of Si(111) 7×7 imaged with different amplitude of drive. Atomic features could be observed at an amplitude of drive of 0.028 nm [33]. With the same mode of operation, metastable surfaces of Si(111) obtained by quenching could be imaged. By changing the level of the frequency shift, 'magic clusters' seen as a cloud on the Si(111) surface could be imaged as a cloud or displaced [34]. Imaging could

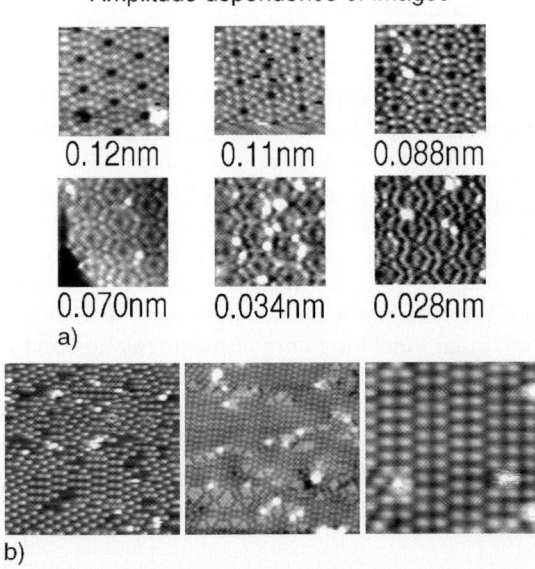

Fig. 5.10. (a) Dynamic force microscopy images of Si(111) 7×7 imaged with heterodyne laser Doppler detection of the cantilever. Atomic features could be observed at 0.028-nm amplitude of drive. **(b)** Various domains observed on quenched Si(111) imaged with the same instrument. The typical amplitude of drive was 0.1 nm

be performed at the third mode of deflection at 4.7 MHz [35]. The distance between the endmost node and the cantilever apex is 16 μm, demonstrating the ability of the optics to use small and stiff cantilevers for imaging. The spring constant is expected to be in the region of 1000 N/m. The optics could be used to implement atomic-resolution DFM with amplitude modulation using the second and third mode of deflection [36].

Fukuma et al. [29] succeeded in acquiring atomic resolution in water with their improved optical lever system. The typical amplitude was around 2 Å, and low amplitude was the key element in achieving atomic resolution, since resolution was lost for a higher amplitude of drive. Hoogenboom et al. [37] and Nishida et al. [3] succeeded in obtaining atomic resolution in water with their Fabry–Perot system with heterodyne doppler interferometry and photothermal excitation.

5.3.2
Manipulation

A low amplitude of drive around 1 Å using the second mode of deflection enables stable manipulation of Si atoms on Si(111) 7 × 7 at room temperature. Si atoms could be fished out vertically in the repulsive mode, as well as moved laterally in the attractive regime [38].

5.3.3
Atomic-Resolution Lateral Force Microscopy

Torsion of the cantilever can be detected either with an optical lever or by other optical detection schemes where the laser spot is positioned on one longitudinal edge of a rectangular cantilever. The first mode of torsion of a dynamic-mode atomic force microscope cantilever lies around 1.6 MHz, so the detection system should have a frequency bandwidth to cover the frequency, and have a low noise floor for sustaining self-excitation. A heterodyne laser Doppler interferometer was used to make a commercial cantilever self-excite at its torsional mode. Figure 5.11a shows a frequency shift to displacement curve of the torsional mode. The sample was Si(111) 7 × 7. An acute frequency shift is observed at the last 0.5 Å before the increase in frequency. Figure 5.11b shows images of Si(111) 7 × 7 acquired with the FM mode [39]. Imaging with tunnelling gap regulation is reported in the literature [40]. Combining tunnelling current gap regulation and FM measurement with heterodyne laser Doppler detection allowed the lateral force gradient to be detected with a lateral amplitude of drive around 10 pm. A positive frequency shift was observed in regions between a pair of neighbouring adatoms, showing a good match between a model where the tip apex is oscillating laterally and subjected to frequency shift by the lateral force gradient [41].

5.3.4
Other Techniques for High Frequency Motion Detection

One emerging technique for high frequency motion detection is the use of electron emission. An electron emitter and oscillator built on the same chip with close

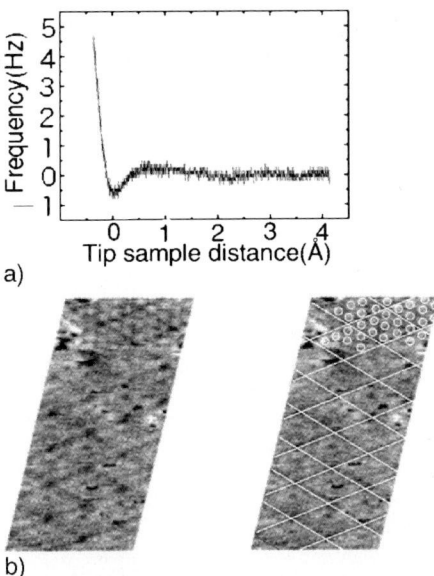

a)

b)

Fig. 5.11. (**a**) Frequency shift to displacement curve of the cantilever which was self-excited at its torsional mode, and (**b**) atomic resolution image acquired by dynamic lateral force microscopy with frequency modulation

separation by a microelectromechanical system fabrication technique is used. The motion of the oscillator is detected as the modulation of the current, since the oscillator serves as the gate [42].

The mix-down technique is a powerful tool for measuring fast phenomena with a relatively slow system. Various groups applied this technique to map surface acoustic waves with scanning probe microscopes that do not have sensitivity at the frequency of the surface acoustic waves [43, 44]. Volodin et al. [45] applied this technique for the measurement of the vibration of carbon nanocoils with the atomic force microscope. Dupas et al. [46] and Fukuma et al. [47] applied the heterodyne optical beam deflection method to allow the use of high-frequency cantilevers and vibration modes.

5.4
Summary and Outlook

This chapter focused on the emerging trend of high-frequency DFM using small cantilevers or higher vibration modes. The key elements were:

1. Focusing of the laser beam to allow the use of micron-sized cantilevers
2. High-frequency detection where the signal-to-noise ratio is obtained by narrow-band homodyne methods or heterodyne methods
3. Photothermal excitation of the cantilever vibration, which is effective above 10 MHz, and is also effective in water for avoiding spurious signals.

4. Cantilever fabrication methods to realize cantilevers with a spring constant in the 100–1000 N/m range, and with a natural frequency in the megahertz range.

The combination of high-frequency DFM and small and stiff cantilevers allows a wider range of imaging parameters to be chosen, resulting in higher resolution and the choice of interaction between the tip and the sample, and represents the future of DFM.

References

1. Binnig G, Quate CF, Gerber C (1986) Atomic force microscope. Phys Rev Lett 56:930
2. Giessibl FJ (1995) Atomic resolution of the silicon(111) 7 × 7 surface by atomic force microscopy. Science 267:68
3. Giessibl FJ (2003) Advances in atomic force microscopy. Rev Mod Phys 75:949
4. Albrecht TR, Grütter P, Horne D, Rugar D (1991) Frequency modulation detection using high-Q cantilevers for enhanced force microscope sensitivity. J Appl Phys 69:668
5. Akamine S, Barrett RC, Quate CF (1990) Improved atomic force microscope images using microcantilevers with sharp tips. Appl Phys Lett 57:316
6. Carr DW, Craighead HG (1997) Fabrication of nanoelectromechanical systems in single crystal silicon using silicon on insulator substrates and electron beam lithography. J Vac Sci Technol B 15(6):2760–2763
7. Waters RL, Aklufi ME (2002) Micromachined Fabry-Perot interferometer for motion detection. Appl Phys Lett 81:3320
8. Hoogenboom BW, Frederix PLTM, Yang JL, Martin S. Pellmont Y, Steinacher M, Zaech S, Langenbach E, Heimbeck H-J, Engel A, Hug HJ (2005) A Fabry-Perot interferometer for micrometer-sized cantilevers. Appl Phys Lett 86:074101
9. Hashiguchi G, Mimaura H (1995) New fabrication method and electrical characteristics of conical silicon field emitters. Jpn J Appl Phys 34:1493
10. Hashiguchi G, Mimaura H (1994) Fabrication of silicon quantum wires using separation by implanted oxygen wafer. Jpn J Appl Phys 33:L1649
11. Kawakatsu H, Saya D, Kato A, Fukushima K, Toshiyoshi H, Fujita H (2002) Millions of cantilevers for atomic force microscopy. Rev Sci Instrum 73:1188
12. Kawakatsu H, Saya D, Fukushima K, Toshiyoshi H, Fujita H (2000) Fabrication of a silicon based nanometric oscillator with a tip form mass for scanning force microscopy operating in the GHz range. J Vac Sci Technol B 16:607
13. Kawakatsu H, Toshiyoshi H, Saya D, Fujita H (1999) A silicon based nanometric oscillator for scanning force microscopy operating in the 100 MHz range. Jpn J Appl Phys B 6:3962
14. Saya D, Fukushima K, Toshiyoshi H, Fujita H, Hashiguchi G, Kawakatsu H (2000) Fabrication of silicon-based filiform-necked nanometric oscillators. Jpn J Appl Phys 39:3793
15. Saya D, Fukushima K, Toshiyoshi H, Hashiguchi G, Fujita H, Kawakatsu H (2002) Fabrication of single-crystal Si cantilever array. Sens Actuators A 95:281
16. Hashiguchi G, Goda T, Hosogi M, Hirano K, Kaji N, Baba Y, Kakushima K, Fujita H (2003) DNA manipulation and retrieval from an aqueous solution with micromachined nanotweezers. Anal Chem 75:4347–4350
17. Kim GM, Kawai S, Nagashio M, Kawakatsu H, Brugger J (2004) Nanomechanical structures with 91 MHz resonance frequency fabricated by local deposition and dry etching. J Vac Sci Technol B 22:1658
18. Yang JL, Despont M, Drechsler U, Hoogenboom BW, Frederix PLTM, Martin S, Engel A, Vettiger P, Hug HJ (2005) Miniaturaized single-crystal silicon cantilevers for scanning force microscopy. Appl Phys Lett 86:134101

19. Ando T, Kodera N, Takai E, Maruyama D, Saito K, Toda A (2001) A high-speed atomic force microscope for studying biological macromolecules. Proc Natl Acad Sci USA 98:12468

20. Olympus Co., Hachioji, Tokyo

21. Binnig G, Rohrer H, Gerber C, Weibel E (1982) Surface studies by scanning tunneling microscopy. Phys Rev Lett 49:57

22. Martin Y, Williams CC, Wickramasinghe HK (1987) Atomic force microscope force mapping and profiling on a sub 100-Å scale. J Appl Phys 61:4723

23. Rugar D, Mamin HJ, Erlandsson R, Stern JE, Terris BD (1988) Force microscope using a fiber-optic displacement sensor. Rev Sci Instrum 59:2337

24. Meyer G, Amer NM (1988) Novel optical approach to atomic force microscopy. Appl Phys Lett 53:1045

25. Kawakatsu H, Saito T (1996) Scanning force microscopy with two optical levers for detection of deformations of the cantilever. J Vac Sci Technol B 14:872–876

26. Kawakatsu H, Kawai S, Saya D, Nagashio M, Kobayashi D, Toshiyoshi H, Fujita H (2002) Towards atomic force microscopy up to 100 MHz. Rev Sci Instrum 73:2317

27. Giessibl FJ, Hembacher S, Bielefeldt H, Mannhart J (2000) Subatomic features on the silicon (111)-(7 × 7) surface observed by atomic force microscopy. Science 289:422

28. Hembacher S, Giessibl FJ, Mannhart J (2004) Force microscopy with Light-Atom probes. Science 305:380

29. Fukuma T Kobayashi K, Matsushige K, Yamada H (2005) True atomic resolution in liquid by frequency-modulation atomic force microscopy. Appl Phys Lett 87:034101

30. Umeda N, Ishizaki S, Uwai H (1991) Scanning attractive force microscope using photothermal vibration. J Vac Sci Technol B 9:1318

31. Ando T, Uchihashi T, Kodera NN, Miyagi A, Nakakita R, Yamashita H, Sakashita M (2006) High-speed atomic force microscopy for studying the dynamic behavior of protein molecules at work. Jpn J Appl Phys 45:1897

32. Kobayashi D, Kawai S, Kawakatsu H (2004) New FM detection techniques for scanning probe microscopy. Jpn J Appl Phys 43:4566

33. Kawai S, Kitamura S, Kobayashi D, Meguro S, Kawakatsu H (2005) An ultra-small amplitude operation of dynamic force microscopy with second flexural mode. Appl Phys Lett 86:193107

34. Kawai S, Rose F, Ishii T, Kawakatsu H, Atomically resolved observation of the quenched Si(111) surface with small amplitude dynamic force microscopy. J Appl Phys 99:104312

35. Kawai S, Kawakatsu H (2006) Atomically resolved dynamic force microscopy operating at 4.7 MHz. Appl Phys Lett 88:133103

36. Kawai S, Kawakatsu H, Atomically resolved amplitude modulation dynamic force microscopy with a high resonance and mechanical quality factor cantilever. Appl Phys Lett 89:013108

37. Hoogenboom BW, Hug HJ, Pellmont Y, Martin S, Frederix PLTM, Fotiadis D, Engel A (2006) Quantitative dynamic-mode scanning force microscopy in liquid. Appl Phys Lett 88:193109

38. Kawai S, Kawakatsu H (2006) Mechanical atom manipulation with small amplitude dynamic force microscopy. Appl Phys Lett 89:023113

39. Kawai S, Kitamura S, Kobayashi D, Kawakatsu H (2005) Dynamic lateral force microscopy with true atomic resolution. Appl Phys Lett 87:173105

40. Giessibl FJ, Herz M, Mannhart J (2002) Friction traced to the single atom. Proc Natl Acad Sci USA 99:12006

41. Kawai S, Sasaki N, Oshima K, Kawakatsu H, Lateral force gradient mapping and its simulation (unpublished)

42. Yamashita K, Sun W, Kakushima K, Fujita H, Toshiyoshi H (2006) RF microelectromechanical system device with a lateral field-emission detector. J Vac Sci Technol B 24:927

43. Strozewski KJ, Mc Bride SE, Wetsel GC (1992) High-frequency surface-displacement detection using an STM as a mixer demdulator. Ultramicroscopy 42:388–392
44. Voigt PU, Koch R (2002) Quantitative geometry of the Rayleigh wave oscillation ellipse by surface acoustic wave scanning tunneling microscopy. J Appl Phys 92:7160
45. Volodin A, Buntinx D, Ahlskog M, Fonseca A, Nagy JB, Van Haesendonck C (2004) Coiled carbon nanotubes as self-sensing mechanical resonators. Nano Lett 4:1775
46. Dupas E, Gremaud G, Kulik A, Loubet J-L (2001) High-frequency mechanical spectroscopy with an atomic force microscope. Rev Sci Instrum 72:3891
47. Fukuma T, Kimura K, Kobayashi K, Matsushige K, Yamada H (2004) Dynamic force microscopy at high cantilever resonance frequencies using heterodyne optical beam deflection method. Appl Phys Lett 85:8287

6 Torsional Resonance Microscopy and Its Applications

Chanmin Su · Lin Huang · Craig B. Prater · Bharat Bhushan

6.1
Introduction to Torsional Resonance Microscopy

Atomic force microscopy has achieved tremendous benefits from a variety of oscillating tip modes of operation, most notably intermittent contact (e.g., TappingMode™) and non-contact mode. In these modes of operation, the AFM cantilever is oscillated typically at its fundamental flexural resonance, the "diving board" resonance. These modes have the advantage that they largely eliminate lateral forces on the sample that tend to damage tips and samples in contact mode. The vertical interaction force is also substantially reduced due to high mechanical Q of the cantilevers, rendering imaging of the delicate soft sample possible.

It has been recognized, however, that AFM cantilevers can oscillate at many different modes [1–17], including higher order flexural modes and torsional (twisting) modes and study tip–surface interaction in different ways. The torsional modes of oscillation have been used, for example, to study friction [2] non-linear tip–surface interactions [5], thin film magnetism [8, 18], nanometer scale dynamic friction [16], surface stiffness [10], surface damping [17], and viscous fluids responses [11].

In 2003, a torsional resonance imaging method called TRmode™ [19] was commercialized. In this imaging mode, lateral forces that act on the tip can cause a change in the torsional resonant frequency, amplitude, and/or phase of the cantilever (see Fig. 6.1) [20]. Images of topographic structure, mechanical properties [21,22], and in-plane force or force gradient fields [18–20] can be constructed from these data.

AFM measurements at torsional resonances have many advantages. These advantages will be outlined briefly in this introductory section and then discussed in more detail later in the chapter.

Among the key advantages are:

– Sensitive to lateral forces on tip
– Able to map lateral stiffness, dissipation, magnetic fields
– Compatibility with other oscillating tip modes
– Maintains tip in near-field with low force
– Wide bandwidth interaction detection
– Specifically sensitive to force/mass near the tip.

Fig. 6.1. Finite element simulation showing the first torsional mode of a rectangular cantilever

TappingMode and non-contact imaging modes were developed in part to eliminate frictional forces between the AFM tip and the sample. This has a substantial advantage for limiting tip/sample wear, but it also comes at a cost. The cost is that the tip is minimally sensitive to lateral forces. Such forces, however, can be of interest. In fact, a variant of contact mode AFM, lateral force microscopy maps both the vertical deflection and the torsional deflection. The flexural motion is an indicator of vertical force and the torsional deflection reflects the lateral forces. These lateral forces have been used, for example, in chemical force microscopy [23] to map the presence of surface chemical species. In traditional TappingMode AFM, however, these lateral forces are not directly observed. Additionally in TappingMode tip position varies relative to the sample by tens of nanometers, rendering measurements that requires a constant force field impossible.

Torsional resonance imaging, however, recovers the ability to achieve low-force imaging and the ability to map in-plane forces. This allows researchers to map lateral stiffness and energy dissipation. It also provides the ability to map in-plane long-range forces, for example electrostatic or magnetic forces. Traditional AFM measurements are sensitive to these long-range forces and force gradient in vertical direction only.

The torsional resonance method is also fundamentally compatible with conventional oscillating tip modes like TappingMode and non-contact mode and contact mode. These methods typically use a piezoelectric oscillator to drive the cantilever into oscillation. In the simplest case, one can switch from a flexural resonance to a torsional resonance by changing the driving frequency of the piezooscillator (a more optimal drive system uses two piezooscillators that can be operated in phase or out of phase is shown in Fig. 6.2 in the next section). The switching between flexural and torsional resonance can be performed as often as desired and commercial systems allow a user to switch between TappingMode and TRmode on alternating scan lines, allowing near simultaneous imaging of the same sample in both imaging modes.

An interesting property of torsional resonance is that it is extremely sensitive to the mass at the end of the tip. The reason for this is that the torsional cantilever resonance frequency is determined in part by the rotational moment of inertia, which contains terms that scale like:

$$\int r_t^2 \, dm \tag{6.1}$$

where dm is the mass distributed at a distance r_t from the central axis of the cantilever. Because of the r_t^2 term, the moment of inertia is heavily weighted towards the mass at the apex of the tip (the largest r_t). For a conventional cantilever, this variation of the torsional resonance frequency allows one to detect a change in mass of the tip with the resolution of 10^{-17} kg/Hz [24]. This characteristic has potential applications related to detecting tip wear and the manipulation of small particles.

Another advantage of torsional resonance is that the tip may be maintained at low force and continuously in the near-field. For traditional oscillating tip modes, the AFM tip may spend up to 99% of its oscillation cycle in a regime with essentially no near-field interaction with the surface. This is advantageous for eliminating tip wear and sample damage, but it is problematic for the measurement of properties that demand near-field interactions. One of the main characteristics of TRmode is its ability to maintain probe tip in the near-field regime at the boundary between direct contact and the long-range force regime. In this transition region, the tip apex has a substantial near-field interaction with the sample, but can provide minimum damage to the sample itself. This transitional regime can be difficult to address in conventional AFM since the cantilever will snap into contact with a sample if the local force gradient exceeds the cantilever spring constant. Torsional resonance provides the ability to stably operate in this near-field region. A specific application of this benefit that will be discussed below is the measurement of local electrical properties of soft samples.

Torsional resonance method also has a wide detection bandwidth. The details depend on cantilever geometry, but the first torsional resonance is often an order of magnitude higher than the first flexural resonance. That means that the cantilever has a roughly flat response bandwidth up to 10 times wider in the torsional direction than the flexural direction. Recently Sahin and colleagues (personal communication, Sahin, Magonov, Su, Quate, Solgaard, 2005) demonstrated the ability to use this wide bandwidth to reconstruct the dynamic forces that occur between a tip and sample. All of the above advantages and the underlying science will be described in the following sections.

6.2
TRmode System Configuration

Schematic diagrams of TRmode AFM are shown in Figs. 6.2 and 6.3. The torsional drive system (Fig. 6.2) consists of two parallel actuators, marked as piezos 1 and 2.

Fig. 6.2. TRmode drive in a nearly proportional assembly drawing: piezoelements (*1, 2*); spacer to the mechanical fixture (*3*); cantilever chip (*4*); rotation center when *1* and *2* are driven asymmetrically of the same amplitude (*5*); cantilever (*6*) ; tip (*7*)

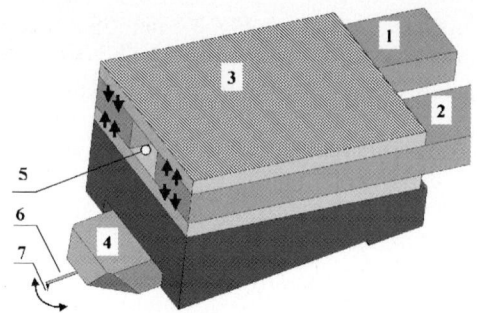

To excite a torsional resonance, drive signals of opposite phase are applied to the two piezos. This actuation will create a rotation of the cantilever holder about the center point 5. Since the tip is far away from the center, the trajectory of the tip is almost a horizontal motion. Due to the asymmetry of the tip mass about the rotation axis, such mass will provide an initial force when the lever is subjected to a translational movement, yielding a torque about the rotation axis. It is obvious that a more massive tip will generate the inertial torque more effectively, therefore driving torsional resonance more efficiently. The situation of the TappingMode excitation is very similar where the piezo pair 1 and 2 moves in phase, causing the tip holder and cantilever chip 4 to move upside and down. The tip and cantilever mass will produce an inertial force to deflect the cantilever rotating about the fixed end of the cantilever. The lateral and torsional cantilever deflections are detected using a four-segment photodetector, as shown in Fig. 6.3.

It is worthwhile to notice that the AFM system shown in Fig. 6.3 measures angular displacement of the cantilever, resulting in a change in the reflected laser beam angle. This provides specific advantages for detection of torsional motion. Consider first the case of flexural motion. For flexural bending of a cantilever, the bend angle θ_f is given approximately by the relationship $\theta_f = \frac{3}{2}\frac{z}{L}$, where z is the vertical tip deflection and L is the length of the cantilever. The cantilever length serves as the lever arm that converts vertical deflection to an angular change. Typical values of cantilever lengths are on the order of $100\,\mu m$. In this case the tip needs to move at least a few nanometers to generate an angular deflection signal enough to be detected.

The lever arm for torsional motion, however, is much shorter, as it corresponds to the tip height plus half the cantilever thickness (this is the distance from the tip apex to the center of torsional rotation). AFM tip heights are typically ~ 10 to 20 times shorter than the cantilever length L. So torsional detection is 10 to 20 times more sensitive to tip motion than flexural detection. As a result it is much easier to detect subnanometer tip motion laterally with the same detection electronics and optics.

The feedback system for TRmode imaging is almost identical to other AFM imaging modes. As shown in Fig. 6.4, the raw deflection signal is sent to the signal processor and then compared with a set point to generate the control error. The error signal is then amplified by a gain control unit, usually consisting of integral and proportional gain, to regulate feedback bandwidth and response speed. I/P gain

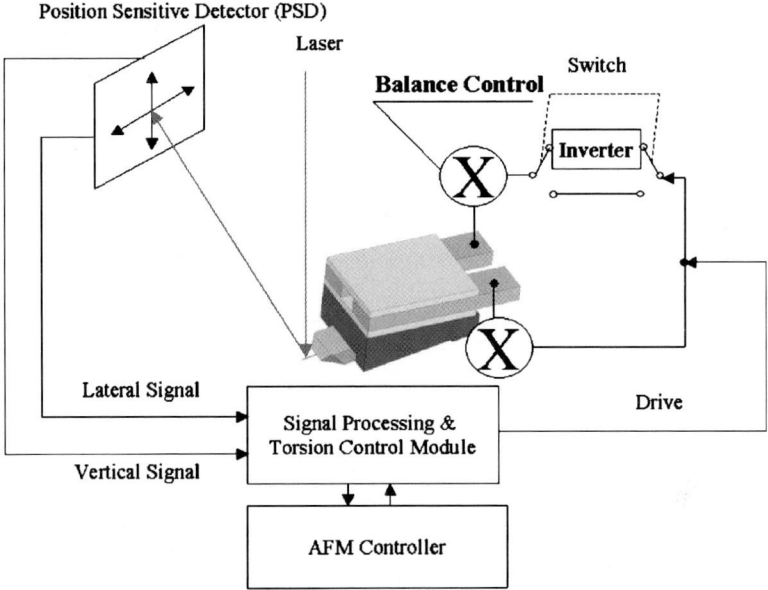

Fig. 6.3. TRmode drive integrated to a four-quadrant detector. Torsional displacement of the cantilever is detected by the lateral component of the detector. On the drive control part a switch that determines symmetry and balance of the piezoactuator pair decides whether to drive the lever into flexural oscillation for TappingMode or torsional oscillation for TRmode

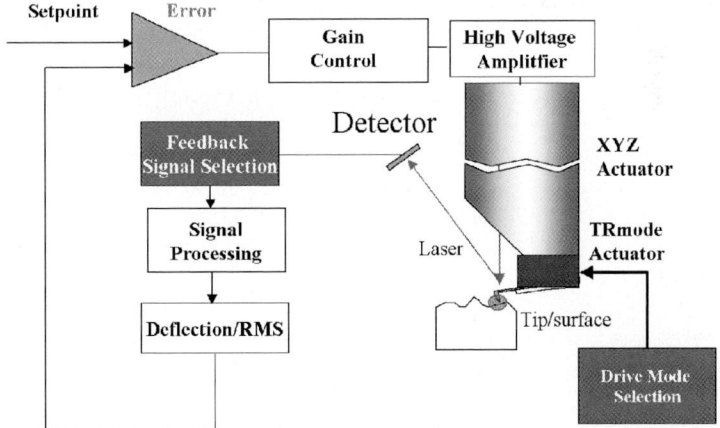

Fig. 6.4. Feedback loop of a torsional resonance imaging system

regulated error signal is further amplified to a high voltage to move the cantilever through an *xyz* translational stage to minimize the magnitude of the error signal.

Although the configuration in Figs. 6.2 and 6.3 are sufficient to provide steady torsional resonance amplitude for a feedback in ambient imaging, excitation of torsional resonance is much more challenging in fluids where quality factor Q is reduced to the order of 1 to 10. The reduction of Q in fluid causes a substantial decrease in the effective mechanical amplification factor observed by the detection system. Parasitic resonance due to acoustic reflection or other mechanical parts may easily dominate the displacement frequency spectra as the piezoactuation amplitude increase by orders of magnitude. In fluids, misalignment of the cantilever torsional axis and the actuator rotation axis is penalized by the presence of the carry on mass of fluid. In this situation precise alignment of the drive rotation axis and cantilever torsional axis will remove the dependency of inertial force and oscillating environment. One of the solutions is to use active piezoelement on the cantilever, as shown in Fig. 6.5. The resulting flexural and torsional responses in air and fluids are shown in Figs. 6.6 and 6.7. In Fig. 6.5 the torsional resonance lever consists of two sections, the active portion and the passive portion. The active arm has two separate elements, which can be actuated in phase or out of phase for flexural and torsional resonances. The active portion is about 340 μm long and the passive portion about 100 μm. The thickness of the Si portion and the piezoelectric portion are equal to about 3.5 μm. Figure 6.6 displays responses of the lever under in phase drive for flexural resonances and out of phase drive for torsional resonances. The complete absence of the flexural response at the torsional resonance peaks, and vice versa, demonstrates pure flexural and torsional oscillation without mode coupling.

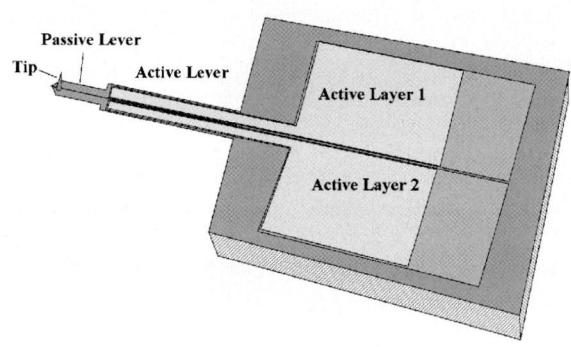

Fig. 6.5. Active torsional cantilever having the drive axis nearly coincide with the cantilever torsional axis

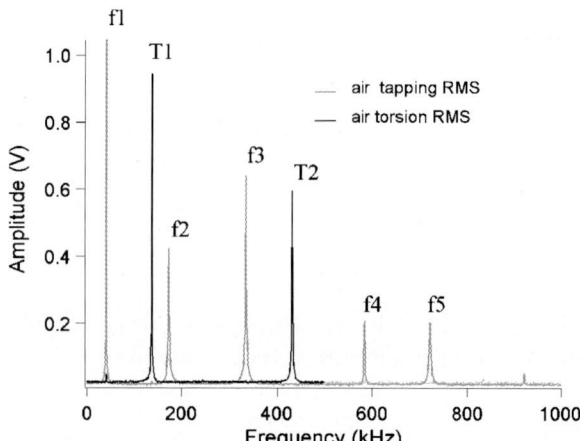

Fig. 6.6. Flexural and torsional spectra of an active torsional lever in air

Fig. 6.7. Torsional responses of the active lever in air and water

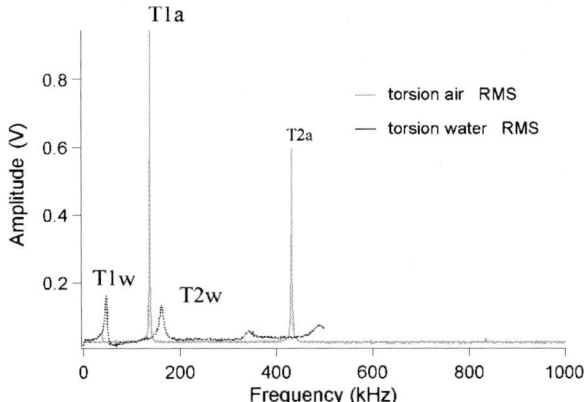

The peaks marked f1, f3, f4 and f5 are flexural resonance of the long active lever at its fundamental resonances and three over tones. The peaks marked f2 and f6 are fundamental and overtone flexural resonance of the small passive cantilever in Fig. 6.5. When the active torsional lever is immersed in water the peaks shifts to low frequencies but with the same clean spectra as in air.

6.3
Torsional Modes of Oscillation

To understand the dynamics of TRmode, it is important to analyze the cantilever's torsional response. Reinstaedtler et al. [22] have analyzed these dynamics recently and concluded that a point-mass model is valid only for weak tip–surface interaction where interaction induced stiffness is less than cantilever stiffness. Consequently we separate torsional responses into several domains, namely free oscillation, weak interaction and contact interaction. For a freely oscillating lever can start with the elastic beam equation:

$$c_T \frac{\partial^2 \theta}{\partial x^2} = \rho J \frac{\partial^2 \theta}{\partial t^2} + \gamma_{air} \frac{\partial \theta}{\partial t} \tag{6.2}$$

where θ is the cantilever rotation and $c_T = wb^3 G/3$ is the torsional stiffness and $J = (w^3 b + wb^3)/12$ is the area moment of inertial about the torsional axis. The terms w, b, L and ρ represent cantilever width, thickness, length and density, respectively. The term γ_{air} is the damping coefficient of air. Solving (6.2) gives one solution for torsional resonant frequencies:

$$\omega_n = \frac{2n-1}{2L} \pi \sqrt{\frac{c_T}{\rho J}} \tag{6.3}$$

where ω_n is the torsional resonance frequency of the nth modes and

$$Q = \omega \frac{\rho J}{\gamma_{air}} \tag{6.4}$$

The AFM cantilever generally carries a tip that may have significant mass compared to the cantilever itself. Therefore it is necessary to correct for the tip mass. The corrected torsional resonance frequencies are shown below:

$$f_{\text{tip,lever}} = \frac{1}{4L} \left(1 - \frac{J_{\text{mass}}}{\frac{\rho w b^3 L}{3} + J_{\text{mass}}} \right) \sqrt{\frac{c_T}{\rho J}} \tag{6.5}$$

where

$$J_{\text{mass}} = m_{\text{tip}} \left(h + \frac{b}{2} \right)^2 \tag{6.6}$$

Combining (6.5) and (6.6), notice the lever mass $m_{\text{lever}} = \rho w b L$, for $h \gg b/2$

$$\frac{f_{\text{lever}} - f_{\text{tip,lever}}}{f_{\text{lever}}} = \frac{1}{1 + \frac{1}{3} \frac{m_{\text{lever}}}{m_{\text{tip}}} \frac{b^2}{h^2}} \tag{6.7}$$

f_{lever} and $f_{\text{tip,lever}}$ represent torsional resonance frequency of a rectangular lever and a rectangular lever with a mass center at h above the lever surface. For a conventional lever with $h \sim 15\,\mu\text{m}$, $b \sim 3\,\mu\text{m}$, the modification of resonance frequency due to the presence of the tip mass can be as much as 45% when the tip mass is only 10% of the lever mass.

Finally additional change of the torsional resonance frequency can be induced by the presence of the warping effect. Such an effect is more significant only if the cantilever width is comparable to its thickness. The warping factor is shown as:

$$\xi = \frac{1}{3} w \cdot b^3 \left(1 - 0.632 \frac{b}{w} \right) \tag{6.8}$$

leading to

$$f_{\text{lever,warp}} = \frac{1}{4L} \sqrt{\frac{\xi G}{\rho J}} \tag{6.9}$$

for tipless lever and

$$f_{\text{tip,lever}} = \frac{1}{4L} \left(1 - \frac{J_{\text{mass}}}{\rho \xi L + J_{\text{mass}}} \right) \sqrt{\frac{\xi G}{\rho J}} \tag{6.10}$$

with the tip mass correction.

The validity of the analytic solutions has been checked out by finite elements analysis for a FESP lever with 235 μm length, 30 μm width and 3 μm thickness, having a pyramid tip of 15 μm high. The results are shown in Table 6.1.

It is important to recognize that there is an additional oscillation mode that can generate lateral motion of the tip. This mode is the lateral bending mode in the plane of the cantilever's width and length (Fig. 6.8). For a cantilever with no

Table 6.1. Cantilever oscillation modes

Mode number	FEA modal analysis (Hz)	Analytical solution (Hz)
1	69345	66105
2	435460	414280
3	817150	793270
4	1.0352E+06	949010
5	1.2209E+06	1.1600E+06

tip, this mode would not be detectable in TRmode because it will not generate angular change in the cantilever surface. The mass of the tip, however, breaks the cantilever symmetry and couples this sideways motion into a small torsional signal. Experimentally the lateral deflection signal can not differentiate pure torsion from tip–coupled lateral bending. A distinct difference is that, while the lateral bending modes resonance frequency is independent of the cantilever thickness, torsional resonance frequency increases linearly with the thickness. In the case of the 225 μm lever there is a transition thickness at about 2 μm above which torsion resonance frequency becomes higher than lateral bending resonance. At the transition thickness it is much more difficult to differentiate the two modes. For this reason it is desirable to choose a cantilever thickness where these two oscillation modes are well separated in frequency.

The contribution of the asymmetric tip mass to TRmode excitation is shown in Fig. 6.9 where a 225 μm long cantilever is actuated by periodic motion at the fixed end of the cantilever. The lower frequency peak (solid curve) is due to lateral bending. The angular change is associated with about 20 times higher lateral displacement of the tip compared to that of the torsional mode at about 10^6 Hz. Therefore it is important to differentiate these two modes experimentally because the bending mode corresponds to much lower imaging resolution due to the large lateral displacement. Figure 6.8 shows the thickness dependence of the bending mode resonance frequency compared to that of the torsional mode.

Another key issue to consider is the effective spring constant of torsion motion versus that of the flexural mode. In a simple rectangular approximation the spring constant of the torsional displacement at the tip apex can be expressed as [25]

$$k_{\text{T}} = \frac{Gwb^3}{3L(h + b/2)^2} \tag{6.11}$$

and for flexural deformation

$$k_{\text{flex}} = \frac{Ewb^3}{4L^3} \tag{6.12}$$

The ratio of the spring constant for flexural and torsion is:

$$\frac{k_{\text{flex}}}{k_{\text{T}}} = \frac{E}{G} \frac{3(h + b/2)^2}{4L^2} \tag{6.13}$$

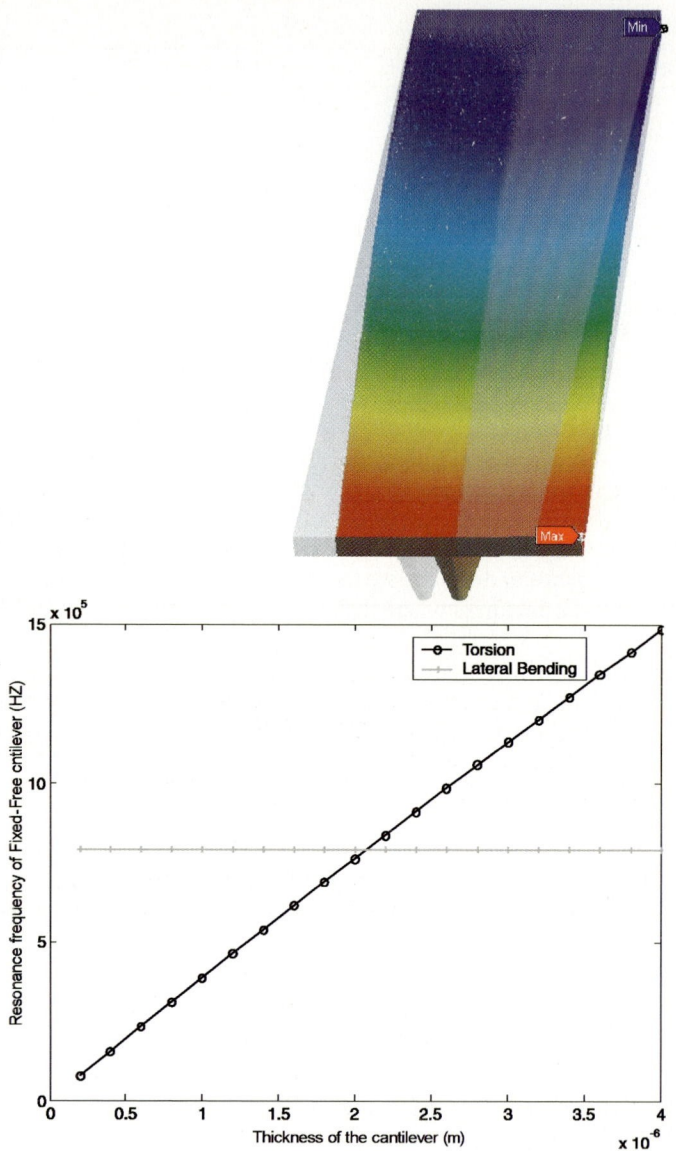

Fig. 6.8. Thickness dependence of the resonance frequencies of bending and torsional modes. The cantilever is 225 μm long, 30 μm wide with varying thickness

For the case of isotropic materials where the tip height h is much larger than the lever half thickness $b/2$ (6.13) reduces to:

$$\frac{k_{\text{flex}}}{k_{\text{T}}} \approx 1.5(2+v)\frac{h^2}{L^2} \tag{6.14}$$

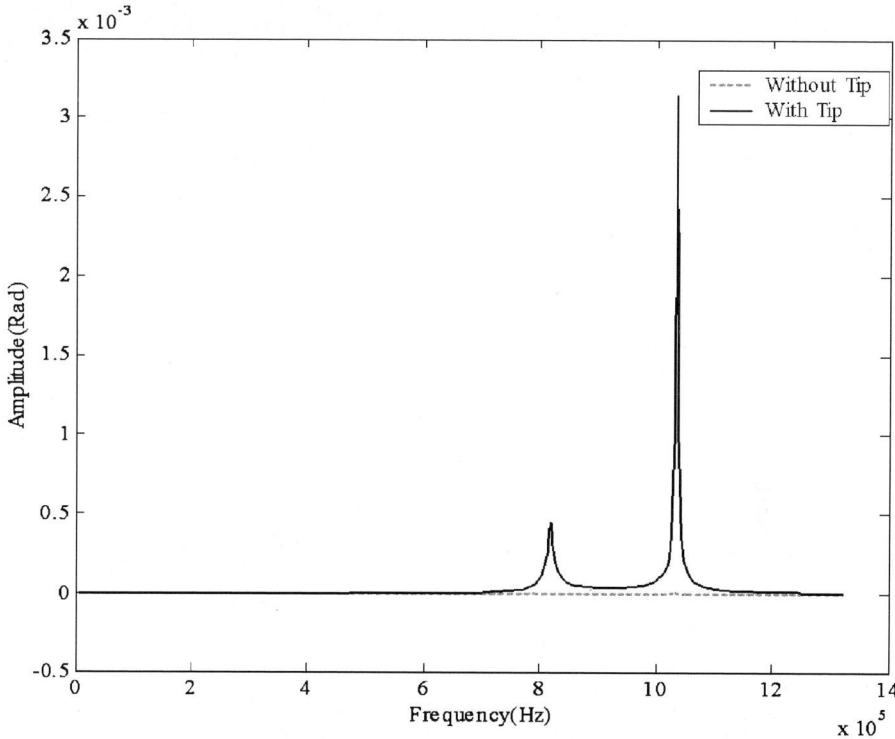

Fig. 6.9. FEA harmonic analysis of an FESP lever tip rotation about the torsional axis. The drive actuation is 1 nm lateral motion at the fixed end of the cantilever

For a tip height of 15 μm and the lever length 225 μm, this ratio is about 0.016, i.e., the torsional stiffness at the tip apex is about 63 times higher than flexural stiffness.

6.4
Imaging and Measurements with TRmode

6.4.1
TRmode in Weakly-Coupled Interaction Region

Similar to tapping the most important control parameter for TRmode AFM feedback is its amplitude change as a function of tip–surface distance. The interaction mechanism of the probe apex with the samples in the case of torsional resonance mode was investigated in great detail by Reistaedtler et al. [22]. Consider the cantilever model in (6.2) with the boundary condition:

$$c_{\mathrm{T}} \frac{\partial \theta}{\partial x} |_{x=L} = h F_{\mathrm{lat}}(z) \tag{6.15}$$

where F_{lat} represents the lateral interaction force that subjects the cantilever to a torque at the tip apex. If one approximates the lateral interaction by a linear spring

K_{lat} in parallel with a damping dash pot η for dissipative interaction (Voigt model), (6.15) becomes:

$$c_{\text{T}}\frac{\partial\theta}{\partial x}\Big|_{x=L} = -h^2\theta(L,t)(K_{\text{Lat}}+i\omega\eta) \tag{6.16}$$

Unlike the TappingMode where the tip apex experiences several regions of interaction, from contact to interaction free state in each cycle, torsional resonant tip only moves slightly at a near constant tip-sample distance z. The dissipating spring Voigt model holds as long as the tip is in small displacement and z remains constant.

The steady state solution of (6.2) and (6.16) can be derived by inserting a periodic solution:

$$\theta(x,t) = \theta_0\,e^{i(kx-\omega t)} \tag{6.17}$$

yielding

$$\theta(x=L) = \frac{\theta_0 kL}{kL\cos kL + ((K_{\text{Lat}}+i\eta\omega)/k_{\text{T}})\sin kL} \tag{6.18}$$

A plot of cantilever torsional response based on (6.18) for varying lateral contact stiffness and damping is shown in Fig. 6.10.

As seen in Fig. 6.10, the torsional resonance shifts to higher frequency as lateral contact stiffness increases. If the drive frequency is fixed, as shown in Fig. 6.10 with the vertical dash line at 204 kHz, the shifting of the resonance frequency will cause amplitude to decrease. This shift is more pronounced as the tip gets closer to the surface, leading to an increase of K_{Lat}. Similarly the amplitude of the TRmode can also decrease as the damping factor $\eta = \eta(z)$ increases. The variable η is a strong function of tip–surface distance since various frictional dissipation depends substantially on the tip–surface distance and contact pressure.

The origin of the K_{Lat} varies depending on tip–surface distance and whether the distance function is approaching or departing the sample surface. When elastic contact dominates the interaction force:

$$K_{\text{Lat}} = 8G'a \tag{6.19}$$

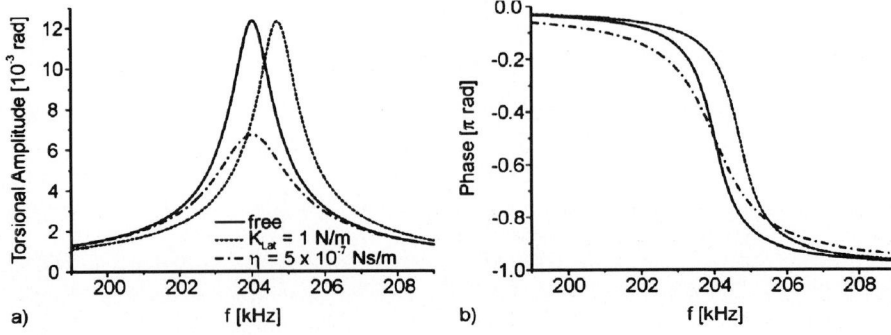

Fig. 6.10. Torsional resonance amplitude as a function of frequency for various lateral stiffness and damping (courtesy of Reinstadtler et al. [22])

where G' is the reduced shear modulus[1] and a is the contact radius. When contact adhesion is considered the loading force is superpositioned with an adhesion force $F_{ad}=3/2\pi\gamma R$ for the JKR model and $F_{ad}=2\pi\gamma R$ for the DMT model. Further details of these models and their transition region can be found in reviews by Dedkov [26] and Maugis [27].

The damping coefficient η is usually attributed to viscoelastic deformation or other dissipative interactions such as capillary force, and adhesion [28]. Oscillating charge [29] induced by moving tip or ruptured bonds [30] can also contribute to energy dissipation significantly.

Experimentally the control of TRmode imaging is similar to TappingMode control where the feedback error is based on the amplitude change due to tip-sample distance change, usually referred to as the amplitude-distance curve. A typical amplitude-distance curve of TRmode on a relatively hard material is shown in Fig. 6.11. The torsional resonance amplitude (upper curve with the right vertical axis) and flexural deflection are measured simultaneously. As marked by the vertical line PP' torsional resonance amplitude starts dropping as soon as the probe detected attractive force and decreased to a very low but non-zero value when the probe is still in the attractive zone. The amplitude change between lines OO' and PP' marks the entire torsional resonance amplitude control region, corresponding to a variation of z in less than 10 nm. During imaging an amplitude set point is selected for feedback reference. Whenever the amplitude increases or decreases as a result of surface topographic variation, the difference between the new amplitude and the set point will produce an error signal for z piezo in Fig. 6.4 to correct tip position in order

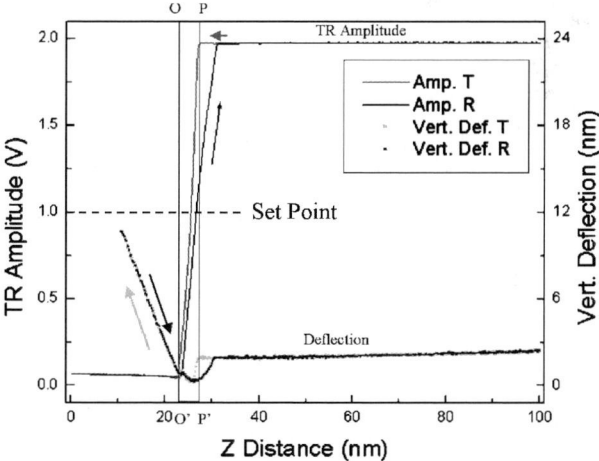

Fig. 6.11. Torsional resonance amplitude as a function of the tip–sample distance on a Si sample. The *horizontal axis* is the z ramp size. The vertical axis on the *left* and *right* represent amplitude and flexural deflection respectively. The line marked TR amplitude represents approaching (trace, marked as T) and retract (retract, marked as R) curves of torsional resonance amplitude. Deflection curves shown are flexural deflection during approaching and retraction

---

[1] $\frac{1}{G'} = \frac{2-v_1^2}{G_1} - \frac{2-v_2^2}{G_2}$, where G_i and v_i are shear modulus and Poisson's ratio of material i

to maintain the amplitude as a constant. Plotting the feedback signal as a function of the *x*-*y* positions creates a topographic image, as long as the feedback signal is properly calibrated or a position sensor is used.

In TappingMode, amplitude versus distance curves are commonly used to study the tip–sample interaction and cantilever response. Torsional amplitude versus distance curves are also illustrative, but there are some striking differences in comparison with tapping. First, the full control range within OO′ and PP′ corresponds to a negative constant flexural deflection. This adhesion region is metastable for both contact mode and tapping operation. Contact mode cannot maintain stable control because of the dual-value of the *z* position for a certain deflection in this adhesive region. TappingMode must also operate with enough kinetic energy and amplitude to pull the tip out of the adhesion zone to avoid feedback loop oscillation. TRmode, however, can keep steady control throughout the imaging process. Secondly the amplitude at full contact does not go to zero, as tapping amplitude does when a repulsive contact is established. The reason is that torsional resonant frequency makes a relatively small shift in contact compared to the shift for TappingMode. Once the tip is in full contact, there is still enough response at the original drive frequency to provide a non-zero amplitude due to the overlap of the responses of TRmode free oscillation and contact oscillation (Fig. 6.12).

Conversely, the contact resonance frequency of flexural mode used by Tapping-Mode is about twice of the free resonance for hard materials. The resonant frequency shift in tapping responses is so substantial that the amplitude at the free tapping drive becomes essentially zero when contact is established.

One of the greatest concerns in imaging is the magnitude of tip–sample interaction force. High interaction force either damage sample or the tip. Imaging control usually prefer low force feedback, as long as the feedback is stable. As demonstrated in the previous section the spring constant of the TRmode is about two orders magnitude higher than flexural spring constant. However, since angular deflection sensitivity of torsional resonance is about 15 times higher than that of tapping. An additional amplification factor (3×) further increases lateral detection sensitivity. For a 225 μm lever tapping operation the voltage of 60 nm vertical displacement

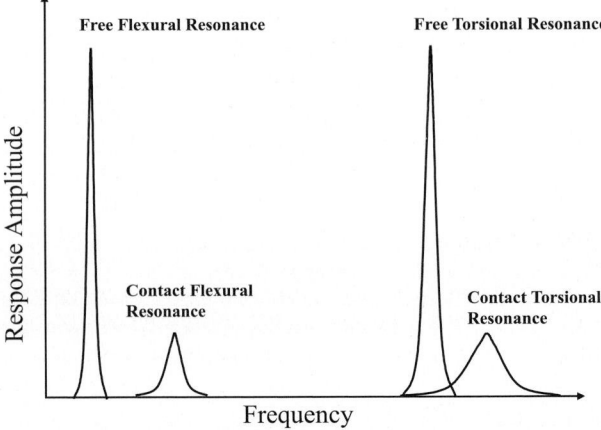

Fig. 6.12. A schematics of the resonance peak shift for flexural and torsional modes after the tip is in constant contact with the surface

corresponds to about 1.3 nm lateral displacement at the probe apex. It is well-known that the tapping force is proportional to the operating amplitude (with the same given set point) [31], lower TRmode displacement amplitude reflects less interaction force. Additionally the Q factor of the TRmode is usually about 3 to 5 times higher than the TappingMode. The two contribution factors, namely displacement amplitude and mechanical Q, should place the TRmode interaction force into the same range of the tapping interaction.

6.4.2
TRmode Imaging and Measurement in Contact Mode

Performing torsional resonance measurements in contact mode has been pioneered by Yamanaka [32] and Rabe [33]. In both cases the sample or the probe was periodically modulated when the cantilever probe is under constant deflection. The frequency sweep was used to determine exact contact resonance frequency to derive lateral contact stiffness or the threshold of stick-slip transition under load. Since a dedicated modulation actuator is needed for sample or tip modulation the measurement system of contact resonance is either in flexural mode or torsion mode. On the other hand if the flexural contact resonance and torsional contact resonance is measured simultaneously derivation of mechanical property of the sample could be much less dependent on the knowledge of the tip shape and contact area. Since TRmode provides the opportunity of concurrent or interleave operation of flexural and torsion modes this section will demonstrate the benefit of contact mode in combination of TRmode and flexural resonance mode.

In contact mode the AFM feedback maintains constant deflection such that the tip is applying a constant force to the sample. The tip-surface interaction will determine cantilever response as in (6.18). By solving this (6.18) explicitly the contact stiffness can be expressed as a function of cantilever property and contact resonance frequency:

$$K_{Lat} = -\frac{c_T}{h^2} 2\pi \sqrt{\frac{\rho J}{c_T}} f_{T,n} \cdot c \tan\left(2\pi \sqrt{\frac{\rho J}{c_T}} f_{T,n} L\right) \tag{6.20}$$

where $f_{T,n}$ refers to the nth torsional contact resonance frequency. All the parameters on the right-hand side of the equation are cantilever geometries, materials data and measured torsional resonance frequencies. K_{Lat} can therefore be considered as a directly measured parameter in the experiment. The practical interest is, however, material property such as elastic modulus. From (6.19) one can see that both reduced modulus and contact area contributes to lateral contact stiffness. Since contact area depends on tip apex shape, sample deformation mode, determination of contact area presents the utmost challenge to probe-based contact mechanics at the nanometer scale. Measurement of K_{Lat} only provides a qualitative clue to materials property.

Since TRmode also provides flexural measurement capacity it is worth noting that flexural contact resonance also depends on the same contact area as the following:

$$K_{flex} = 2E'a \tag{6.21}$$

where E' is the reduced Young's modulus[2], and K_{flex} is the contact stiffness of the flexural mode. The flexural contact stiffness has been solved in terms of the measurement parameters by Rabe et al. [34] for a full solution of the flexural contact resonance:

$$K_{\text{flex}} = k_{\text{flex}} \frac{(k_n L)^3}{3} \frac{1 + \cos k_n L \cosh k_n L}{\sinh k_n L \cos k_n L - \sin k_n L \cosh k_n L} \tag{6.22}$$

where k_{flex} is the spring constant of the flexural lever shown in (6.12), k_n is the wave number of the nth mode, and $k_n L$ is determined by the dispersion equation:

$$k_n^2 L^2 = 4\pi^2 \frac{12\rho L^2}{b^2 E'} f_{\text{flex},n} \tag{6.23}$$

For point-mass model (6.20) and (6.22) reduce to:

$$K_{\text{Lat}} = 8G'a = \frac{wb^3 G}{3h^2 L} \left(1 - \frac{f_{\text{T,cnt}}^2}{f_{\text{T,free}}^2}\right) \tag{6.24}$$

and

$$K_{\text{flex}} = 2E'a = \frac{wb^3 E}{4L} \left(1 - \frac{f_{\text{flex,cnt}}^2}{f_{\text{flex,free}}^2}\right) \tag{6.25}$$

where E and G are the cantilever Young's and shear modulus, respectively. The subscripts for frequency f are: T for torsion, $flex$ for flexural, cnt for contact and $free$ for free resonance. Since contact resonances of flexural and torsional modes are measured independently for the same contact radius a, (6.24) and (6.25) reduce to:

$$\frac{4G'}{E'} = \frac{k_{\text{T}} \left(1 - \frac{f_{\text{T,cnt}}^2}{f_{\text{T,free}}^2}\right)}{k_{\text{flex}} \left(1 - \frac{f_{\text{flex,cnt}}^2}{f_{\text{flex,free}}^2}\right)} \tag{6.26}$$

where the spring constant of torsion k_{T} and flexural k_{flex} can be measured through thermal tune. All the frequencies of different resonances in (6.26) can be determined experimentally. Equation (6.26) therefore provides a direct measurement of elastic anisotropy in and out of the surface plane. For an isotropic material, the equation can give Poison's ratio without the knowledge of the contact area.

Another complimentary dynamic imaging mode is the so-called lateral excitation (LE) mode. LE mode refers to the AFM measurement techniques in which the cantilever is driven by the lateral oscillation of sample surfaces through tip–sample interaction, such as lateral force modulation AFM (LM-AFM), acoustic friction force microscopy (AFFM), and lateral atomic force acoustic microscopy (lateral AFAM). The excitation frequency of sample surfaces can be within a wide range. In

---

[2] $\frac{1}{E'} = \frac{1-v_1^2}{E_1} - \frac{1-v_2^2}{E_2}$, where E_i and v_i are Young's modulus and Poisson's ratio of material i, representing tip or sample

LM-AFM, the sample is laterally vibrated at a frequency ($\sim$16 KHz) well below the cantilever torsional/lateral bending resonance frequency. The torsional amplitude and phase are employed for friction imaging. In AFFM and lateral AFAM, the sample is oscillating laterally at megahertz frequencies (up to 3 MHz) to excite the cantilever vibrating in torsional or lateral bending resonance. The torsional amplitude and contact-resonance spectra are used for friction imaging.

6.5
Applications of TRmode Imaging

6.5.1
High-Resolution Imaging Application

TRmode uses near-field interaction to control feedback making the imaging control particularly sensitive to variation of adhesive forces at the nanometer scale. Condensation on the ledges of atomic steps of HOPG has been studied by other methods such as friction force microscopy [35], lateral modulation in contact mode [36] and dissipation study in UHV [37]. TRmode presents the opportunity to interact with the tip at the adhesive zone and control the interaction using feedback to maintain the set point.

To determine if TRmode can provide higher resolution and sensitivity than other imaging modes, we focused on the single atomic layer steps using a small scan size. The step height was confirmed to be a single atomic step of HOPG. The height image in Fig. 6.13a reveals four atomic layers in this area marked by 1 through 4. The tapping phase image (Fig. 6.13b) shows only two step edges. Note, that the data scale of the tapping phase image was reduced to two degrees to enhance the contrast of the image. The TR phase image shows detailed structures around the step edge with high lateral resolution and high sensitivity. As marked by the white arrow, in the area we didn't see any contrast in both tapping height and phase, we observed TR phase contrast. This indicates that TR phase data is extremely sensitive to surface ledges of single or subatomic steps. The TR phase data scale is set to twenty degrees, which is an order of magnitude larger than the tapping phase data scale.

To understand the origin of the TR phase contrast, we measured the TR amplitude change as a function of the tip–surface separation at different sample areas, on a HOPG terrace and at a step edge. In this experiment, the feedback loop was turned off and x-y position fixed. The Z piezo was used to push the cantilever toward the surface until an amplitude drop was detected. The Z piezo was then retracted to let the cantilever oscillation fully recover. The travel range of the Z piezo is 150 nm with a repetition rate of 2 Hz. The two sets of force curves are shown in Fig. 6.13 (P1 and P2). We found a drastic difference between the two kinds of amplitude versus z curves. The amplitude versus z curve on the terrace shown in Fig. 6.13 (P1) has small hysteresis and shallow slope. In contrast, the amplitude versus z curve taken from the step edge area showed a sudden drop of the amplitude and a large hysteresis between trace and retrace curves. From previous shear force microscopy and TDFM results, we have learned that a thin water layer adsorbed on the sample surface can introduce sudden drops of the lateral oscillation amplitude of the probe [38–40]. The authors attributed the observed phenomena to the solid-like properties of the water

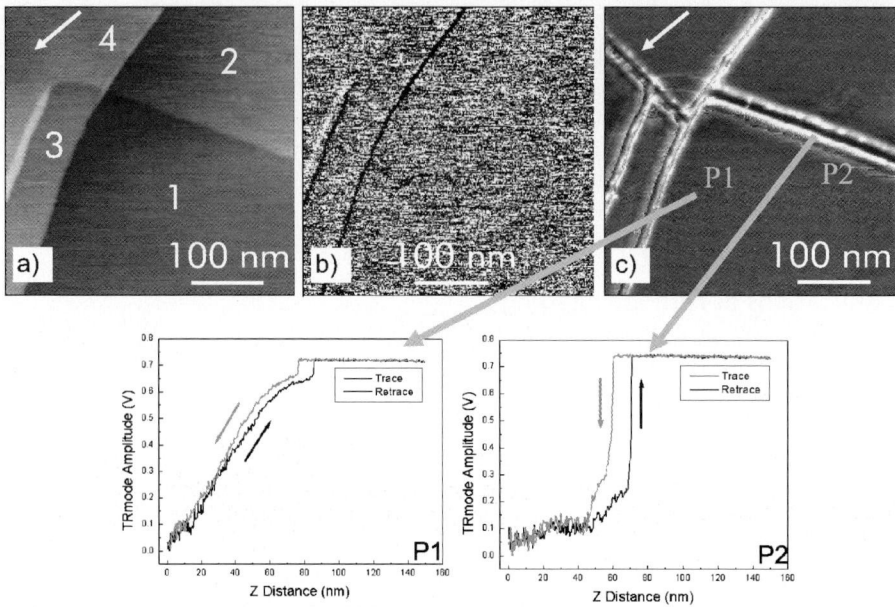

Fig. 6.13. HOPG surface imaged with interleaved TappingMode and TRmode. (**a**) Tapping height with data scale of 5 nm, the *numbers 1–4* mark four single atomic layers in this area. (**b**) Tapping phase with data scale of 2° and (**c**) TR phase with data scale of 20°. The *white arrows* in (**a**) and (**c**) mark an area showing TR phase contrast without height variations. The scan size is 400 nm with a scan rate of 1 Hz. TRmode amplitude change as a function of tip–surface distance without feedback loop control on a HOPG terrace marked P1; curve taken at a step edge showing high TR phase contrast marked P2

thin film when it is confined between the probe and a solid surface. It is believed that the initial adsorption of the water on the HOPG surface concentrates on the surface defects like step edges. The large hysteresis in force curves and sudden drop of the amplitude can be considered as a signature of a water layer presence.

The capillary force-induced amplitude change can also cause artifacts in height measurement. At the points of strong adhesive force the amplitude drop will force feedback to pull the tip away and interpret such incident as an increase of topographic height. Such a *height* profile is particularly clear at the HOPG atomic ledges, as shown in Fig. 6.14. In Fig. 6.14a, a height standard with a chemically homogeneous surface was imaged by both TappingMode and TRmode at high speed. TRmode has much better surface tracking, indicated by very small control error and the near vertical wall angle of the height standard cross section data. On the other hand if TRmode feedback is used to image HOPG surfaces with capillary adhesive forces at atomic ledges an overshoot of height at the ledge is observed (Fig. 6.14b). In this case TRmode translates chemical force into the height data.

The strategy of dealing with such a situation is to combine TappingMode feedback with TRmode. In this case the height control is based on TappingMode for feedback. At the end of each height scan line the probe is put at proximate contact position where the adhesion force is maximized and follows the trajectory determined

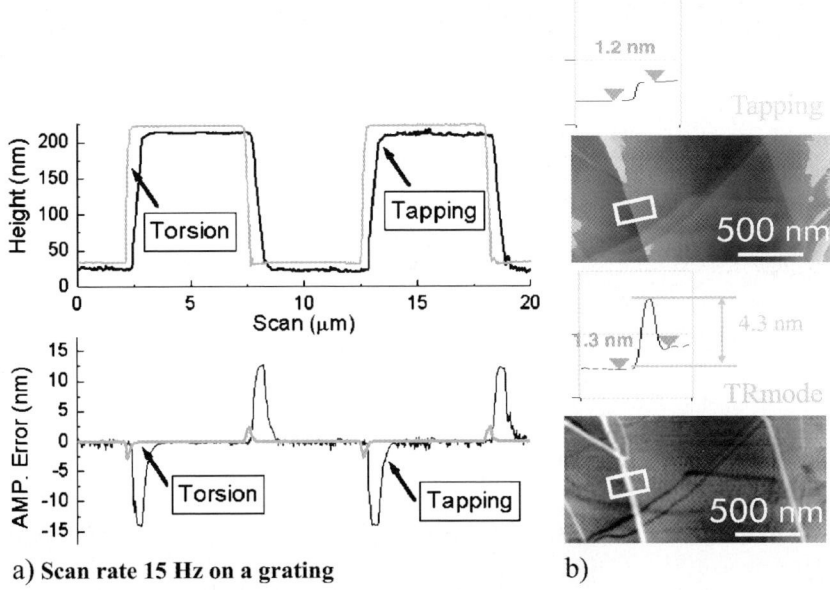

a) **Scan rate 15 Hz on a grating**   b)

Fig. 6.14. Comparative study of height standard and HOPG topographic steps. (**a**) A 180 nm height standard scanned by TappingMode and TRmode at a high scanning rate. TappingMode height data distorsion is shown in the *lower part* of curves represented by control errors. (**b**) The same height standard scanned by TRmode feedback (*upper red*). The control error is shown as the *lower red curve*

by the previous height scan without feedback. TRmode only performs measurements at a constant tip–surface offset. This concept is shown schematically in Fig. 6.15.

The sensitivity of TRmode to adhesion forces also leads to interesting applications as show in Fig. 6.16. The images are TappingMode and TRmode height images captured from a Si wafer deposited with carbon nanotubes. Depending on the nanotubes orientation, we observed different contrast in the TRmode height image. Whenever the nanotube is perpendicular to TRmode oscillating direction nanotube height in TRmode displayed a correct profile above the Si surface, as marked by white arrows in Fig. 6.16b. When nanotubes are parallel to the tip lateral oscillation direction the height profile becomes negative compared to the Si surface, as marked by black arrows in the same figure. In TappingMode height image all the carbon nanotubes displayed a positive height profile. The cause of this discrepancy is adhesive force. Since carbon nanotubes can exhibit strong hydrophobicity [41], the capillary induced amplitude decrease will be absent if the tip motion is entirely above the carbon nanotube. When the displacement of the tip is perpendicular to the nanotubes (in the area pointed by white arrows in Fig. 6.16b), the large tip size could always capture some adhesive interaction in the course of the oscillation. The near-field interaction environment resembles the other part of the sample and the feedback only responds to the relative height change. On the other hand, when the tip oscillates parallel to the carbon nanotube, which may be hydrophobic in this case, the absence of the capillary force locally will cause amplitude increase and

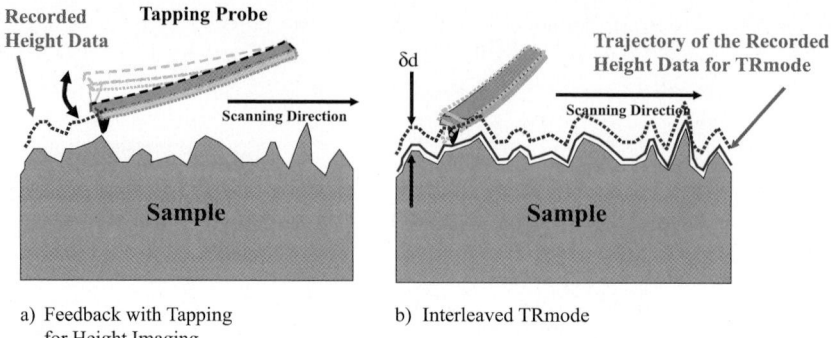

a) Feedback with Tapping   b) Interleaved TRmode
   for Height Imaging

Fig. 6.15. Schematics of interleave scanning of TappingMode and TRmode. (**a**) Topographic scanning by tapping using feedback. The height data is defined by the average position of the tip during tapping as shown in the *dotted line*. (**b**) The height trajectory in the *dotted line* is negtively offset by δd to bring the tip closer to the surface and the scanning is guided by the trejactory without feedback

Fig. 6.16. TappingMode and TRmode height images on nanotubes deposited on a Si wafer. The scan size is 500 nm for both images

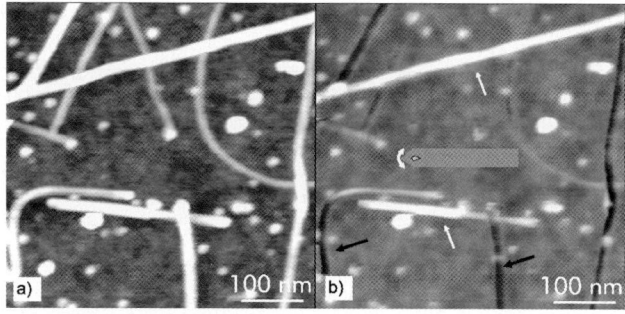

feedback will respond as the occurrence of a lower profile, leading to the negative height features indicated by black arrows in Fig. 6.16b. Again the combination of the TappingMode and TRmode will provide correct height profiles, as well as adhesive chemical property maps.

The unique combination of TappingMode and TRmode is impossible for shear force microscopy due to the absence of the vertical interaction for precise topographic information. Contact mode in conjunction with friction force, however, will perturb the sample so much that the experiment in Fig. 6.16 is unfeasible.

6.5.2
Electric Measurements Under Controlled Proximity by TRmode

In the field of material science and nanotechnology, it is very important to measure the electrical properties of various samples on the nanometer scale. Several techniques have been developed for this task. Among these techniques, scanning tunneling microscopy (STM), conductive/tunneling AFM and scanning capacitance microscopy (SCM) are widely used. In STM, a sharp metal probe is brought close to a surface, with a bias voltage applied between the tip and the surface. There is some finite probability that electrons will tunnel through the gap between the tip and the sample when the separation is small. This tunneling current is used as a feedback signal to control the tip–surface distance to maintain a constant current at a set point as the tip is scanned across the sample. STM can be used to measure properties of metals, semiconductors and other materials with high to medium conductivities. STM has a significant drawback. Since it uses the tunneling current as the feedback signal, the sample area being scanned needs to have some conductivity to allow the feedback loop to work throughout the scan. In general, STM cannot be used to scan an insulating sample with conductive patches. To overcome this problem, a tunneling atomic force microscope has been developed (US Patent 5,874,734). In using a conductive probe running in contact mode, the feedback system maintains a constant force between the tip and sample as in the standard AFM. During scanning, a constant or variable bias voltage may be applied between the tip and sample and the current distribution may be measured simultaneously with the topographic image. The advantage of this technique is using the deflection force between probe and surface as the feedback signal to control the tip–surface distance and force. The technique works on hard insulating samples with conductive patches and ultralow conductivity samples. However, there are several problems associated with using contact mode to image some samples. First, the feedback can only maintain a constant force between the tip and the surface in the vertical direction. When the tip scans across the surface, there is generally a large shear force present, and this high lateral force can easily damage both the tip and the sample. Next, the sensitivity of contact mode is limited because feedback is based on a static signal, as opposed to a dynamic signal. Static signals are more susceptible to thermal drift and charging, and thus sensitivity is compromised. For these reasons, contact mode is also not preferred when imaging soft and delicate samples.

The ideal solution would also reduce tip wear and increase throughput for measurements on electrical properties of dielectric and insulating films. The results described in this section demonstrate that TRmode is the best candidate for this task because the amplitude of the torsional resonance of the probe is used to control the probe so that it remains in close proximity to the sample surface, thus allowing precise measurement of the current between the tip and surface at nanometer scale tip–sample separations. Meanwhile, TRmode is a dynamic imaging mode that can be used to image soft and delicate samples similar to TappingMode.

All experiments were performed with a Dimension 3100 AFM and a NSIV controller (Veeco Instruments, Santa Barbara, CA). A TUNA application module was used to measure the current distribution. A negative sample bias will result in a positive (bright) current in the image display. The probes used for this study are FESP type and LTESP type coated with Pt/Ti thin film on the tip side (Veeco Instruments, Santa Barbara, CA). The FESP are single-beam etched Si probes with a fundamental flexural mode frequency of 60 kHz and a fundamental torsional resonance frequency of 550 kHz with a quality factor around 900. The dimension of the cantilever is typically 225 μm × 30 μm × 3 μm with a flexural spring constant of 1 to about 5 N/m, and a torsional spring constant estimated to be 30 to 150 N/m. We have also used the NSC 14/W2C/15 probes (MikroMasch, Estonia) for some of the TR-TUNA measurements. To prove that TRmode has advantages over contact mode on scanning the soft samples, we performed force-curve measurements on a PDMS sample. In this experiment, the feedback loop was turned off and x-y position fixed. The Z piezo was used to push the cantilever toward the surface until an amplitude drop was detected. The Z piezo was then retracted to let the cantilever oscillation fully recover. In the plot shown in Fig. 6.17, the TR amplitude and the cantilever vertical deflection were plotted as a function of tip travel in the z-direction. When tip approaches to the surface, TR amplitude drops from 2 to $\sim$ 0.2 V (region indicated by the shadow box). In this region, the cantilever vertical deflection changes from $\sim$ 0.01 to 0.002 V, which shows the downward bending of the lever due to net attractive forces. With Z piezo pushing further into the surface, the TR amplitude drops slowly close to zero. We observed the vertical deflection of the cantilever increased gradually and changed from 0.002 to 0.1 V at the final z position. This indicates that the cantilever went through from bending downward by net attractive forces to upward by the repulsive forces in the vertical direction.

When the Z piezo retracts from the surface, the TR amplitude recovers to 2 V with some hysteresis between 1 and 2 V. Meanwhile, the cantilever bending changes from upward to downward and finally to its neutral position. The significance of the result is that the TRmode operates at the vertical attractive force regions under normal conditions, for example, set point to free air amplitude ratio ranging from 95% to 10%.

The schematic in Fig. 6.18 illustrates the working principle of TRmode AFM or tunneling AFM. A conductive AFM probe is brought into contact with a sample surface. We can choose to control the tip–sample interactions by contact mode or in our case by TRmode. The sample can have either a thin non-conductive layer or conductive regions at different places. A DC constant or variable bias voltage

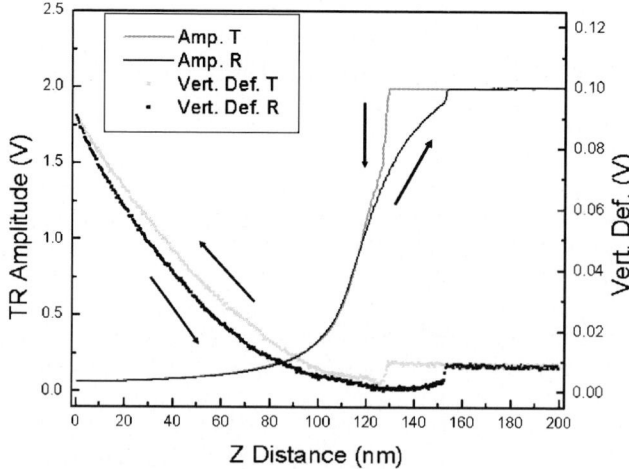

Fig. 6.17. A force curve plot shows TR amplitude and the cantilever vertical deflection as a function of tip–sample separation on a soft sample PDMS. The area with *lower vertical deflection* comparing free cantilever marks that the cantilever bends downward to the surface in the normal TRmode operation region. That indicates the total net forces between the probe and the sample in the *vertical direction* is attractive. The ramp rate is 1 Hz with a ramp size of 200 nm. The free air TR amplitude is 2 V

is applied between the probe and the sample. When the tip scans over the sample, the feedback loop keeps a constant TR amplitude and the conductive or tunneling current can be measured at each point simultaneously on the surface. The current signal can be displayed together with the topographic image of the sample.

As discussed earlier, TappingMode is capable of imaging soft materials. However, the tip spends only a very short period of time on or close to the surface, which cannot meet the requirements of electrical measurement. To clearly demonstrate our point, we have performed an experiment on the carbon black filled vulcanizate. The current image in Fig. 6.19a was acquired with TappingMode at the sample bias voltage of 10 V. There is no detectable current in any area of this image. In contrast, we observed a lot of bright spots in the current image Fig.19b, which was scanned on the same area of sample with the same bias voltage but with TRmode. The current

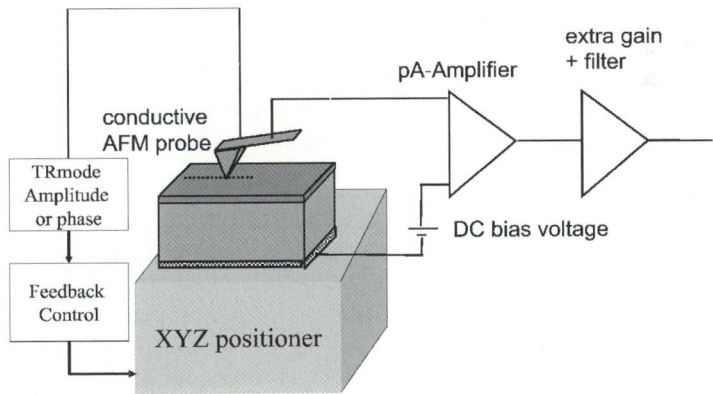

Fig. 6.18. Schematics of TRmode electric measurement. While a feedback loop maintains constant TRmode amplitude, tunneling current is measured by the right hand parts with sub-pico ample sensitivity

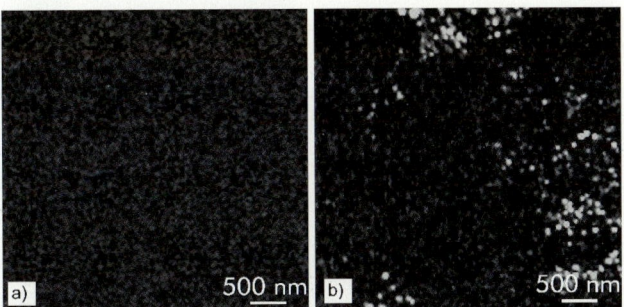

Fig. 6.19. Tunneling current images taken of a carbon black sample filled vulcanizate sample at a sample bias voltage of -10 V. (**a**) The image was acquired simultaneously with TappingMode; there is no any current contrast in this area. (**b**) The image was acquired simultaneously with TRmode; we observed many bright spots in the same sample area. Both images have 4.3 μm scan, and 1 pA current scale

scale for both images is 1 pA, with the scan size of 4.3 μm. The bright spots in Fig. 6.19b represent the areas where there are tunneling currents between tip and sample through the conductive paths.

After optimizing the experimental control parameters, we observed that the tunneling current could appear at relative low bias voltage on some areas. Shown in Fig. 6.20a is a topographic image, 10 μm scan, of this carbon black filled polymer sample. The current images in Fig. 6.20b–d were obtained at the same area with different sample bias voltage from −0.5 to −2 V. There are only a few bright spots in Fig. 6.20b, which indicates the conductive areas at this sample bias voltage. With increasing of the sample bias voltage, it reduces the effective barrier for tunneling. Therefore, we observed that tunneling currents appeared in more areas. To calculate local conductivity more knowledge, such as tip geometry and tip–sample distance, is needed.

The unique strength of using TRmode to control the tip–sample distance for current measurements has been demonstrated not only on soft samples, but also on delicate samples. The following example shows the application of TR-TUNA for the fundamental task in the nanoelectronics field: conductivity test in the nanometer scale. The topographic images in Fig. 6.21a,b were captured from a special sample, where the conductive nanowires were deposited on to a sili-

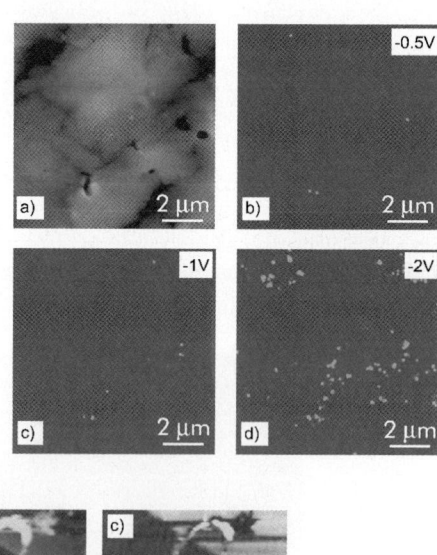

Fig. 6.20. Topography (**a**) and simultaneously obtained tunneling current images (**b**)–(**d**) from a carbon black filled vulcanizate sample, 10 μm scan. The current images were obtained at the same area with different sample bias voltage from −0.5 to −2 V, 1 pA current scale

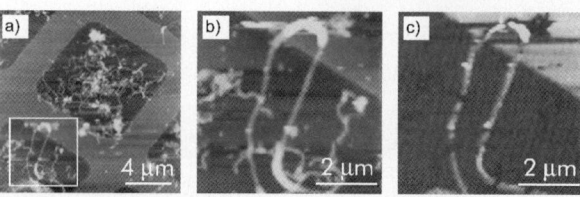

Fig. 6.21. (**a**) Topography of nanowires on silicon surface with deposited conductive gold grids, 16 μm scan, 120 nm height scale. (**b**) A zoom-in image of (**a**) of 6 μm scan size. (**c**) A tunneling current image (contrast reversed) taken at the sample bias of 4 V, 5 pA current scale. Only the nanowires connected to the gold grid show up in this current image

con wafer with conductive gold girds. The square gold grid can be seen clearly in Fig. 6.21a, while the zoom-in image shown in Fig. 6.21b reveals more details in this area. Since the nanowires are only loosely bound to the surface, we could not use contact mode to scan this sample and to obtain decent images. With either TappingMode or TRmode, the nanowires can be stably imaged without being pushed away by the tip. The current image in Fig. 6.21c provided clear evidence that only the nanowires that have a connection to the gold grid exhibit contrast to the tunneling current image. In other words, only the nanowires electrically connected to the gold grid can form a conductive path to show current contrast.

Carbon nanotube is one of the fastest growing areas of nanomaterials. But there are only limited methods of performing the electrical measurements at this very small scale. Since the nanotubes are weakly bonded to the substrate in most cases, using contact mode for the measurement will push the nanotubes around on the surface. The following two examples demonstrated that TR-TUNA could be used to study the nanotube conductivity on various substrates. Figure 6.22a,c is height images obtained with TRmode from a sample that the nanotubes were deposited on a treated silicon wafer. It shows that nanotubes can be repeatedly and stably imaged in great detail by TRmode. The current images (Fig. 6.22b,d) were acquired simultaneously with the height image at the same sample location but with different sample bias voltages. With a negative 1 V of sample bias (Fig. 6.22b), the nanotubes appear brighter than the silicon substrate. When the sample bias was reversed to positive 0.5 V, the nanotubes appear darker than the substrate, which indicates the tunneling current flows from tip to sample. In both cases, the contrast of nanotubes in the current images reveals the better conductivity of nanotubes over the silicon substrate. When we measured the background current level on the Si part with the contact mode TUNA, it is about 50 pA with the bias voltage of 1 V. However, we have measured as much as 7 nA current on the nanotubes at the maximum current locations. One of the possible explanations is that the current is drawn from the conducting nanotube network, which effectively has a much larger contact area than the tip apex size.

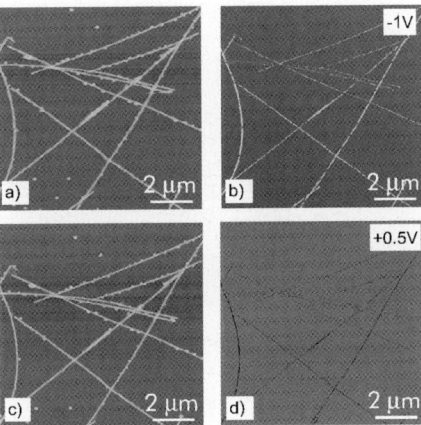

Fig. 6.22. (**a,c**) Topography of carbon nanotubes deposited on silicon, 10 μm scan, 100 nm height scale. (**b**) The corresponding current image of (**a**) taken at a sample bias voltage of −1 V, 10 nA current scale. (**d**) The current image taken at a sample bias voltage of +0.5 V, 10 nA current scale

6.5.3
In-Plane Anisotropy

TRmode is inherently capable of studying in-plane anisotropic properties of samples. To demonstrate this capability, we used the TRmode to scan a polymer sample (polydiacetylene, PDA single crystal). Polydiacetylene single crystal consists of a one-dimensional molecule chain hundreds of micrometers in length. The images shown in Fig. 6.23a,b is TR phase data with different sample orientation. During the scanning, we kept the same scanning angle and other control parameters, but rotated the sample by 90°. Previous studies showed that this sample has strong anisotropy because of the aligned polymer backbone structure [42]. The image in Fig. 6.23a shows minimum phase contrast because the tip oscillation direction is aligned with the sample so that the tip is sliding almost parallel to the backbones. There is minimum friction interaction between the tip and the structure in this configuration. With the sample rotated at 90°, the tip oscillation direction is now perpendicular to the backbone structures and a strong phase contrast is observed (Fig. 6.23b). In this situation, the tip has to climb over the backbone structure and therefore there is a much larger friction force between the tip and sample. The images shown in Fig. 6.23a,b is not at the exact same sample area, however, the backbone structure alignment directions can also be observed under the optical microscope and marked by the white arrows in the images.

Another example of in-plane anisotropy is the magnetic force microscope (MFM). MFM provides a unique way to measure near-surface stray field of magnetic domain in materials or recording heads [43]. While there is no doubt of the importance of MFM in nanometer scale magnetic studies of materials and devices, most measurements are qualitative, providing spatial variation of the force gradient in the direction perpendicular to the samples through force gradient contrast, i.e., dF/dz [44]. A quantitative derivation of the three-dimensional source field is not possible from a one-dimensional measurement of the force gradient. A solution has been sought to tackle this problem by tilting the tip relative to the sample surface. However such a measurement again only proved to be difficult [45]. Efforts have been

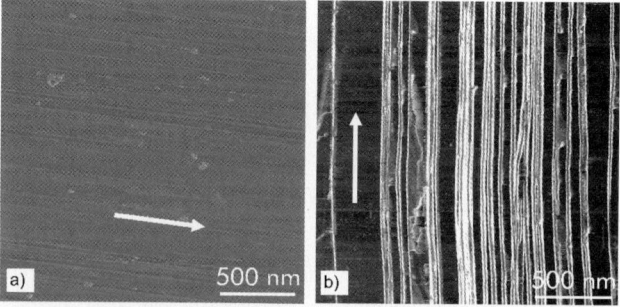

Fig. 6.23. TRmode phase data on PDA samples with different sample orientation. The *white arrows* in the images mark the direction of the backbone structure. The scan size for both images is 2 μm with a phase data scale of 30°. (**a**),(**b**) Both images are of the sample parallel or perpendicular to the tip oscillation directions

devoted to approach the lateral force gradient by using specially configured probes with tips oscillating parallel to surfaces. Such measurements again only provide the single component of the force gradient [46]. The complication of determining other force gradient components in the measurement instrument may easily hinder the possibility of measurement at exactly the same spatial position in the nanometer scale. A method that can measure force or force gradient concurrently or sequentially at the same nanometer scale feature is still needed. With TRmode the force gradient components normal and parallel to the sample can be determined and interleaved at the same location. Such measurements may provide a new dimension of information on the force gradient, leading to a better understanding of the force and its gradient source. We have chosen well-known magnetic samples, namely magnetic recording disks, and compared the measurement results with simulations employing mature finite element tools. It was anticipated that with the well-known magnetic field components of a disk one might be able to derive nominal tip magnetization direction, which would otherwise be impossible with the single component of force gradient measurements.

Figure 6.24 shows the same measurement of a magnetic hard disk of relative high aerial density. The tip lateral motion is perpendicular to the longitudinal bit direction. Again, the location of these two images is confirmed by topographic image check. The bottom left of Fig. 6.24 shows the cross section data of the reference lines indicated in the images above. Except for some details the results of tapping and torsional resonance phase shift features are opposite to each other, as if the contrast is reversed. Notice the data scales of the two curves are different by a factor of 5. After

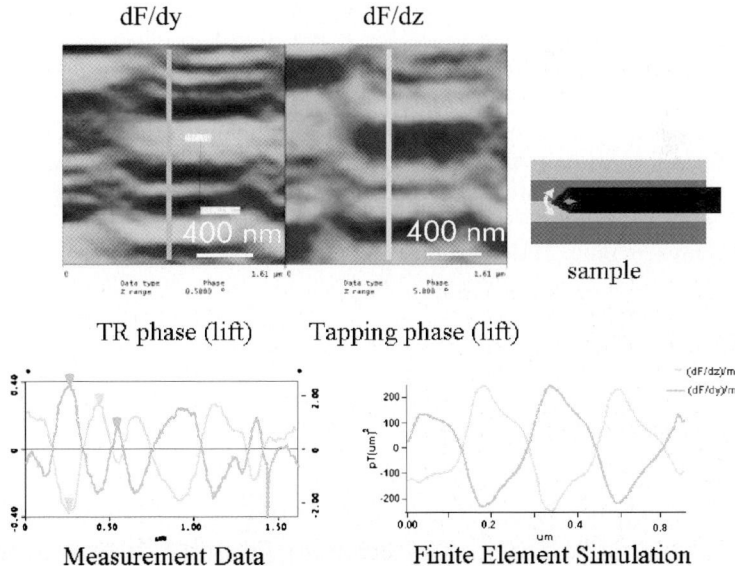

Fig. 6.24. Force gradient measurement by TRmode (*top left*) and TappingMode (*top middle*). The *top right* image shows the relative direction of tip rotation relative to the domain orientation. From the *bottom left* one can see the cross section data as indicated in the lines of the top images. The *bottom right* shows the finite element calculation data plotted for the case of tip magnetization rotated 37° relative to the cantilever surface normal

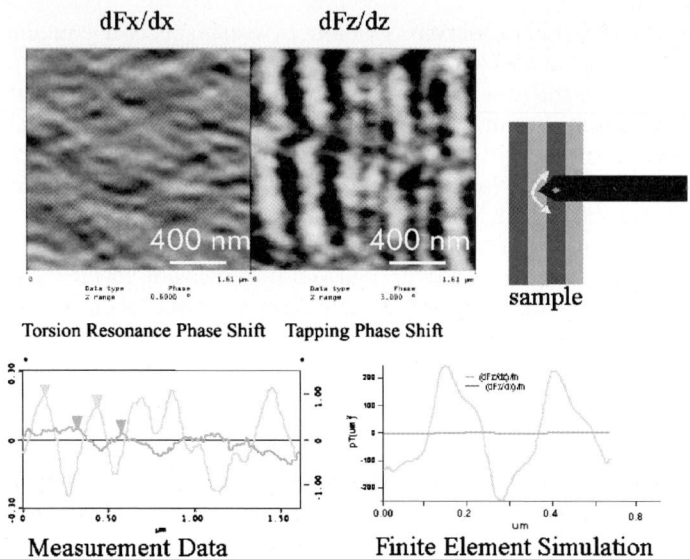

Fig. 6.25. Sample rotated 90° relative to Fig. 6.24 with force gradient measurement by TRmode (*top left*) and TappingMode (*top middle*). The *top right* shows the relative direction of tip rotation relative to the domain orientation. The *bottom left* shows the cross section data as indicated in the lines of the top images. The *bottom right* shows the finite element calculation data plotted for the case of tip magnetization with an angle of 37° relative to the cantilever surface normal

the sample was rotated 90°, the imaging data of both TRmode and TappingMode are shown in Fig. 6.25. The tapping data shows similar domain structure as Fig. 6.24 in the same imaging mode, though with some undulation. On the other hand, the TRmode image shows dramatically different behavior compared to the same imaging mode before the sample was rotated. The bit direction in reference tip lateral motion is shown in the top right of Figs. 6.24 and 6.25. The bottom right charts in Figs. 6.24 and 6.25 represent finite element simulation results. Since the magnetism of the hard drive is known from manufacturer specifications, the simulation takes magnetization of the tip as the variable and rotates its direction until the force gradient data complies with the measurements of force gradient in all the directions. These experiments demonstrate that, with additional in-plane force gradient information from TRmode, the deduction of 3D magnetic vector from a know tip or the tip magnetization from a known sample is feasible.

6.6
Torsional Tapping Harmonics for Mechanical Property Characterization

When an oscillating probe tip periodically contacts a sample surface, a complex dynamic process occurs as the tip passes through various interaction regimes. For example, a probe tip may pass in and out of long-range electrostatic forces, the onset of fluid meniscus forces, elastic contact and adhesion. These forces occur on time scales that are not readily observed during normal AFM imaging, as the cantilever oscillation is usually averaged over many cycles by the detection electronics.

It has been recognized, however, that each impact event acts as a force impulse that excites a range of modes of cantilever oscillation [5–7]. These excited modes can provide a window to examine dynamic processes normally not available in a conventional AFM. This section will examine how these time-dependent forces can be observed and discuss the application of this technique to mechanical property measurements.

From Fourier theory we know that any periodic signal can be constructed by summing appropriate harmonic components. So the periodic tip–sample force in TappingMode may be considered as the sum of Fourier components $F_n(n\omega_0)$. The tapping impulse force $F_i(t)$ can be represented as:

$$F_i(t) = \sum_n F_n(n\omega_0)\, e^{in\omega_0 t} \tag{6.27}$$

where $\omega_0 = 2\pi f_0$ and f_0 is the tapping frequency. The tip and sample feel forces at the fundamental tapping frequency and each integer multiple Fourier frequency $n\omega$ that makes up the pulse. In principle, if one were to measure the amplitudes of all the modes excited, one could accurately reconstruct the time-dependent interaction force.

In practice, conventional AFMs have been able to capture the amplitudes of only a small number of higher order vibration modes. The reason is that the amplitude of higher frequency motion is heavily attenuated above the cantilever's fundamental flexural frequency. Figure 6.26 shows a plot of the transfer function of a cantilever oscillated in the flexural direction. The transfer function shows the cantilever amplitude versus frequency, normalized against the fundamental flexural frequency. Note

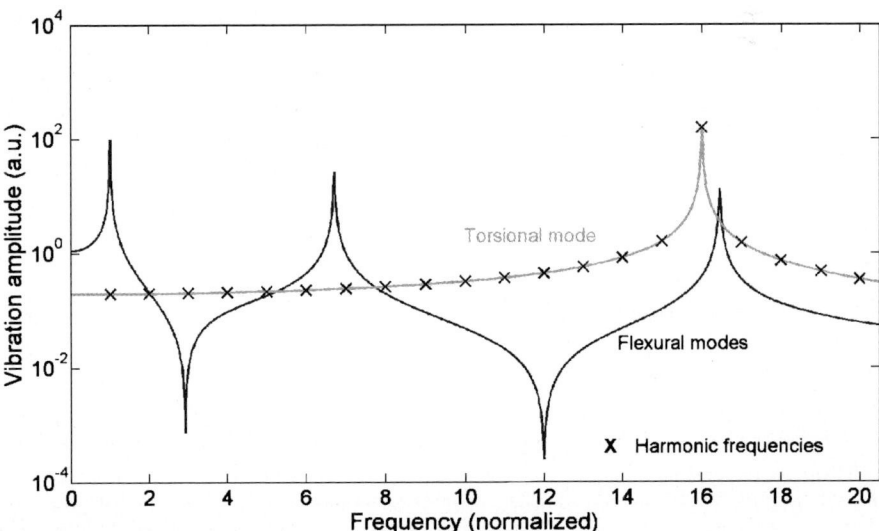

Fig. 6.26. Flexural and torsional transfer functions of an AFM cantilever (personal communication, Sahin, Magonov, Su, Quate, Solgaard, 2005)

that the cantilever's amplitude response is five orders of magnitude lower than the fundamental at only three times the fundamental resonant frequency. This puts many of the higher Fourier components below the detection limit of the AFM.

6.6.1
Detecting Cantilever Harmonics Through Torsional Detection

Sahin and colleagues (personal communication, Sahin, Magonov, Su, Quate, Solgaard, 2005) however, have recently demonstrated a technique to recover more of the higher Fourier components by taking advantage of the wider bandwidth available with torsional oscillations. The authors achieved the coupling of tapping forces to torsional motion by offsetting the AFM tip from the center line of the cantilever, as shown in Fig. 6.27. The offset cantilever generates a torque that excites twisting motion of the cantilever with each tap. Figure 6.28 also shows the response characteristics of the torsional motion of the same cantilever. Note for the case shown that the fundamental and the next 14 higher Fourier components of the flexural frequency are below the first flexural resonance. Hence the torsional response provides a wide bandwidth with relative uniform amplitude.

In this case each time the tip taps a sample surface, it excites torsional motions *integer multiples* of the tapping frequency. It is worth pointing out that these are not higher flexural order modes of oscillation, but harmonics of the tapping frequency.

Fig. 6.27. Offset tip cantilever for measurement of torsional tapping harmonics (personal communication, Sahin, Magonov, Su, Quate, Solgaard, 2005)

6.6.2
Reconstruction of Time-Resolved Forces

Time-resolved forces can be reconstructed in the following way:

1. Measure the torsional transfer function $G(\omega)$, i.e., the conversion between a known force and a known torsional amplitude. A representative transfer function is shown in Fig. 6.26.
2. Record the torsional deflection signal $T(\omega)$ with sufficient bandwidth that is at least ten times the fundamental flexural frequency.
3. Measure the complex amplitudes $T_n(n\omega_0)$ at integer multiples n of the fundamental flexural frequency ω_0, i.e., the tapping frequency.

Fig. 6.28. Time-varying tip-sample force reconstructed from torsional tapping harmonics. (**a**) Oscilloscope traces of the periodic torsional vibration signals at the position sensitive detector obtained on graphite. (**b**) Time resolved tip–sample force measurements obtained on graphite (**c**) torsional harmonic spectra (personal communication, Sahin, Magonov, Su, Quate, Solgaard, 2005)

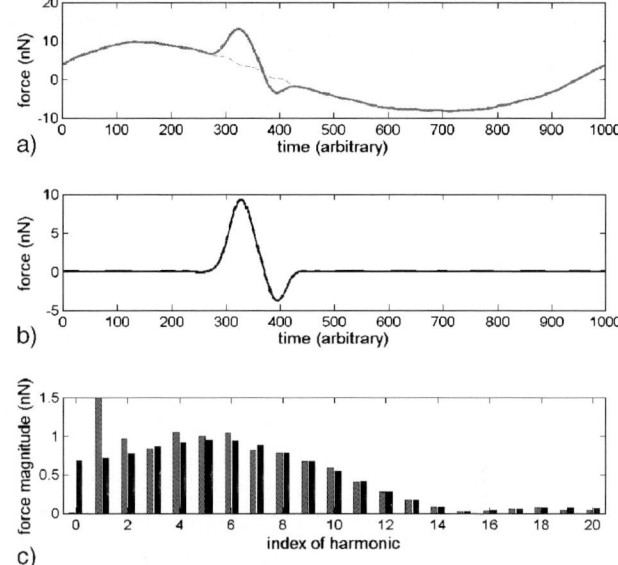

4. Convert torsional amplitudes $T_n(n\omega_0)$ into torsional forces $F_n(n\omega_0)$, using the inverse transform function $G^{-1}(n\omega_0)$ of the cantilever's torsional motion. That is,

$$F_n(n\omega_0) = G^{-1}(n\omega_0)T_n(n\omega_0) \tag{6.28}$$

5. Convert back to the time domain by summing the Fourier series:

$$F_i(t) = \sum_n F_n(n\omega_0)e^{in\omega_0 t} \tag{6.29}$$

An example of a time-varying force is shown in Fig. 6.28. Note that the tip-surface interaction goes through several regimes including initial attraction, repulsive contact, and adhesion, before breaking free from the surface. These regimes are entirely analogous to those seen in quasistatic AFM force curves, except that the forces are being measured on time scales thousands of times faster.

6.6.3
Force-Versus-Distance Curves

Once the time-resolved force measurements are made, it is relatively straight forward to convert this into force-versus-distance plots. We start with the following knowledge:

– $F_i(t)$: The time-varying force, calculated above
– $z(t)$: The cantilever's vertical deflection as a function of time

The signal $z(t)$ tracks the cantilever's flexural oscillation at the tapping frequency. It is a largely sinusoidal signal, with small harmonic distortions from the tip–sample interactions and the cantilever dynamics.

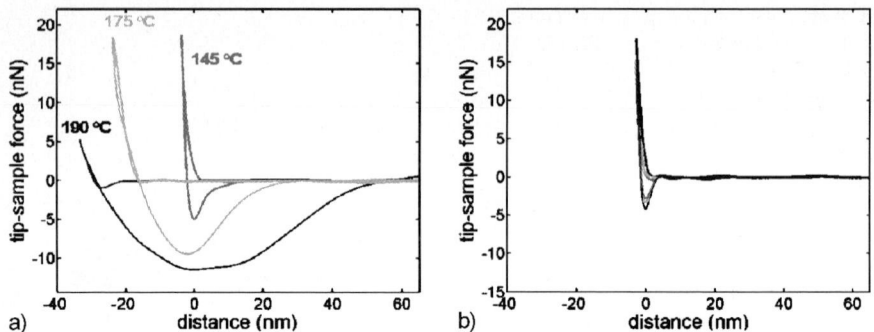

Fig. 6.29. High speed force versus distance curve reconstructed for torsional tapping harmonics (**a**) polystyrene above glassy transition at different temperatures. (**b**) PMMA at the same temperatures but below glassy transition (personal communication, Sahin, Magonov, Su, Quate, Solgaard, 2005)

This signal $z(t)$ is then inverted numerically to obtain $t(z)$, which is plugged into the force function to get $F(z)$. The resulting plot charts the force the cantilever feels at each distance from the sample surface. An example force versus distance plot is shown in Fig. 6.29.

It is important to contrast this reconstructed force curve with typical quasistatic AFM force curves. Typical quasistatic AFM force measurements complete each force cycle on time scales around a second. The tapping harmonic technique can in principal generate a high-speed force curve every tapping cycle, thousands to millions of times per second. This provides several important advantages. First, it allows many more interaction experiments in the same period of time. Second, it opens the door to high-speed mapping of material properties. Additionally, it allows the exploration of mechanical properties at many higher orders of magnitude of frequencies than quasistatic tests.

6.6.4
Mechanical Property Measurements and Compositional Mapping

It has long been established that it is possible to extract localized mechanical properties from AFM force curves [47]. The basic approach is to analyze the load-displacement curve against a physical model that is representative of the system under study. Many different theories have been used to model elastic contact, including Hertzian theory, Sneddon, DMT, and Maugis [48]. These models predict the loading force required to achieve a specific elastic deformation (displacement) of the sample. With knowledge of the AFM tip radius, it is possible to extract a measurement of elastic modulus which provides the best fit to the load-displacement curve.

It is also possible to use torsional tapping harmonics to create high-sensitivity compositional maps of heterogeneous materials. Materials with different mechanical properties will have different tip–sample contact profiles. Specifically, the time-varying force $F_i(t)$ will be different. As we saw in (6.29), this time-varying force is made up of a sum of Fourier components $F_n(n\omega_0)$. It is possible to image the

Fig. 6.30. Composition and material property mapping. Images of the twelfth harmonic of a PMMA/polystyrene blend at various temperatures. *Top*: imaging when both polystyrene and PMMA are glassy state. *Middle*: polystyrene transformed into rubbery state while PMMA remains glassy. *Bottom*: polystyrene completes glassy transition, as shown in the *dark contrast zone*, indicating softer elastic properties. The torsional harmonic clearly distinguishes between the two materials and provides relative measurements of surface stiffness (personal communication, Sahin, Magonov, Su, Quate, Solgaard, 2005)

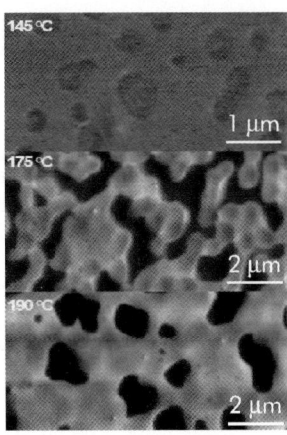

amplitude and/or phase of any of these Fourier components in real time by using a lock-in amplifier tuned to the appropriate reference frequency $f_r = n\omega_0/2\pi$.

This technique then provides simultaneous topographic images and relative compositional and material property maps, an example of which is shown in Fig. 6.30. The compositional maps are somewhat analogous to those created by phase imaging [49], except that it is possible to optimize the contrast by appropriate selection of the reference frequency. Specifically, the harmonic multiplier n is chosen to achieve the best discrimination between multiple materials under study. Additionally, the contrast in phase imaging is largely due to differences in material dissipation. The ability to image both amplitude and phase of tapping torsional harmonics allows separate identification of elastic and dissipative material properties.

6.7
Conclusion

Torsional resonance microscope provides dynamic information of the tip–sample interaction in lateral dimensions. Since the tip displacement amplitude can be as low as subnanometers, it reflects near-field interaction due to different mechanisms, such as mechanical, electrical and magnetic interactions. When TRmode is applied in combination with flexural interaction such as TappingMode or contact mode it is possible to study the interaction force and force gradient in both vertical and lateral directions concurrently. TRmode also provides a unique opportunity for electric, magnetic or near-field optical control where the tip needs to stay in close proximity of the surface. When TRmode is applied together with the contact mode one may be able to derive vertical and shear contact stiffness simultaneously. One of the benefits of performing multiple dimensional measurements was to derive elastic anisotropy, independent of the contact area.

The dynamic responses of TRmode were proven to be valuable in determining non-linear harmonics of the TappingMode, dynamic friction and other surface mechanical properties.

References

1. Neumeister JM, Ducker WA (1994) Lateral, Normal and Longitudinal Spring Constants of Atomic-Force Microscopy Cantilevers. Rev Sci Instrum 65:2527–2531
2. Warmack RJ, Zheng XY, Thundat T, Allison DP (1994) Friction Effects in the Deflection of Atomic-Force Microscope Cantilevers. Rev Sci Instrum 65:394–399
3. Turner J, Hirsekron S, Rabe U, Arnold W (1997) High-Frequency Response of Atomic-Force Microscope Cantilevers. J Appl Phys 82:966–979
4. Rabe U, Turner J, Arnold W (1998) Analysis of the High-Frequency Response of Atomic Force Microscope Cantilevers. Appl Phys A 66:S277–S282
5. Stark M, Stark R, Heckl WA, Guckenberger R (2000) Spectroscopy of the Anharmonic Cantilever Oscillations In TappingMode Atomic-Force Microscopy. Appl Phys Lett 77:3293–3295
6. Stark R, Heckl W (2000) Fourier Transformed Atomic Force Microscopy: TappingMode Atomic Force Microscopy Beyond The Hookian. Surf Sci 457:219–228
7. Hillenbrand R, Stark M, Guckenberger P (2000) Higher-Harmonics Generation in Tapping-Mode Atomic-Force Microscopy: Insights into the Tip-Sample Interaction. Appl Phys Lett 76:3478–3480
8. Löhndorf M, Moreland J, Kabos P (2000) Microcantilever Torque Magnetometry of Thin Magnetic Films. J Appl Phys 87:5995–5997
9. Löhndorf M, Moreland J, Kabos P (2000) Ferromagnetic Resonance Detection with a Torsion-Mode Atomic-Force Microscope. Appl Phys Lett 76:1176–1178
10. Turner JA, Wiehn JS (2001) Sensitivity of flexural and torsional vibration modes of atomic force microscope cantilevers to surface stiffness variations. Nanotechnology 12:322–330
11. Green CP, Sader JE (2002) Torsional Frequency Response of Cantilever Beams Immersed in Viscous Fluids with Applications to the Atomic Force Microscope. J Appl Phys 92:6262–6274
12. Yamanaka K, Takano H, Tomita E, Fujihira M (1996) Lateral Force Modulation Atomic Force Microscopy of Langmuir-Blodgett Film in Water. Jpn J Appl Phys Part 1 Reg Papers Short Notes Rev Papers 35(10):5421–5425
13. Reinstädtler M, Rabe U, Scherer V, Turner JA, Arnold W (2003) Imaging of Flexural and Torsional Resonance Modes of Atomic Force Microscopy Cantilevers Using Optical Interferometry. Surf Sci 532:1152–1158
14. Caron A, Rabe U, Reinstadtler M, Turner J, Arnold W (2004) Imaging Using Lateral Bending Modes of Atomic Force Microscope Cantilevers. Appl Phys Lett 85:6398–6400
15. Yamanaka K, Tomita K (1995) Lateral Force Modulation Atomic-Force Microscope for Selective Imaging of Friction Forces. Jpn J Appl Phys Part 1 Reg Papers Short Notes Rev Papers 34(5B):2879–2882
16. Reinstadtler M, Rabe U, Scherer V, Hartmann U, Goldade A, Bhushan B, Arnold W (2003) On the Nanoscale Measurement of Friction Using Atomic-Force Microscope Cantilever Torsional Resonances. Appl Phys Lett 82:2604–2606
17. Chang W, Fang T, Chou H (2003) Effect of interactive damping on sensitivity of vibration modes of rectangular AFM cantilevers. Phys Lett A 312:158–165
18. Su C, Huang L, Neilson P, Kelley V (2003) In-situ measurement of in-plane and out-of-plane force gradient with a torsional resonance mode AFM. In: Koenraad PM, Kemerink M (eds.) Scanning Tunneling Microscopy/Spectroscopy And Related Techniques: 12th International Conference, CP696. AIP, Melville, New York, pp. 349–356
19. Su C, Babcock K, Huang L (2005) US Patent 6,945,099
20. Huang L, Su C (2004) Torsional resonance mode imaging for high-speed atomic force microscopy. In: Koenraad PM, Kemerink M (eds.) Scanning Tunneling Mi-

croscopy/Spectroscopy And Related Techniques: 12th International Conference, CP696. AIP, Melville, New York, p. 357; (2005) A Torsional Resonance Mode Afm For In-Plane Tip Surface Interactions. Ultramicscopy 100:277–285

21. Kasai T, Bhushan B, Huang L, Su C (2004) Topography and Phase Imaging Using the Torsional Resonance Mode. Nanotechnol 15:731–742
22. Reinstädtler M, Kasai T, Rabe U, Bhushan B, Arnold W (2005) Imaging and Measurement of Elasticity and Friction Using the TRmode. J Phys D Appl Phys 38:R269–R282
23. Frisbie C, Rozsnyai A, Noy A, Wrighton M, Lieber C (1994) Functional-Group Imaging by Chemical Force Microscopy. Science 265:2071–2074
24. Sharos LB, Raman A, Crittenden S, Reifenberger R (2004) Enhanced Mass Sensing Using Torsional and Lateral Resonances in Microcantilevers. Appl Phys Lett 84:4638–4640
25. Oshea S, Welland M, Wong T (1993) Influence of Frictional Forces on Atomic-Force Microscope Images. Ultramicroscopy 52:55–64
26. Dedkov GV (2000) Experimental and Theoretical Aspects of the Modern Nanotribology. Phys Status Solid A Appl Res 179:3–75
27. Maugis D (1992) Adhesion of Spheres: The Jkr-Dmt Transition Using a Dugdale Model. J Colloid Interface Sci 150:243–269
28. Hoffmann P, Jeffery S, Pethica J, Ozer H, Oral A (2001) Energy Dissipation in Atomic Force Microscopy and Atomic Loss Processes. Phys Rev Lett 87:265502-1 to 265502-4
29. Volokitin AI, Persson BNJ (2005) Adsorbate-Induced Enhancement of Electrostatic Non-contact Friction. Phys Rev Lett 94:086104-1 to 086104-4
30. Baljon ARC, Robbins MO (1996) Energy Dissipation During Rupture of Adhesive Bonds. Science 271(5248):482–484
31. Su CM, Huang L, Kjoller K (2004) Direct Measurement of Tapping Force with a Cantilever Deflection Force Sensor. Ultramicroscopy 100(3–4):233–239
32. Yamanaka K, Nakano S (1998) Quantitative Elasticity Evaluation by Contact Resonance in an Atomic Force Microscope. Appl Phys A Mater Sci Process 66:S313–S317
33. Rabe U, Kopycinska M, Hirsekorn S, Arnold W (2002) Evaluation of the Contact Resonance Frequencies in Atomic Force Microscopy as a Method for Surface Characterisation (Invited). Ultrasonics 40(1–8):49–54
34. Rabe U, Amelio S, Kester E, Scherer V, Hirsekorn S, Arnold W (2000) Quantitative Determination of Contact Stiffness Using Atomic Force Acoustic Microscopy. Ultrasonics 38(1–8):430–437
35. Annis BK, Pedraza DF (1993) Effect of Friction on Atomic-Force Microscopy of Ion-Implanted Highly Oriented Pyrolytic-Graphite. J Vac Sci Technol B 11(5):1759–1765
36. Kawagishi T, Kato A, Hoshi Y, Kawakatsu H (2002) Mapping of Lateral Vibration of the Tip in Atomic Force Microscopy at the Torsional Resonance of the Cantilever. Ultramicroscopy 91:37–48
37. Spychalski-Merle A, Krischker K, Göddenhenrich T, Heiden C (2000) Friction Contrast in Resonant Cantilever Vibration Mode. Appl Phys Lett 77:501–503
38. Antognozzi M, Humphris ADL, Miles MJ (2001) Observation of Molecular Layering in a Confined Water Film and Study of the Layers Viscoelastic Properties. Appl Phys Lett 78:300–302
39. Brunner R, Marti O, Hollricher O (1999) Influence of Environmental Conditions on Shear-Force Distance Control in Near-Field Optical Microscopy. J Appl Phys 86:7100–7106
40. Vaccaro L, Bernal MP, Marguis-Weible F, Duschl C (2000) Shear Force Surface Contrast on Self-Assembly Monolayers. Appl Phys Lett 77:3110–3112
41. Li SH, Li HJ, Wang XB, Song YL, Liu YQ, Jiang L, Zhu DB (2002) Super-Hydrophobicity of Large-Area Honeycomb-Like Aligned Carbon Nanotubes. J Phys Chem B 106(36):9274–9276

42. Marcus MS, Carpick RW, Sasaki DY, Eriksson MA (2002) Material Anisotropy Revealed by Phase Contrast in Intermittent Contact Atomic Force Microscopy. Phys Rev Lett 88:226103, 1–4

43. Grutter P, Meyer E, Heinzelmann H, Rosenthaler L, Hidber H, Guntherodt HJ (1988) Application of Atomic Force Microscopy to Magnetic-Materials. J Vac Sci Technol A Vac Surf Films 6:279–282

44. Wadas A, Grutter P (1989) Theoretical Approach to Magnetic Force Microscopy. Phys Rev B 39:12013–12017; Wadas A, Hug HJ (1992) Models for the Stray Field from Magnetic Tips Used in Magnetic Force Microscopy. J Appl Phys 72:203–206

45. Wadas A, Grutter P, Guntherodt HJ (1990) Analysis of In-plane Bit Structure by Magnetic Force Microscopy. J Appl Phys 67:3462–3467

46. Antognozzi M, Haschke H, Miles MJ (2001) STM'01 Abstract, Vancouver, Canada, 15–20 July 2001, p 439

47. Burnham NA, Colton RJ (1989) Measuring the Nanomechanical Properties and Surface Forces of Materials Using an Atomic Force Microscope. J Vac Sci Tech A 7(4):2906–2913

48. Israelachvili JN (1992) Intermolecular and Surface Forces: With Applications to Colloidal and Biological Systems, 2nd edn. Academic, London

49. Magonov SN, Elings V, Whangbo MH (1997) Phase Imaging and Stiffness in TappingMode Atomic Force Microscopy. Surf Sci 375:L385–L391

7 Modeling of Tip-Cantilever Dynamics in Atomic Force Microscopy

Yaxin Song · Bharat Bhushan

Abstract Atomic force microscopy (AFM) is commonly used for atomic and nanoscale surface measurements. Two operational modes of AFM exist: static mode and dynamic mode. In dynamic AFM mode, a cantilever is driven to vibrate by its holder or the sample. The changes of cantilever vibration parameters (amplitude, resonance frequency, and phase angle) due to tip–sample interaction are used to reveal surface properties. Analytical and numerical models that can accurately simulate surface-coupled cantilever dynamics are essential for explaining AFM scanning images and evaluating the sample's material properties. In this chapter, the existing dynamic modes of AFM are categorized in terms of cantilever deflection and excitation mechanism. Cantilever models for cantilever response simulation are summarized. Using these models, the important relations of cantilever frequency shift, vibration amplitude and phase angle with tip–sample interaction in various dynamic modes are derived, with an emphasis on newly-developed torsional resonance (TR) mode and lateral excitation (LE) mode. Some specific issues, such as the excitation of higher-order vibration modes in TappingMode (TM), the effects of tip eccentricity on cantilever responses in TR and LE modes, and how the cantilever dynamics affects the atomic-scale topographic and friction maps obtained in friction force microscopy (FFM) measurements, are investigated. Based on the derived relations between cantilever responses and tip–sample interaction, methods for quantitative evaluation of the sample's mechanical parameters are described.

Nomenclature

a_0	Intermolecular distance
a_c	Contact radius
a_g	Amplitude of sample harmonic motion
a_{lat}, a_n, a_t	Sample motions in lateral, normal, tangential directions
a_{lb}, a_{tr}, a_{vb}	Eigenvalues of cantilever in lateral bending, torsion, vertical bending
A	Area of cross section
A_0, A_c, A_t	Free vibration amplitude, amplitude under interaction, transient amplitude
A_θ, A_θ^0	Torsional amplitude with and without interaction
b	Width of cross section
c	Damping coefficient
c_e	Effective lateral stiffness of cantilever
c_{lat}, c_n, c_t	Contact damping coefficients in lateral, normal, tangential directions

c_{lb}, c_{tr}, c_{vb}	Damping coefficients of cantilever in lateral bending, torsion, vertical bending
C_0, C_1, C_2, C_3, C_4	Constants
d	Closest tip–sample separation
$d_n, \dot{d}_n$	Tip–sample separation, time differential of tip–sample separation
d_x, d_y, d_z	Displacements along the x, y, z-axes
$d_{x1}, d_{y1}, d_{z1}, d_{x2}, d_{y2}, d_{z2}$	Nodal displacements along the x, y, z-axes of the beam element
d_{lat}^P, d_n^P, d_t^P	Displacements at tip end (point P) in lateral, normal, tangential directions
d_x^C, d_y^C, d_z^C	Nodal displacements along the x, y, z-axes at node C
D	Equilibrium tip–sample separation
D_1, D_2, D_3, D_4	Constants
e	Mathematical constant
e_t	Tip eccentricity
E	Elastic modulus
E^*	Effective elastic modulus
E_{dis}	Energy dissipation due to tip–sample interaction
E_s, E_t	Elastic moduli of sample, tip
f_c	Normal contact force
f_{lat}, f_n, f_t	Interaction forces in lateral, normal, tangential directions
f_x, f_y, f_z	Interaction forces in the x, y, z-directions
f_x^C, f_y^C, f_z^C	Nodal forces along the x, y, z-axes at node C
F_{lat}, F_n, F_t	Amplitudes of interaction forces in lateral, normal, tangential directions
F_x, F_y, F_z	Amplitudes of interaction forces in the x, y, z-directions
$F_{x1}, F_{y1}, F_{z1}, F_{x2}, F_{y2}, F_{z2}$	Nodal forces along the x, y, z-axes at nodes 1 and 2 of beam element
$g_z = h_g \cos(\Omega t)$	Holder harmonic motion in the z-direction
$g_\theta = \theta_g e^{i\Omega t}$	Holder torsional motion
G	Shear modulus
G_s, G_t	Shear moduli of sample, tip
G^*	Effective shear modulus
h	Thickness of cantilever cross section
h_g	Amplitude of holder harmonic motion in the z-direction
h_G	Vertical distance between tip mass center and central line of cantilever
H	Hamakar constant
$H_\theta(\Omega), H_\theta^0(\Omega)$	Frequency response functions with and without interaction in TRmode
i_1, i_2	Arbitrary integers

I_a	Imaginary part of a_{tr}
$I_p = I_y + I_z$	Polar area moment of inertia
I_y, I_z	Moments of inertia about the y- and z-axes
J	Torsion constant
k	Spring constant
k_n, k_{lat}, k_t	Contact stiffness in normal, lateral, tangential directions
k_n^{ref}	Contact stiffness of reference sample
l, L, L_e	Tip length, cantilever length, length of beam element
L_h	Constant
m	Effective mass
$M_{x1}, M_{y1}, M_{z1}, M_{x2}, M_{y2}, M_{z2}$	Nodal moments about the x, y, z-axes at nodes 1 and 2 of the beam element
M_x^C, M_y^C, M_z^C	Nodal moments along the x, y, z-axes at node C
n	Number of degrees of freedom
Q	Quality factor
r_{ki}	Distance between the kth atom on the tip and the ith atom on the surface
R	Tip radius
R_a	Real part of a_{tr}
s_1	$= \cos(\phi + I_a L)$
s_2	$= \cos(\phi - I_a L)$
s_3	$= \sin(\phi + I_a L)$
s_4	$= \sin(\phi - I_a L)$
$s_{\text{lat}}, s_n, s_\theta$	Cantilever lateral, normal, twist stiffness
t	Time
T_{lat}	$= f_{\text{lat}} l$
T_t	Kinetic energy of the tip
T^*	Magnitude of sinusoidal lateral force
T_c^*	Critical lateral force magnitude at which stick-slip occurs
$v(x, t), w(x, t)$	Cantilever displacements in the y-directions
x, y, z	Coordinates
x_1, x_2	Coordinates
x_h, y_h, z_h	Coordinates of the holder
x_t, y_t, z_t	Tip coordinates
x', y'	Coordinates of graphite surface
V_{ts}	Interaction potential
W	Normal load
$Z, \dot{Z}, \ddot{Z}$	Normal deflection, velocity, acceleration
$\boldsymbol{a}_1, \boldsymbol{a}_2$	Unit lattice vectors
$\boldsymbol{a}^\alpha = \{a_t, a_{\text{lat}}, a_n\}^T$	Displacement vector of sample surface
$\boldsymbol{A}_{C\alpha}$	Transfer matrix
$\boldsymbol{c}_{\text{ts}}$	Equivalent damping matrix due to tip–sample interaction
$\boldsymbol{c}_{\text{ts}}^\alpha = \text{diag}(c_t, c_{\text{lat}}, c_n)$	Damping matrix of tip–sample interaction

C	System damping matrix
C_{ts}	Global damping matrix due to tip–sample interaction
C_Λ	Normalized damping matrix
$d, \dot{d}, \ddot{d}$	System displacement, velocity and acceleration vectors
d^e	Nodal displacement vector of beam element
d^C	Nodal displacement vector at node C
$d^\alpha = \{d_t^P, d_{lat}^P, d_n^P\}^T$	Displacement vector at tip end (point P)
f^e	Nodal force vector of beam element
$f_{ts}^C = \{f_x^C, f_y^C, f_z^C, M_x^C, M_y^C, M_z^C\}^T$	Nodal force vector at node C
$f_{ts}^P = \{f_x, f_y, f_z\}^T$	Interaction force vector at the tip end (point P)
$f_{ts}^\alpha = \{f_t, f_{lat}, f_n\}^T$	Interaction force vector on the sample surface
F_{ext}	External force vector
F_{ts}	Interaction force vector
$g_h = \{x_h, y_h, z_h\}^T$	Vector of holder motion
G_t	Kronecker matrix for position information at node C
$H_{lat}(\Omega)$	Frequency response function vector in LE mode
$H_z(\Omega)$	Frequency response function vector for free vibration in TM
$H_z^{ts}(\Omega)$	Frequency response function vector under interaction in TM
$H_\theta(\Omega)$	Frequency response function vector of free vibration in TRmode
$H_\theta^{ts}(\Omega)$	Frequency response function vector under interaction in TRmode
I	Identity matrix
k^e	Element stiffness matrix
k_{ts}	Equivalent stiffness matrix due to tip–sample interaction
$k_{ts}^\alpha = \text{diag}(k_t, k_{lat}, k_n)$	Stiffness matrix of tip–sample interaction
K	System stiffness matrix
K_{ts}	Global stiffness matrix due to interaction
l_z, l_θ	Positioning vectors for vertical, torsional motion of the holder
m^e	Element mass matrix
m_t	Mass matrix of tip
M	System mass matrix
r_t	Tip position vector
$u(t)$	Cantilever deflection relative to the holder
u_{tol}	Total displacement vector
U	Amplitude of cantilever deflection relative to the holder
V_G	Velocity vector of cantilever tip at its mass center
α	Tilting angle of cantilever to sample surface

$\alpha_0, \alpha_e, \alpha_s$	Damping factor of cantilever, effective damping factor, damping factor of interaction
β	Angle from the x axis to the x' axis
β_w, β_θ	Eigenvalues of cantilever in vertical bending, torsion, in absence of damping
β_w^{ref}	Eigenvalues of cantilever in vertical bending with reference sample, in absence of damping
γ	$= \Omega/\omega_{tr}$
γ_c	Coupling factor
$\delta_{lat}, \dot{\delta}_{lat}$	Lateral relative displacement, velocity
$\Delta_{lat}, \Delta_n, \Delta_t$	Amplitudes of relative displacements in lateral, normal, tangential directions
$\Delta_x, \Delta_y, \Delta_z$	Amplitudes of relative displacements in the x, y, z-directions
ε_{ts}	Parameter of Lennard–Jones potential
η_c	$= \frac{c_{tr}}{GJ}\Omega$
η_θ	$= \sqrt{\frac{\rho I_p}{GJ}}\Omega$
$\eta_{lat}, \eta_n, \eta_t$	Contact viscosity in lateral, normal, tangential directions
$\theta(x, t)$	Twist angle of cantilever
$\tilde{\theta}, \tilde{\theta}_0$	Angles
θ_g	Amplitude of holder torsional motion
$\theta_x, \theta_y, \theta_z$	Rotations about the x, y, z-axes
$\theta_{x1}, \theta_{y1}, \theta_{z1}, \theta_{x2}, \theta_{y2}, \theta_{z2}$	Nodal rotations about the x, y, z-axes of beam element
$\theta_x^C, \theta_y^C, \theta_z^C$	Rotations about the x, y, z-axes at node C
$\Theta(x), \Theta_1(x), \Theta_2(x)$	Modal shape functions of torsion
Θ_c, Θ_p	Torsional amplitudes in coupled and pure torsional analysis
λ	Tip location coefficient
λ_l	Periodic lattice spacing of sample surface
Π	Characteristic function
ρ	Mass density
σ	Parameter of Lennard–Jones potential
$\varsigma_i (i = 1, 2, \ldots n)$	The ith damping ratio
ν_s, ν_t	Poisson's ratios of sample, tip
$\varphi, \varphi_1, \varphi_2, \varphi_t$	Phase angles
ϕ, ϕ_0	Torsional phase with and without interaction
$\phi_v(x), \phi_w(x)$	Modal shape functions of lateral bending, vertical bending
$\Phi(x)$	Auxiliary function
χ_k	Relative lateral contact stiffness
χ_η	Relative lateral contact viscosity
ω	Circular frequency
ω_0	Fundamental resonance frequency in radian
ω_c	Contact resonance frequency in radian

$\omega_{ci} (i = 1, 2, \ldots n)$	The ith circular frequency under interaction
$\tilde{\omega}_e$	Normalized effective resonance frequency
ω_{free}	Free resonance frequency in radian
$\omega_i (i = 1, 2, \ldots n)$	The ith circular frequency
$\omega_{\text{lb}}, \omega_{\text{tr}}$	Resonance frequencies of lateral bending, torsion, in radian
$\Delta\omega$	Circular frequency shift
Ω	Driving frequency in radian
$\tilde{\Omega}$	Normalized driving frequency
Γ	Position matrix in FFM
Γ_{lat}	Positioning vector in LE mode
Γ_z	$= \boldsymbol{M} \boldsymbol{l}_z$
Γ_θ	$= \boldsymbol{M} \boldsymbol{l}_\theta$
Λ^2	Eigenvalue matrix
Λ_c^2	Eigenvalue matrix under interaction
$\boldsymbol{\varphi}_i (i = 1, 2, \ldots n)$	The ith normalized eigenvector
$\boldsymbol{\varphi}_{ci} (i = 1, 2, \cdots n)$	The ith normalized eigenvector under interaction
Φ	$= [\boldsymbol{\varphi}_1, \boldsymbol{\varphi}_2, \ldots, \boldsymbol{\varphi}_n]$, normalized eigenvector matrix
Φ_c	$= [\boldsymbol{\varphi}_{c1}, \boldsymbol{\varphi}_{c2}, \ldots, \boldsymbol{\varphi}_{cn}]$, normalized eigenvector matrix under interaction

Abbreviations

AFAM	Atomic force acoustic microscopy
AFFM	Acoustic friction force microscopy
AFM	Atomic force microscopy
AM-AFM	Amplitude modulation atomic force microscopy
DOF	Degrees of freedom
EVI	Elastic-viscous index
FE	Finite element
FFM	Friction force microscopy
FM-AFM	Frequency modulation atomic force microscopy
FMM	Force modulation mode
FRF	Frequency response function
HOPG	Highly-oriented pyrolytic graphite
LE mode	Lateral excitation mode
LFM	Lateral force microscopy
LM-AFM	Lateral force modulation atomic force microscopy
MP	Metal particle
NC-AFM	Non-contact atomic force microscopy
STM	Scanning tunnelling microscopy
TM	TappingMode
TRmode	Torsional resonance mode
UAFM	Ultrasonic atomic force microscopy

7.1
Introduction

Atomic force microscopy (AFM) is commonly used for atomic and nanoscale measurement of various properties, including surface topography, friction, adhesion, and viscoelasticity [5,6]. During measurement, a cantilever is scanned over a sample surface. Surface properties are revealed by observing the cantilever deflections or dynamic changes of vibration parameters (amplitude, resonance frequency, and phase angle) due to tip–sample interaction. Figure 7.1 shows the schematic diagram of an AFM tip-cantilever assembly interacting with a sample surface. The scanning is implemented by the motion of a cylindrical piezoelectric tube, which can act as the holder of either the cantilever or the sample. The deflection of the cantilever is measured using the optical lever method. A laser beam is projected on the upper surface of the cantilever close to the tip (point C). The reflected beam is led by a mirror into a four-segment photodiode. The flexural angle (θ_y^C) and twist of the cantilever (θ_x^C) are obtained by calibrating the vertical and lateral voltage output of the photo diode, respectively.

AFM cantilever deflection and vibration information under tip–sample interaction are utilized for surface topography, friction and material property imaging. For quantitative explanation of these images and evaluation of material mechanical properties, the relationship between cantilever response and tip–sample interaction needs to be established. Analytical and numerical models that can accurately simulate the surface-coupled dynamics of the cantilever are essential for this purpose. In addition, a thorough understanding of cantilever dynamics is helpful for development of AFM measurement techniques.

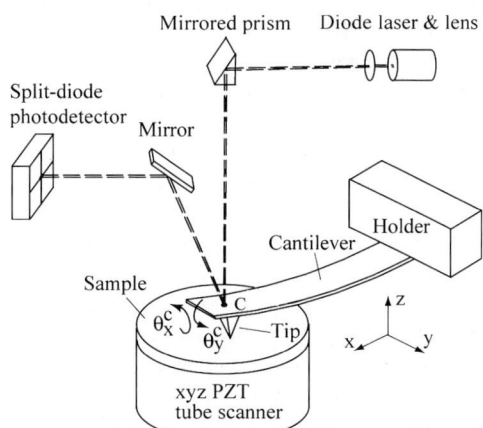

Fig. 7.1. Schematic diagram of an AFM tip-cantilever assembly interacting with a sample surface. The four-segment photodiode measures the flexural angle θ_y^C and twist angle θ_x^C of the cantilever

7.1.1
Various AFM Modes and Measurement Techniques

A rectangular cantilever in AFM can be modeled as a three-dimensional (3D) beam with the clamped-free boundary conditions. As shown in Fig. 7.2, the cantilever

Fig. 7.2. Four deformation shapes of a rectangular cantilever with free-clamped boundary conditions. Vertical bending is related to flexural angle θ_y and normal deflection d_z. Lateral bending is related to rotation angle θ_z and lateral deflection d_y. Twist angle θ_x is due to cantilever torsion. Extension causes the cantilever longitudinal displacement d_x

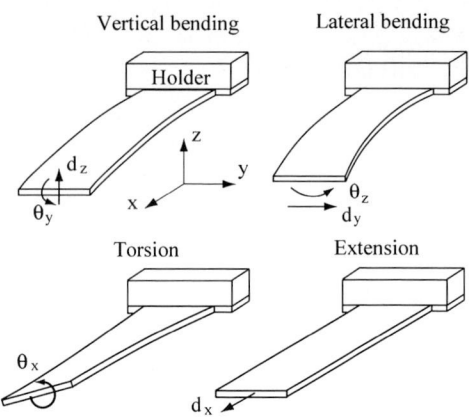

has four deformation shapes: vertical bending (bending about the y-axis), lateral bending (bending about the z-axis), torsion (about the x-axis), and extension (along the x-axis). Vertical bending is related to flexural angle θ_y and normal deflection d_z. Lateral bending is related to rotation angle θ_z and lateral deflection d_y. Twist angle θ_x is due to cantilever torsion. Extension causes the cantilever longitudinal displacement d_x. In terms of the cantilever state of motion during measurement, two basic types of AFM modes exist: static mode and dynamic mode. In static AFM modes, the cantilever is in quasistatic motion. In dynamic AFM modes, the cantilever is driven to vibrate near or at its resonance frequency and then the cantilever tip is brought to the proximity of a sample surface for imaging. Compared to static AFM, dynamic AFM can provide a better signal-to-noise ratio and higher resolution in measurement of material and surface properties [16, 18, 62].

Table 7.1 summarizes the static and dynamic AFM modes. Static AFM modes include contact mode and friction force microscopy (FFM, or lateral force microscopy, LFM). In contact mode [39], the cantilever tip is in constant contact with the sample surface. The cantilever normal deflection is monitored from the flexural angle θ_y^C. The normal tip–sample interaction force is calculated as the product of cantilever spring constant (stiffness of vertical bending) and normal deflection. By keeping a constant normal deflection through the z-motion of the piezotube, the surface topography is tracked. If the cantilever is brought to the sample surface, pressed down and then pulled away, a force-distance curve can be obtained for adhesion measurement. FFM [34, 36, 37] are commonly used for friction measurement. In constant-force mode of FFM, a constant normal load is maintained and the cantilever tip is scanned over sample surfaces. The scan direction is perpendicular to the longitudinal direction of the cantilever and the friction force along that direction is obtained by measuring the cantilever twist angle.

Dynamic AFM modes considered here are categorized as the following types in terms of cantilever deflection and excitation mechanism: (1) TappingMode (TM), non-contact AFM (NC-AFM), (2) force modulation mode (FMM), atomic force acoustic microscopy (AFAM) mode (or ultrasonic atomic force microscopy, UAFM), (3) torsional resonance (TR) mode, (4) lateral excitation (LE) mode, and (5) combined normal and lateral excitation mode.

Table 7.1. Summary of AFM modes

	Static modes		
Mode	Cantilever deflection	Output	Detected surface properties
Contact mode	Vertical bending	Normal deflection	Topography, adhesion
Friction force microscopy	Vertical bending, torsion and lateral bending	Normal deflection and twist angle	Topography, friction

		Dynamic modes				
Mode	Schematics	Cantilever deflection	Excitation source	Output / Driving frequency	Output	Detected surface properties
Tapping mode, non-contact AFM,	~10–100 nm	Vertical bending	Holder	Fundamental flexural resonance frequency	Normal deflection amplitude, phase and frequency shift	Topography and normal viscosity
Force modulation mode, atomic force acoustic microscopy mode		Vertical bending	Sample surface or holder	Fundamental and higher order flexural resonance frequency	Normal deflection amplitude and resonance frequency	Normal stiffness
Torsional resonance mode	~0.3–2 nm	Torsion and lateral bending	Holder	Torsional resonance frequency	Torsional amplitude, phase and resonance frequency	Topography, lateral stiffness and viscosity
Lateral excitation mode	~1 nm	Torsion and lateral bending	Sample surface	In a wide range, from very low (~ 20 kHz) to very high (up to 3 MHz)	Torsional amplitude, phase and resonance frequency	Topography, friction, lateral stiffness and viscosity
Combined normal and lateral excitation mode	~0.5 nm, ~1 nm	Vertical bending, torsion and lateral bending	Sample surface	In vertical direction higher than first flexural resonance frequency; in lateral direction much lower than torsional resonance frequency	Normal deflection amplitude and phase, torsional amplitude, phase and resonance frequency	Normal stiffness and lateral stiffness

In TM and NC-AFM, vertical bending dominates the cantilever deflection and the cantilever is excited by the vertical harmonic motion of its holder. TM is also named as amplitude modulation AFM (AM-AFM). In TM, the cantilever is driven at a fixed frequency close or equal to the fundamental resonance frequency of vertical bending. During measurement, the vibrating tip touches the sample surface intermittently. The vibration amplitude is compared to the set point and the difference is used as a feedback parameter to track sample topography. The phase can be used for material viscosity imaging. Figure 7.3 shows the TM topography and phase angle images for a metal particle (MP) tape [8]. The topography and phase images show different characteristics. The phase image is correlated to the viscoelastic properties of the MP tape.

NC-AFM is also called frequency modulation AFM (FM-AFM) [1]. In NC-AFM, the cantilever is always oscillated at its resonance frequency with a constant amplitude. During measurement, the vibrating tip does not touch the sample surface. Under the tip–sample interaction, the cantilever resonance frequency and oscillation amplitude are changing. A feedback loop detects the cantilever oscillation signal, shifts it by 90°, and uses it as the excitation signal so that the cantilever is always excited in resonance. Another feedback loop adjusts the excitation amplitude to keep the cantilever oscillating at a constant amplitude. In NC-AFM, the excitation signal is self-driven by the cantilever oscillation, which is dramatically different from the constant excitation used in TM. The spatial difference of frequency shift due to the tip–sample interaction can be used for contrast. The topography images are obtained by varying the tip–sample distance during the scan to keep a constant frequency shift. Compared to TM, NC-AFM can improve the imaging resolution dramatically by using a very high quality factor Q. Using NC-AFM, atomic resolution can be obtained by reducing the tip–sample distance and working in vacuum [19, 31].

The dominant cantilever deflection in FMM and AFAM is vertical bending. In FFM and AFAM, the cantilever is driven to vibrate by the vertical motion of either the sample surface or the holder. During measurement, the cantilever tip is in constant contact with the sample surface and driven to vibrate vertically. The amplitude of the tip is kept as small as possible in order that the linear approximation of tip–sample

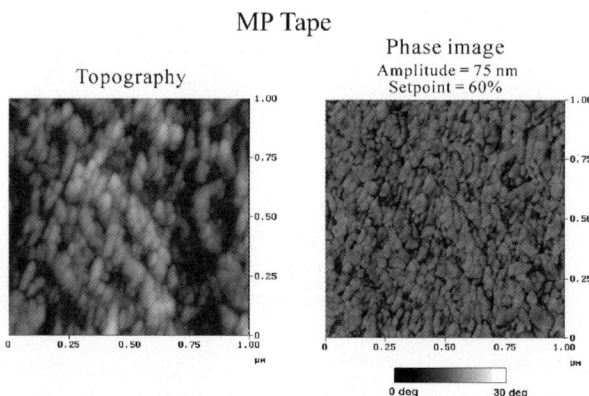

Fig. 7.3. TM topography and phase angle images for MP tape. The set point is defined as the ratio of the cantilever vibration amplitudes after engagement to the free amplitude in air [8]

forces is allowed, and to avoid lift-off. In FMM [33, 52], topography information is first obtained during primary scanning using TM. During interleave scanning, the cantilever is moved up and down at the resonance frequency of the holder's bimorph (below the fundamental resonance of the cantilever). The z-direction feedback control is deactivated and the topography information from the primary scan is used to maintain a constant lift scan height. The cantilever vibration amplitude can be used to image local stiffness. FMM can be used for elasticity contrast for soft materials such as polymers. For stiffer materials such as metals and ceramics, the contact stiffness between tip and surface is much higher than the spring constant of the cantilever (ranging from 0.01 to 80 N/m). The samples do not deform and thus the contrast due to elasticity becomes very low. In AFAM [3, 26, 40–44, 60, 62, 66], the cantilever is driven at ultrasonic frequencies at the fundamental contact resonance and several other higher-order contact resonances. At higher-order modes, the effective stiffness of the cantilever is enhanced to deform the samples and sample elasticity can be evaluated.

In TRmode [7,25,30,47], two piezoelectric elements are attached to the cantilever holder and vibrate out-of-phase to drive the cantilever into torsional oscillation. Under lateral (in-plane) tip–sample interaction, a lateral force and a torque are exerted on the cantilever, causing it to deflect in a combination of torsion and lateral bending. In LE mode [11, 46, 47, 51, 68], the cantilever is driven to vibrate by the lateral oscillation of sample surfaces in a direction perpendicular to the longitudinal axis of the cantilever. As in TRmode, the cantilever in LE mode deflects in both torsion and lateral bending. Compared to other AFM modes such as TM, in which the tip–sample interaction is mainly in normal direction, TR and LE modes were developed for in-plane surface property measurement. The properties of materials like thin films can be measured more readily with TR and LE modes. In addition, TR and LE modes have some inherent advantages in surface property imaging. Firstly, in TR and LE modes, the cantilever tip vibrates laterally (parallel) to the sample surface. During measurements, the tip remains close to the sample surface, ensuring more intensive tip–sample interaction and more surface material properties-related information [30]. Secondly, the torsional/lateral bending stiffness of a cantilever is typically two orders of magnitude higher than that of vertical bending. Therefore, most of the deformation in TR and LE modes occurs in the sample. TR and LE modes can be used to measure stiff and hard samples [30].

Two operation modes are possible for TRmode. In one mode, the cantilever is excited into torsional vibration and then approached to the sample surface. By keeping a constant torsional amplitude, surface topography can be measured from the z piezo motion [25, 30]. The then-obtained torsional phase data can be used for imaging of material viscoelasticity [30]. In the second mode, the cantilever is pressed on the sample surface with a constant normal load and then driven to vibrate at a frequency equal or close to the torsional contact resonance frequency. The variation of resonance frequency, torsional amplitude, and phase angle are used for mapping of material properties [12, 45]. Figure 7.4 shows the images of a MP tape using TM, TRmode I (constant amplitude) and TRmode II (constant deflection) [12]. Compared to the TM phase angle image, better contrast resulting from variations in viscoelasticity can be seen in the phase angle image using TRmode I. Amplitude and phase angle images using TRmode II have the largest contrast.

TM surface height, TM phase angle and TR mode I phase angle images of MP tape
(Setpoint = 60%, TM amplitude in air = 20 nm, TR amplitude in air = 1 V)

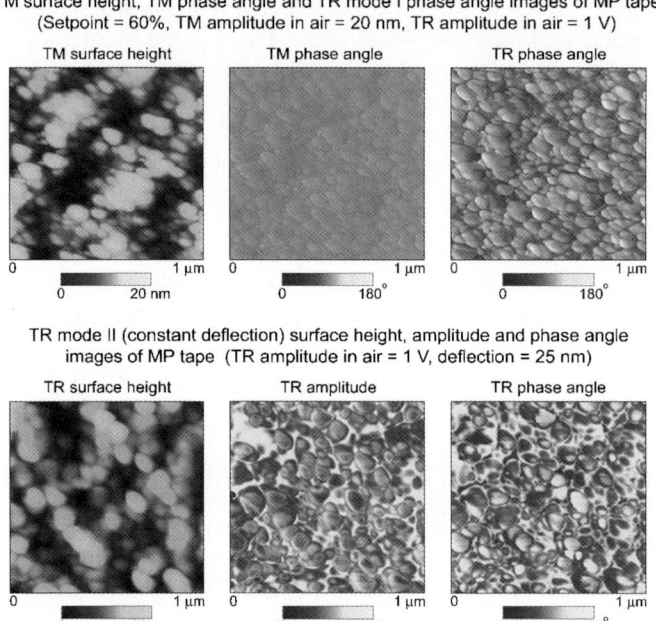

Fig. 7.4. Images of MP tape using TM, TRmode I (constant amplitude) and TRmode II (constant deflection) [12]

LE mode refers to the AFM measurement techniques in which the cantilever is driven by the lateral oscillation of sample surfaces through tip–sample interaction, such as lateral force modulation AFM (LM-AFM) [68], acoustic friction force microscopy (AFFM) [51], and lateral atomic force acoustic microscopy (lateral AFAM) [11, 46, 47]. The excitation frequency of sample surfaces could be in a wide range. In LM-AFM, the sample is laterally vibrated at a frequency (∼16 kHz) well below the cantilever torsional/lateral bending resonance frequency. The torsional amplitude and phase are employed for friction imaging. In AFFM and lateral AFAM, the sample is oscillating laterally at megahertz frequencies (up to 3 MHz) to excite the cantilever vibrating in torsional or lateral bending resonance. The torsional amplitude and contact-resonance spectra are used for friction imaging. Compared with conventional FFM, the advantages of friction measurement using LE mode are that the topography-induced friction can be reduced significantly and friction measurement can be operated at higher relative velocities (on the order of 1 mm/s).

In combined normal and lateral excitation mode, which can be viewed as a combination of AFAM and LM-AFM, the cantilever is vertically excited by the sample surface at a frequency much higher than the flexural fundamental resonance frequency so that the tip is cyclically indented into the sample [69]. At the same time, lateral oscillation of the surface at a frequency much lower than the cantilever torsional/lateral bending resonance frequency induces the cantilever vibration in torsional and lateral bending. By modulation of the flexural vibration amplitude, subsurface features of normal stiffness can be imaged. The lateral stiffness of the subsurface can be imaged from the torsional responses of the cantilever.

7.1.2
Models for AFM Cantilevers

Analytical and numerical models have been developed for dynamic simulation of AFM cantilevers with and without tip–sample interaction. Table 7.2 summarizes the AFM cantilever models for dynamic modeling.

Due to the existence of attractive and repulsive interaction regimes and the non-linear nature of the normal tip–sample interaction forces, the dynamic behavior of the cantilever in TM and NC-AFM is very complicated. Point-mass models are employed in the investigation of cantilever dynamics in TM and NC-AFM. Point-mass models approximate the dynamics of the distributed-parameter cantilever system by the motion equation of a lumped mass. Using the point-mass models, researchers have obtained the analytical descriptions of non-linear cantilever dynamics and provided insightful understanding of physical factors governing the motion of the cantilever [13, 20, 49, 63, 64].

Table 7.2. Summary of cantilever models for dynamic modeling

Model	Schematics	Deflection modeled	Applications
1D point-mass model		Vertical bending	Non-contact mode, Tapping mode
1D beam model		Vertical bending	Tapping mode, atomic force acoustic microscopy
Torsional model		Torsion	Torsional resonance mode, lateral excitation mode
Coupled torsional-bending model		Torsion and lateral bending	Torsional resonance mode, lateral excitation mode
3D point-mass model		3D translational displacements	Profiling process of friction force microscopy
3D Finite element beam model		Vertical bending, torsion and lateral bending	Profiling process of friction force microscopy, all dynamic AFM modes

It is recognized that in experiments, higher-order flexural modes of the cantilever are often excited. Point-mass models cannot simulate the cantilever dynamics involving higher-order modes other than the fundamental one [41,58,62]. Neither can point-mass models account for the effects of the geometry and location of the tip on the cantilever dynamics. Furthermore, point-mass models provide solutions corresponding to the cantilever vertical displacement while in experiment the detecting system of AFM measures the rotation angle of the cantilever. This could give rise to some inaccuracy in data explanation since the vertical displacement and rotation angle of a vibrating cantilever do not have a one-to-one relation when higher modes of the cantilever are involved. One-dimensional (1D) beam models [10,32,41,42,58] have been employed to investigate the cantilever response in TM and AFAM. Using the 1D beam models, analytical modal analyses of the tip-cantilever system were performed by representing the tip–sample interaction by a linear spring and dashpot [16,41,62,65]. This linearization is valid only if the tip oscillates around an equilibrium position with very small amplitudes. Considering the non-linear Hertzian contact boundary conditions, the non-linear amplitude-frequency relation for various flexural modes were obtained using the method of multiscales [60]. In many cases, numerical methods, e.g., finite element (FE) method [4, 55] or mode superposition method [32, 58], are employed to simulate the non-linear dynamics of a 1D beam.

In operation of TR and LE modes, the lateral oscillation of the cantilever tip over the sample surface is quite small (0.3–2 nm for TRmode and $\sim$ 1 nm for LE mode). The tip-surface distance (therefore normal tip–sample force) remains almost constant and the vertical deflection of the cantilever is uncoupled with the torsion and lateral bending. The dynamic response of the cantilever in TR and LE modes has been modeled as the pure torsional vibration of a shaft. Torsional modal analyses of tip-cantilever system were performed with linear elastic tip–sample interaction [67]. The relation between the torsional amplitude/phase shift and lateral contact stiffness/viscosity in TRmode was derived from a forced torsional analysis [53, 54]. The coupled torsional-bending model, which considers both the torsion and lateral bending of the cantilever in TR and LE modes, was developed recently [56].

A 3D point-mass model [22–24] was developed to simulate the tip motion during the profiling process of FFM. In the 3D point-mass model, the tip-cantilever system is represented by three masses connected by elastic springs to its holder. The point-mass model has three uncoupled translational degrees of freedom (DOFs). In each translational direction, the motion of the point-mass is described as a single DOF oscillator. The friction force in that direction is obtained as the product of the translational displacement (relative to the holder) of the mass and the spring stiffness. As a mathematical approximation of the real tip-cantilever system, this model's parameters (effective masses and spring stiffnesses) can only be obtained by estimation and the simulated responses are translational displacements instead of the rotation angles detected in FFM. The 3D point-mass model also neglects the coupling between the lateral bending and torsion of the cantilever.

The 3D FE beam model of tip-cantilever systems [55] was developed for numerical simulation of free (without tip–sample interaction) and surface-coupled (with tip–sample interaction) dynamics of AFM cantilevers in various dynamic modes. Representing the cantilever by 3D beam elements, this versatile model can address the exact excitation mechanisms, tip geometry/location, tilting of the cantilever to the sample surface, and all the possible couplings among the different deflections of the AFM cantilever. The FE model's parameters can be determined from the cantilever geometry and material properties. Translational displacements, as well as flexural and twist angles are the simulated cantilever responses. The 3D FE beam model was also used in simulation of the profiling process of FFM for investigation of the effects of cantilever dynamics on atomic-scale topographic and friction maps obtained in FFM measurements [57].

7.1.3
Outline

In Sect. 7.2, tip–sample interaction forces are described first. Point-mass model and 1D beam model are then introduced briefly for vertical bending modeling of AFM cantilevers in TM, NC-AFM, FMM and AFAM. Pure torsional analysis is carried out to study the cantilever dynamics in TRmode. Coupled torsional-bending analysis is performed to obtain the cantilever response in TR and LE modes. Parametric analyses are performed to investigate under what condition the pure torsional analysis is acceptable. The differences between TR and LE modes are demonstrated and explained. In Sect. 7.3, a 3D FE beam model is introduced for numerical simulation of free and surface-coupled dynamics of a tip-cantilever system. FE formulations are derived for modeling of TM, TR and LE modes. Numerical simulations of TM, TR and LE modes are performed in frequency and time domains. The excitation of higher-order flexural modes in TM and the effects of tip eccentricity on cantilever response in TR and LE modes, are studied. In Sect. 7.4, the 3D FE beam model is employed to simulate the atomic-scale topographic and lateral force profiling process in FFM. The way to calculate the interatomic forces between the tip and graphite surface is described. The simulated results are discussed in comparison of the experimental results and the possible methods to detect the full atomic structures of surfaces are discussed. In Sect. 7.5, methods are described to evaluate the sample's mechanical properties from the measured contact stiffness/viscosity between the tip and sample surface. Section 7.6 summarizes this chapter.

7.2
Modeling of AFM Tip-Cantilever Systems in AFM

Different models have been developed for modeling of AFM cantilever dynamics with and without tip–sample interaction. These models are used for various AFM modes with different cantilever deflections and excitation mechanisms. Cantilever responses including contact resonance frequency, vibration amplitude and phase angle have been used for material property imaging. The purpose of dynamic modeling of tip-cantilever systems is to investigate the relations between the cantilever responses and the tip–sample interaction. The tip–sample interaction is related to the sample's material properties and the cantilever's geometry and material properties. These derived relations between cantilever responses and tip–sample interaction can be used for quantitative explanation of AFM images and evaluation of the sample's material properties. Modeling of cantilever dynamics also helps us to understand the difference of cantilever behaviors and determine the application conditions of different dynamic modes.

7.2.1
Tip–Sample Interaction

To investigate the cantilever response during measurement, tip–sample interaction needs to be described first. The interaction between a cantilever tip and sample can be modeled as the interaction between a sphere and a flat surface. Two different interaction regimes, attractive and repulsive, are distinguished in the normal direction of the cantilever. A van der Waals force is widely used to describe the long-range attractive force. Neglecting the energy dissipation in tip–sample contact, the short-range repulsive force in the normal direction can be calculated using the JKR [28] or DMT [15] model. The JKR model is suitable for soft, compliant materials with high adhesion forces and large tip radii, while the DMT model is suitable to describe the contact forces of hard, stiff materials with low adhesion forces and small tip radii. Adopting the DMT model, the normal interaction force can be described as

$$
f_n = \begin{cases} -\dfrac{HR}{6d_n^2}, & d_n > a_0 \\[2mm] -\dfrac{HR}{6a_0^2} + \dfrac{4}{3}E^*\sqrt{R}(a_0 - d_n)^{3/2}, & d_n \le a_0 \end{cases} \tag{7.1}
$$

Here, d_n is the transient tip–sample separation, H is Hamakar constant, R is the tip radius, a_0 is the intermolecular distance, E^* is the effective elastic modulus given by $E^* = [(1 - v_t^2)/E_t + (1 - v_s^2)/E_s]^{-1}$, where E_t, E_s, v_t, and v_s are the elastic moduli and Poisson's ratios of the tip and sample, respectively. For the contact in which the energy dissipation due to the tip–sample interaction is not negligible, a modified viscoelastic form of (7.1) is

$$
f_n = \begin{cases} -\dfrac{HR}{6d_n^2}, & d_n > a_0 \\[2mm] -\dfrac{HR}{6a_0^2} + \dfrac{4}{3}E^*\sqrt{R}(a_0 - d_n)^{3/2} - \eta_n(a_0 - d_n)^{1/2}\dot{d}_n, & d_n \le a_0 \end{cases} \tag{7.2}
$$

where η_n is the viscosity of the tip–sample contact in the normal direction, and (˙) represents the differential with respect to time t.

According to the Hertzian contact theory [27], the tip–sample interaction force at the lateral direction is the functions of repulsive contact force, i.e.,

$$
f_{lat} = \begin{cases} 0, & d_n > a_0 \\[2mm] -8G^*\left(\dfrac{3Rf_c}{4E^*}\right)^{1/3}\delta_{lat}, & d_n \le a_0 \end{cases} \tag{7.3}
$$

where $f_c = \frac{4}{3}E^*\sqrt{R}(a_0 - d_n)^{3/2}$ is the normal contact force, the effective shear modulus G^* is given by $G^* = [(2 - v_t)/G_t + (2 - v_s)/G_s]^{-1}$, G_t and G_s are

become close to zero; (3) torsional amplitude strongly depends on γ. If the driving frequency is equal to or lower than the designated torsional resonance frequency ($\gamma \leq 1.0$), in the region where χ_k and χ_η are moderate, torsional amplitude decreases as χ_k and/or χ_η increases. Under the same tip–sample interaction, torsional amplitude increases as γ increases toward 1.0 and could never be larger than A_θ^0. If the driving frequency is higher than the designated torsional resonance frequency ($\gamma > 1.0$), the situation becomes complicated. Torsional amplitude depends on how close the driving frequency is to the contact resonance frequency. A peak is seen in the region where the contact resonance happens to be very close to the driving frequency. The torsional amplitude could be larger than A_θ^0 (not shown in Fig. 7.11, though).

Regarding the phase shift, the observations are the following:

1. If tip–sample interaction is weak, the phase shift is close to $-\phi_0$.

2. In the case of small χ_η, the sample can be viewed as a pure elastic material. If χ_k is also small, the phase shift is $-\phi_0$; if χ_k is large, the cantilever is clamped to the sample surface and the phase shift is $0°$. Basically, the phase shift decreases with the increasing of χ_k. When χ_k increases from small to large values, the phase shift changes from $-\phi_0$ to $0°$.

3. In the case of small χ_k, the sample can be viewed as a pure viscous material. As χ_η increases from small to large values, the phase shift approaches from $-\phi_0$ to $90°$.

4. If only χ_k is large, the phase shift keeps close to $0°$; if only χ_η is large, the phase shift keeps a value of $90°$. The phase shift depends strongly on the elastic-viscous index (EVI) that we define as

$$\delta = \chi_k/\chi_\eta = k_{\text{lat}}/(\eta_{\text{lat}}\Omega) \qquad (7.67)$$

   If $\delta > 1$, the sample behaves more like an elastic material. Or else, the sample shows more characteristics as a viscous material. When neither of χ_k and χ_η is small, the phase shift has the biggest sensitivity in a domain around $\delta = 1$. In that domain, the phase shift increases as χ_η increases and/or χ_k decreases.

5. The phase shift of the cantilever without tip–sample interaction, i.e., $-\phi_0$, is very sensitive to γ. For $\gamma < 1.0$, $-\phi_0$ is in the range of $(0°, 90°)$ and increases with the increasing of the system damping (c increases); for $\gamma = 1.0$, $-\phi_0 = 90°$; if $\gamma > 1.0$, $-\phi_0$ is in the range of $(90°, 180°)$ and decreases as the system damping increases.

6. For any range of γ, phase shift increase as γ increases.

7.2.5
Coupled Torsional-Bending Analysis

7.2.5.1
Forced Vibration Analysis in TR and LE Modes

As shown in Fig. 7.12, in TR and LE modes, the lateral force at the tip end causes the lateral bending and torsion of the cantilever. The torsion of the cantilever is

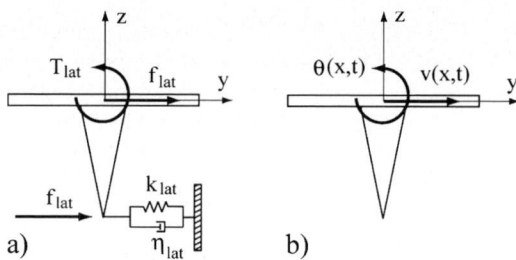

Fig. 7.12. Schematic diagram of the cantilever tip under lateral tip–sample interaction. (**a**) The interaction is linear viscoelastic. A lateral force f_{lat} and a torque $T_{lat} = f_{lat}l$ are exerted on the cantilever. (**b**) The cantilever is under the deflections of lateral bending (lateral displacement $v(x, t)$) and torsion (twist angle $\theta(x, t)$)

governed by (7.39). The lateral bending is governed by the following ordinary differential equations

$$EI_z \frac{\partial^4 v(x, t)}{\partial x^4} + \rho A \frac{\partial^2 v(x, t)}{\partial t^2} + c_{lb} \frac{\partial v(x, t)}{\partial t} = 0 \tag{7.68}$$

where $v(x, t)$ is the lateral displacement along the y-axis, I_z is the moments of inertia about the z-axis, $c_{lb} = \omega_{lb}\rho A/Q$ are the damping coefficients for the lateral bending when the cantilever is vibrating in air. Here, ω_{lb} is the resonance frequency corresponding to lateral bending. For a cantilever with a rectangular cross section, $I_z = hb^3/12$.

In TRmode, assuming a harmonic motion of the holder as $g_\theta(t) = \theta_g e^{i\Omega t}$, one has $\theta(x, t) = \Theta(x) e^{i\Omega t}$ and $v(x, t) = \phi_v(x) e^{i\Omega t}$. Similarly, the lateral force and the relative displacement can be represented as $f_{lat} = F_{lat} e^{i\Omega t} \delta_{lat} = \Delta_{lat} e^{i\Omega t}$. Substituting the solutions into the governing equations for torsion and lateral bending, one has

$$\frac{d^2\Theta(x)}{dx^2} - a_{tr}^2 \Theta(x) = 0 \tag{7.69}$$

$$\frac{d^4\phi_v(x)}{dx^4} - a_{lb}^4 \phi_v(x) = 0 \tag{7.70}$$

where $a_{tr}^2 = -\frac{\rho I_p}{GJ}\Omega^2 + i\frac{c_{tr}}{GJ}\Omega$, $a_{lb}^4 = \frac{\rho A}{EI_z}\Omega^2 - i\frac{c_{lb}}{EI_z}\Omega$. The determination of the response functions $\Theta(x)$ and $\phi_v(x)$ requires two torsional-related and four lateral bending-related boundary conditions. As shown in Fig. 7.13, the amplitude of relative displacement $\Delta_{lat} = l \Theta|_{x=L} + \phi_v|_{x=L}$. The six boundary conditions are

$$\Theta|_{x=0} = \theta_g, \quad GJ\Theta'\big|_{x=L} = -(k_{lat} + i\eta_{lat}\Omega)(l \Theta|_{x=L} + \phi_v|_{x=L}) l \tag{7.71}$$

$$\phi_v|_{x=0} = 0, \quad \phi_v'\big|_{x=0} = 0, \quad EI_z\phi_v''\big|_{x=L} = 0,$$
$$EI_z\phi_v'''\big|_{x=L} = (k_{lat} + i\eta_{lat}\Omega)(l \Theta|_{x=L} + \phi_v|_{x=L}) \tag{7.72}$$

The response functions $\Theta(x)$ and $\phi_v(x)$ can be obtained by solving (7.69) to (7.72). The twist angle and lateral displacement at the end of the cantilever are [56]

Fig. 7.13. Schematic diagram of a cantilever deflection in TRmode. The cantilever is excited by the holder

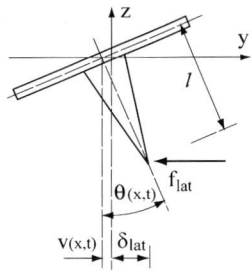

$$\Theta_c \equiv \Theta(x = L) = \frac{2a_{tr}e^{a_{tr}L}GJ}{a_{tr}(e^{2a_{tr}L} + 1)GJ + (e^{2a_{tr}L} - 1)l^2(k_{lat} + i\eta_{lat}\Omega)\gamma_c}\theta_g$$

(7.73)

$$\phi_v(x = L) = \frac{-2a_{tr}e^{a_{tr}L}GJ(1 + i)(k_{lat} + i\eta_{lat}\Omega)\pi_2 l}{a_{lb}^3(e^{2a_{tr}L} - 1)EI_zl^2(k_{lat} + i\eta_{lat}\Omega)\pi_1 + a_{tr}(e^{2a_{tr}L} + 1)GJ}\theta_g$$
$$\times[a_{lb}^3 EI_z\pi_1 + (1 + i)(k_{lat} + i\eta_{lat}\Omega)\pi_2]$$

(7.74)

in which

$$\gamma_c = \frac{a_{lb}^3 EI_z\pi_1}{a_{lb}^3 EI_z\pi_1 + (1 + i)(k_{lat} + i\eta_{lat}\Omega)\pi_2}$$

(7.75)

$$\pi_1 = 1 + e^{2ia_{lb}L} + 4e^{(1+i)a_{lb}L} + e^{2a_{lb}L} + e^{2(1+i)a_{lb}L}$$

(7.76)

$$\pi_2 = 1 - ie^{2ia_{lb}L} + ie^{2a_{lb}L} - e^{2(1+i)a_{lb}L}$$

(7.77)

With the lateral bending being neglected, the cantilever would be in pure torsion and the twist angle at the end of the cantilever is

$$\Theta_p \equiv \Theta(x = L) = \frac{2a_{tr}e^{a_{tr}L}GJ}{a_{tr}(e^{2a_{tr}L} + 1)GJ + (e^{2a_{tr}L} - 1)l^2(k_{lat} + i\eta_{lat}\Omega)}\theta_g$$ (7.78)

Comparing Θ_c with Θ_p, we can see that Θ_c equals Θ_p only if the coupling coefficient $\gamma_c = 1$. That means that the lateral bending of the cantilever can be neglected if $\pi_1 \rightarrow \infty$, or $\pi_2 \rightarrow 0$, or the tip–sample interaction is very weak ($k_{lat} \rightarrow 0$ and $\eta_{lat} \rightarrow 0$).

In LE mode, the sample surface is oscillating in a harmonic motion expressed as $a_{lat}(t) = a_ge^{i\Omega t}$. Compared to TRmode in which the driving frequency has to be at or close to the cantilever torsional resonance frequency, in LE mode, driving frequency could be a value in a wide range. It could be much lower than the torsional/lateral bending resonance frequency of the cantilever, or very high around the aforementioned resonance frequency. As in TRmode, the cantilever undergoes a combination of torsion and lateral bending governed by (7.39) and (7.68). By assuming $\theta(x, t) = \Theta(x)e^{i\Omega t}$ and $v(x, t) = \phi_v(x)e^{i\Omega t}$, we have (7.69) and (7.70) to obtain the response functions $\Theta(x)$ and $\phi_v(x)$. As shown in Fig. 7.14, the relative

Fig. 7.14. Schematic diagram of a cantilever deflection in LE mode. The cantilever is excited to vibrate by the harmonic oscillation of the sample surface through the lateral tip–sample interaction

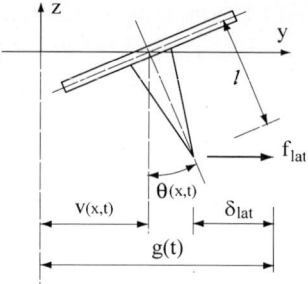

displacement amplitude $\Delta_{\text{lat}} = l\,\Theta|_{x=L} + \phi_v\,|_{x=L} - a_g$. The six boundary conditions are

$$\Theta|_{x=0} = 0\,, \quad GJ\Theta'\big|_{x=L} = -(k_{\text{lat}} + i\eta_{\text{lat}}\Omega)(l\,\Theta|_{x=L} + \phi_v\,|_{x=L} - a_g)\,l \tag{7.79}$$

$$\phi_v|_{x=0} = 0\,, \quad \phi_v'\big|_{x=0} = 0\,, \quad EI_z\phi_v''\big|_{x=L} = 0\,,$$
$$EI_z\phi_v'''\big|_{x=L} = (k_{\text{lat}} + i\eta_{\text{lat}}\Omega)(l\,\Theta|_{x=L} + \phi_v\,|_{x=L} - a_g) \tag{7.80}$$

The twist angle and lateral displacement at the end of the cantilever are [56]

$$\Theta_c \equiv \Theta(x = L) = \frac{a_g}{l + \dfrac{a_t(e^{2a_tL}+1)GJ}{(e^{2a_tL}-1)(k_{\text{lat}}+i\eta_{\text{lat}}\Omega)l}\dfrac{1}{\gamma_c}} \tag{7.81}$$

$$\phi_v(x = L) = \frac{a_{\text{tr}}(1 + e^{2a_tL})GJ(1 + i)(k_{\text{lat}} + i\eta_{\text{lat}}\Omega)\pi_2 l}{a_{\text{lb}}^3(e^{2a_{\text{tr}}L} - 1)EI_zl^2(k_{\text{lat}} + i\eta_{\text{lat}}\Omega)\pi_1 + a_{\text{tr}}(e^{2a_{\text{tr}}L} + 1)GJ} a_g$$
$$\times[a_{\text{lb}}^3 EI_z\pi_1 + (1 + i)(k_{\text{lat}} + i\eta_{\text{lat}}\Omega)\pi_2] \tag{7.82}$$

The pure torsional analysis of LE mode can be obtained by setting $\phi_v = 0$ and the twist angle at the end of the cantilever are

$$\Theta_p \equiv \Theta(x = L) = \frac{a_g}{l + \dfrac{a_{\text{tr}}(e^{2a_{\text{tr}}L}+1)GJ}{(e^{2a_{\text{tr}}L}-1)(k_{\text{lat}}+i\eta_{\text{lat}}\Omega)l}} \tag{7.83}$$

Comparing Θ_c with Θ_p, we can see that as in TRmode, Θ_c equals Θ_p only if the coupling coefficient $\gamma_c = 1$.

7.2.5.2
Parameter Analysis and Discussion

Parameter analyses are performed for the same cantilever as that in Sect. 7.2.4. The first resonance frequency for torsion and lateral bending modes are $\omega_{\text{tr}} = 5.15 \times 10^6$ rad/s (819.4 kHz) and $\omega_{\text{lb}} = 5.97 \times 10^6$ rad/s (949.6 kHz). The cantilever responses in TR and LE modes are calculated as the functions of χ_k and χ_η.

Two dimensionless parameters χ_k and χ_η represent the relative lateral contact stiffness and viscosity to the torsional stiffness of the cantilever. Actually, since the lateral bending stiffness of the cantilever has a proportional relation with its torsional stiffness, χ_k and χ_η also reflect the relative lateral contact stiffness and viscosity to the lateral bending stiffness of the cantilever. If we define the relative lateral contact stiffness and viscosity to the lateral bending stiffness of the cantilever as $\widetilde{\chi}_k = k_{lat}L^3/EI_z$ and $\widetilde{\chi}_\eta = \eta_{lat}\Omega L^3/EI_z$, then for a rectangular cantilever

$$\frac{\chi_k}{\widetilde{\chi}_k} = \frac{\chi_\eta}{\widetilde{\chi}_\eta} = \frac{EI_z l^2}{GJL^2} \approx \frac{(1+\nu)}{2}\left(\frac{b}{h}\right)^2\left(\frac{l}{L}\right)^2 \tag{7.84}$$

where ν is the Poisson's ratio.

Figure 7.15a–f gives, respectively, the normalized amplitude and phase shift of Θ_c, the normalized amplitude and phase shift of $\phi_v(x = L)$, the amplitude of the ratio Θ_p/Θ_c, and the phase difference between Θ_c and Θ_p [56]. It can be seen that the twist of the cantilever has the maximum for very weak interaction, decreases as the interaction increases, and drops to a small value for very strong interaction. The opposite trend is observed for the lateral displacement $\phi_v(x = L)$: for weak interaction, there is negligible lateral displacement; as the interaction increases, lateral displacement also increases and approaches to its maximum for very strong interaction. The different trends shown for the amplitudes of Θ_c and $\phi_v(x = L)$ can be explained as following. For weak interaction, the sample surface cannot really restrain the motion of the tip and the cantilever deflects in pure torsion, as vibrating far away from the sample surface. When the interaction gets very strong, the cantilever tip can be viewed as being clamped to the sample surface. Under excitation, the cantilever tip cannot move and the cantilever has to move laterally back and forth. In this case, we have $|\Theta_c| = \max[|\phi_v(x = l)|]/l$, which is several orders smaller than $\max(|\Theta_c|)$. In our simulation, $\Omega = \omega_{tr}$ and $\omega_{tr} < \omega_{lb}$. Since under positive lateral tip–sample interaction, the resonance frequencies of the system all shift to larger values, the deflection of the cantilever is dominated by the torsional mode. The phase shifts of Θ_c and $\phi_v(x = L)$ in Fig. 7.15b,d indicates that there is 180° to 270° phase difference between the torsion and lateral displacement of the cantilever. Basically, when the cantilever twists positively with the holder, the cantilever also moves laterally to the negative direction of the y-axis due to the negative lateral interaction force. It is concluded from Fig. 7.15e,f that in TRmode the lateral bending has to be considered in cantilever response analysis if the lateral interaction is relatively strong compared to the cantilever torsional or lateral bending stiffness. Figure 7.15f demonstrates the phase difference between Θ_c and Θ_p. In an experiment, this means that pure torsional analysis may lead to a misreading of the phase shift information in TRmode.

In TRmode, sometimes the driving frequency Ω is chosen at the torsional contact resonance frequency, which is always larger than ω_{tr}. Similar parameter analysis for the case of $\Omega > \omega_{tr}$ also shows that the pure torsional analysis approximates well the coupling analysis only if lateral interaction is relatively weak.

The cantilever responses to unit excitation amplitude g_0 in LE mode are shown in Figs. 7.16 and 7.17, for the cases when $\Omega = \omega_{tr}$ and $\Omega = \omega_{lb}$, respectively [56]. In each, Figs. 7.16a–f and 7.17a–f give, respectively, the normalized amplitude and

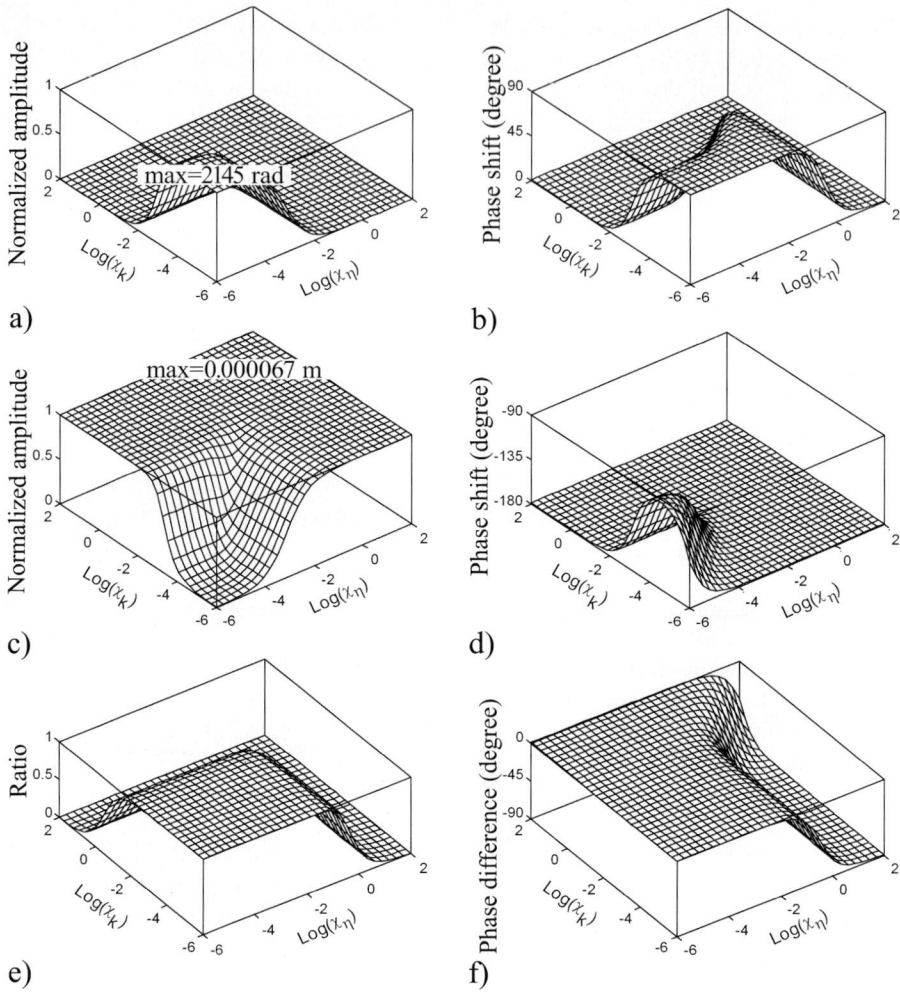

Fig. 7.15. Parameter analysis of cantilever responses to unit excitation amplitude θ_g in TRmode, when $\Omega = \omega_{tr} = 5.15 \times 10^6$ rad/s (819.4 kHz). (**a**) Normalized amplitude of Θ_c. (**b**) Phase shift of Θ_c. (**c**) Normalized amplitude of $\phi_v(x = L)$. (**d**) Phase shift of $\phi_v(x = L)$. (**e**) Amplitude of Θ_p/Θ_c. (**f**) Phase difference between Θ_c and Θ_p [56]

phase shift of Θ_c, the normalized amplitude and phase shift of $\phi_v(x = L)$, the amplitude of the ratio Θ_c/Θ_p, and the phase difference between Θ_c and Θ_p. In spite of the different values, the amplitudes of Θ_c and $\phi_v(x = L)$ show the same trend for two different driving frequencies: the amplitudes of Θ_c and $\phi_v(x = L)$ start from very small values for weak interaction, increase as the interaction increases, and approach their maximum for very strong interaction. Recall that in TRmode, as the interaction increases, $|\Theta_c|$ decreases while $|\phi_v(x = L)|$ increases. The differences in TR and LE modes result from the different excitation modes. For weak interaction, in TRmode, the tip is vibrating without the sample surface and the cantilever deflects

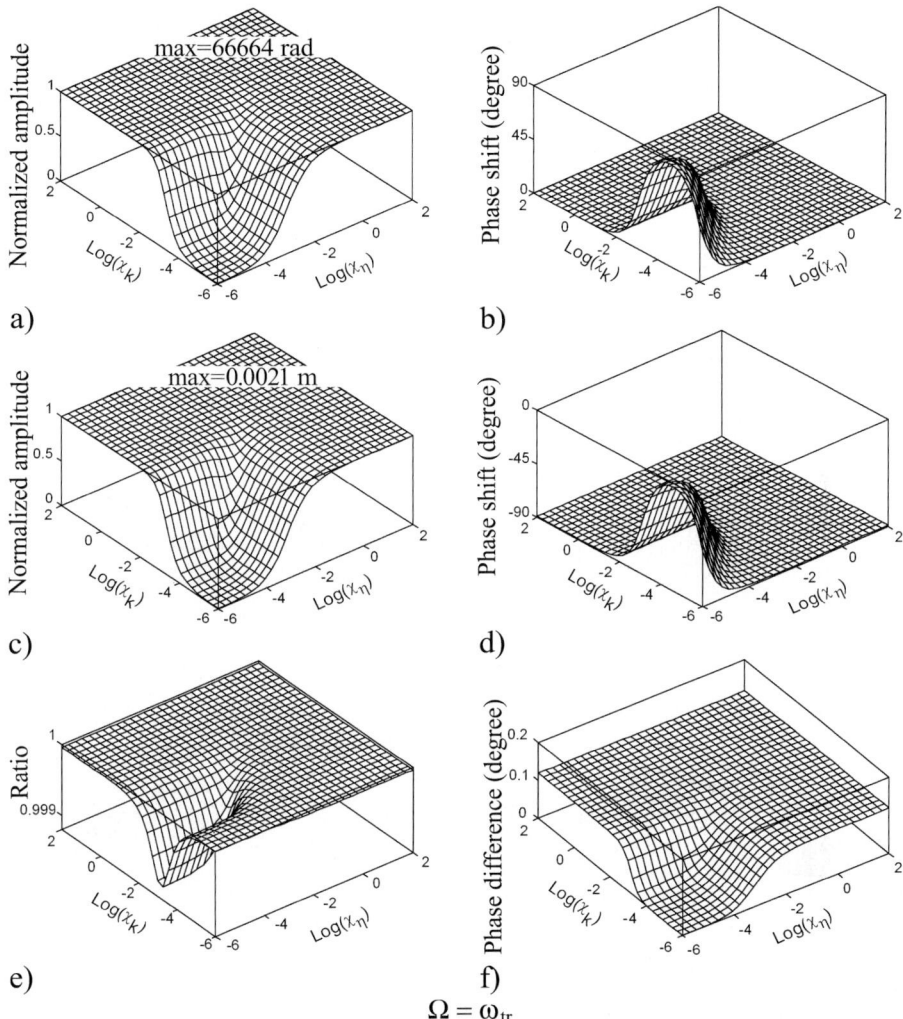

Fig. 7.16. Parameter analysis of cantilever responses to unit excitation amplitude a_g in LE mode, when $\Omega = \omega_{tr} = 5.15 \times 10^6$ rad/s (819.4 kHz). (**a**) Normalized amplitude of Θ_c. (**b**) Phase shift of Θ_c. (**c**) Normalized amplitude of $\phi_v(x = L)$. (**d**) Phase shift of $\phi_v(x = L)$. (**e**) Amplitude of Θ_c/Θ_p. (**f**) Phase difference between Θ_c and Θ_p [56]

in pure torsion, while in LE mode, the lateral oscillation of the sample surface can hardly excite the cantilever and both torsion and lateral displacement are small. When the interaction is very strong, in TRmode, the cantilever tip cannot move and the cantilever move laterally back and forth; in LE mode, the cantilever tip follows the oscillation of the sample surface, causing large lateral motion and torsion of the cantilever. Table 7.3 summarizes the cantilever responses to different tip–sample interaction in TR and LE modes. The phase shifts of Θ_c and $\phi_v(x = L)$ in Figs. 7.16b,d and 7.17b,d show 0°–90° phase difference between the torsion

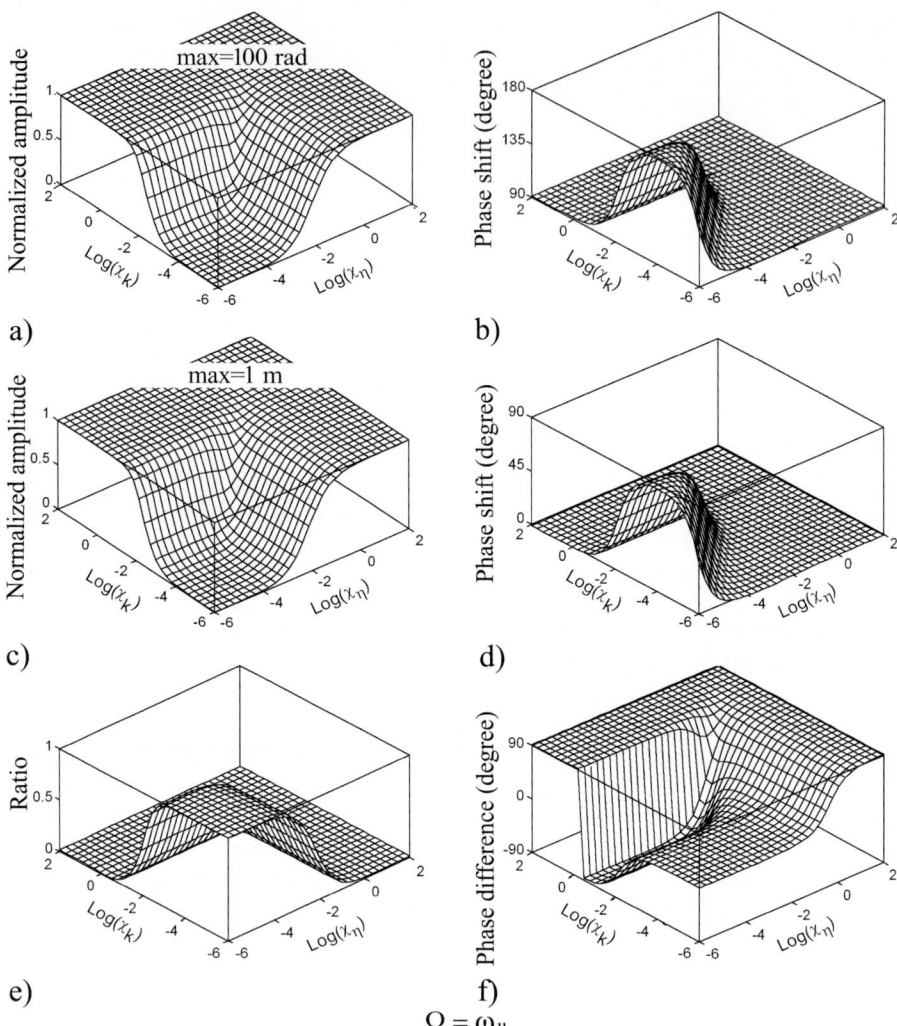

Fig. 7.17. Parameter analysis of cantilever responses to unit excitation amplitude a_g in LE mode, when $\Omega = \omega_{lb} = 5.97 \times 10^6$ rad/s (949.6 kHz). (**a**) Normalized amplitude of Θ_c. (**b**) Phase shift of Θ_c. (**c**) Normalized amplitude of $\phi_v(x = L)$. (**d**) Phase shift of $\phi_v(x = L)$. (**e**) Amplitude of Θ_c/Θ_p. (**f**) Phase difference between Θ_c and Θ_p [56]

and lateral displacement of the cantilever. Basically, in LE mode, when the sample surface moves in positive direction, the cantilever follows and has both positive lateral displacement and twist angle.

Figures 7.16e,f and 7.17e,f show that as in TRmode, pure torsional analysis can provide accurate cantilever torsional response in LE mode if the lateral interaction is relatively weak compared to the cantilever torsional or lateral bending stiffness. Driving frequency has significant effects on the range of χ_k and χ_η in which the pure torsional analysis is valid. The valid range for $\Omega = \omega_{tr}$ is much larger than

Table 7.3. Relations of cantilever responses and tip–sample interaction in TR and LE modes

| | TRmode (Excitation from the holder) | | LE mode (Excitation from the sample) | |
	Twist angle	Lateral displacement	Twist angle	Lateral displacement
Weak interaction	Free amplitude	Zero	Zero	Zero
Increasing interaction	Decreasing	Increasing	Increasing	Increasing
Strong interaction	Zero	Maximum	Maximum	Maximum

that for $\Omega = \omega_{lb}$. This is because when $\Omega = \omega_{tr}$, torsional mode dominates the cantilever deflection while in the case of $\Omega = \omega_{lb}$, more lateral bending deformation gets involved. In pure torsional analysis, for $\Omega = \omega_{lb}$, torsional amplitude peaks at the places where the driving frequencies are close to the contact torsional resonance frequencies. Note that the results in Figs. 7.16 and 7.17 are for a cantilever whose torsional resonance frequency is lower than that of lateral bending ($\omega_{tr} < \omega_{lb}$). Different results should be obtained otherwise. For instance, if $\omega_{lb} < \omega_{tr}$, there would be no peaks shown in the pure torsional amplitudes when $\Omega = \omega_{lb}$. While for $\Omega = \omega_{tr}$, the valid range of χ_k and χ_η for pure torsional analysis would not be as large as that shown in Fig. 7.16e. The lateral bending would contribute much to the cantilever deflection when the contact resonance frequencies of lateral bending become close to the driving frequency. However, the statement is always true that in LE mode, the pure torsional analysis is valid if the lateral interaction is relatively weak.

Due to the different excitation mechanisms, the cantilever response shows different characteristics in TR and LE modes. In TRmode, excitation frequency is close to either the torsional resonance frequency or the torsional contact resonance frequency of the cantilever. In some cases of LE mode, such as AFFM and lateral AFAM, excitation frequency is around the cantilever torsional and lateral bending resonance frequencies, while in others, such as LM-AFM, very low frequency lateral force modulation is employed. The FRFs of the torsional angle Θ_c in TR and LE modes are calculated from (7.73) and (7.81) by setting the excitation amplitude $\theta_g = 1$ and $a_g = 1$, respectively. For comparison, the FRFs of Θ_p are also obtained in the same way using (7.78) and (7.83). The results are shown in Fig. 7.18a,b, respectively [56]. For TRmode, two cases are shown: without interaction and when $\chi_k = 0.1$. For LE modes, results are shown for $\chi_k = 0.01$ and $\chi_k = 1.0$. In all calculation, the lateral viscosity is neglected ($\chi_\eta = 0$) for clear and easy illustration.

In Fig. 7.18a, it is clear that with no interaction, the cantilever is in pure torsional vibration. With lateral tip–sample interaction, lateral bending is coupled with the torsion and the FRF resonances shift to the right. If a very low excitation frequency is used in TRmode, torsional vibration could hardly be excited. If the excitation frequency Ω is around the torsional resonance frequency ω_{tr}, even a medium lateral tip–sample interaction would result in a dramatic decrease in torsional amplitude. Therefore, $\Omega = \omega_{tr}$ is only suitable for measurement with relatively weak interaction. For medium lateral interaction, one could choose an excitation frequency that is close to the torsional contact resonance frequency so that a significant torsional angle can be measured. However, if the interaction is relatively strong, the cantilever has approximately clamped–clamped boundary conditions. Even at the torsional contact

Fig. 7.18. Frequency response functions of torsional angles Θ_c and Θ_p in TR mode (**a**) and in LE mode (**b**). Lateral viscosity η_{lat} is neglected [56]

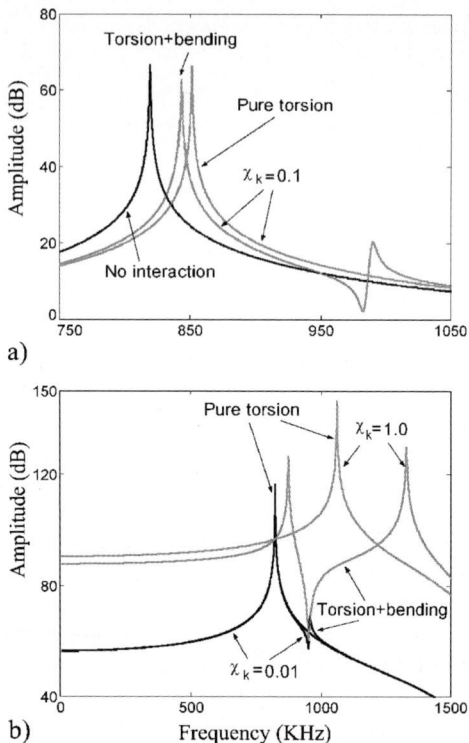

a)

b)

resonance, the torsional amplitude will not be big enough. Generally, the torsional amplitude corresponding to the lateral bending mode is rather small (see the small peak around 970 kHz in Fig. 7.18a). This means the contact-resonance frequency of lateral bending may not be used as the excitation frequency in TRmode. To sum up, TRmode might only be used for measurement of sample surface with relatively weak tip–sample interaction.

In Fig. 7.18b, FRFs for LE mode are very different from those for TRmode. In LE mode, lateral tip–sample interaction always exists. As the interaction becomes stronger, the FRF resonance amplitude increases. If the excitation frequency is much lower than the torsional/lateral bending resonance frequency, torsional vibration can be significant only if the interaction is relatively strong, i.e., low frequency LM-AFM may be used only under relatively strong lateral tip–sample interaction. If the excitation frequency is around the torsional and lateral bending resonance frequencies, the torsional amplitude is increased. When the lateral interaction is relatively small, the torsional amplitude corresponding to lateral bending mode is quite small but when the interaction becomes stronger, it increases rapidly to the same order as that of torsional mode. Therefore, for relatively weak tip–sample interaction, the excitation frequency should be around the torsional resonance frequency, and for relatively strong tip–sample interaction, both torsional and lateral bending modes may be excited for measurement. Table 7.4 summarizes the application conditions for TR and LE mode.

Table 7.4. Summary of application conditions for TR and LE modes

	Driving frequency	Applicable for
TRmode	Free torsional resonance	Weak interaction
	Contact torsional resonance	Weak or medium interaction
LE mode	Very low	Strong interaction
	Contact torsional resonance	Any interaction
	Contact resonance of lateral bending	Medium or strong interaction

In the above analyses for TR and LE modes, it is assumed that there is no slip occurring between the tip and sample surface during measurement, i.e., the lateral interaction force does not exceed the critical friction force. This is the condition that must be satisfied so that the linear viscoelastic interaction model is valid. It is found that in TR and LE modes, under a certain applied normal load, a critical excitation amplitude, which indicates the onset of sliding between the tip and sample, can be determined by observing the shape of the resonance curves [46]. Torsional resonance curves measured with different amplitudes of excitation are shown in Fig. 7.19 [46]. At low excitation amplitudes the shape of the resonance curve is Lorentzian. With the increasing of excitation amplitude, deviations from the Lorentzian shape appear. Above the critical excitation amplitude, the resonance curve flattens out and the frequency-span of the flattened part increases with the excitation amplitude. By choosing an excitation amplitude smaller than the critical one, the non-slip condition can be satisfied.

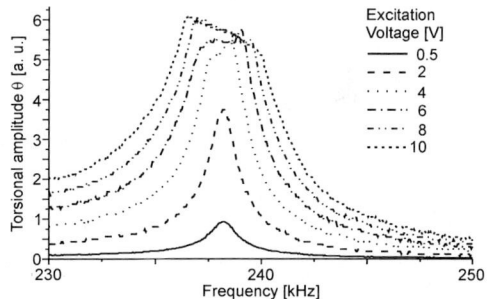

Fig. 7.19. Torsional resonance curves of a cantilever measured on fused silica with different amplitudes of excitations [47]

7.3
Finite Element Modeling of Tip-Cantilever Systems

FE methods are powerful and versatile tools for dynamic analysis of complex mechanical systems. The free and surface-coupled dynamics of AFM cantilevers in various dynamic modes can be simulated by finite element models. This section introduces a 3D FE beam model of tip-cantilever systems [55]. Representing the cantilever by 3D beam elements, this model addresses all the complexities in cantilever dynamic modeling, arising from the excitation mechanism, tip geometry/location, tilting of the cantilever to the sample surface, and the couplings among the different deflections of the AFM cantilever. FE formulations are given for TM, TR and LE modes and are summarized in Table 7.5.

7.3.1
Finite Element Beam Model of Tip-Cantilever Systems

The FE model of the tip-cantilever system is illustrated in Fig. 7.20. The center of the tip bottom is position at point A, whose coordinates are $x = \lambda L(0 < \lambda \leq 1)$, $y = e_t$, and $z = 0$. The distance between the mass center of the tip (point G) and point A is h_G. The cantilever is discretized by 3D beam elements and the tip is modeled as a rigid bar. The tip–sample interaction occurring at the end of the tip (point P) is transferred to the cantilever through the point C. For convenience, the point C shall be a node of two adjoining beam elements.

Figure 7.21 shows a 3D beam element. Note that the local coordinate system of the beam elements employed to discretize the cantilever is coincident with the global coordinate system of the tip-cantilever assembly in Fig. 7.20. At each node of a 3D beam element, there are six DOFs, three translation displacements and three rotations. The element nodal displacement vector is

$$\boldsymbol{d}^{\mathrm{e}} = \{d_{x1}, d_{y1}, d_{z1}, \theta_{x1}, \theta_{y1}, \theta_{z1}, d_{x2}, d_{y2}, d_{z2}, \theta_{x2}, \theta_{y2}, \theta_{z2}\}^{\mathrm{T}} \qquad (7.85)$$

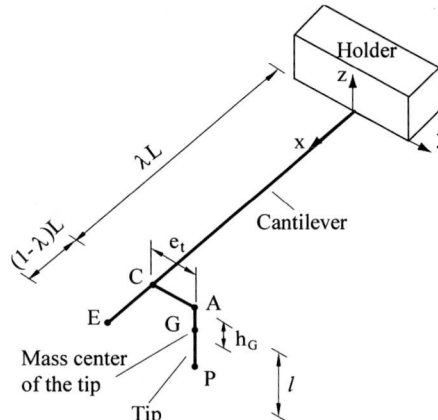

Fig. 7.20. Schematic diagram of tip-cantilever system. The cantilever is represented by 3D beam elements and the tip by a rigid bar. The tip is not positioned perfectly on the central line of the cantilever

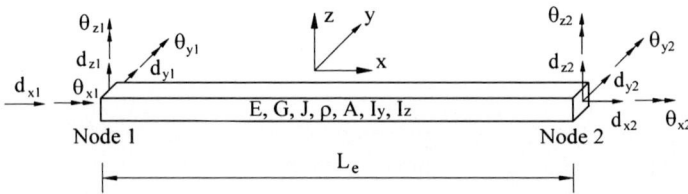

Fig. 7.21. Three-dimensional beam element. At each node of the element, there are three translational and three rotational displacements

Table 7.5. Summary of the FE formulations for TM, TR and LE modes

	TM	TRmode	LE mode
Excitation	Holder: $g_z(t) = h_g e^{i\Omega t}$	Holder: $g_\theta(t) = \theta_g e^{i\Omega t}$	Sample: $a_{\mathrm{lat}} = a_g e^{i\Omega t}$
Total displacement	$d(t) = u(t) + l_z g_z(t)$	$d(t) = u(t) + l_\theta g_\theta(t)$	$d(t)$
Motion equation	$M\ddot{u} + C\dot{u} + Ku = -\Gamma_z \ddot{g}_z(t) + F_{\mathrm{ts}}(u, \dot{u})$	$M\ddot{u} + C\dot{u} + Ku = -\Gamma_\theta \ddot{g}_\theta(t) + F_{\mathrm{ts}}(u, \dot{u})$	$M\ddot{d} + C\dot{d} + Kd = F_{\mathrm{ts}}(d, \dot{d})$
Frequency response function (no interaction)	$H_z(\Omega) = \dfrac{U}{h_g} = [K + i\Omega C - \Omega^2 M]^{-1}\Gamma_z \Omega^2$	$H_\theta(\Omega) = \dfrac{U}{\theta_g} = [K + i\Omega C - \Omega^2 M]^{-1}\Gamma_\theta \Omega^2$	N/A
Frequency response function (under interaction)	$H_z^{\mathrm{ts}}(\Omega) = \dfrac{U}{h_g}$ $= [(K + K_{\mathrm{ts}}) + i\Omega(C + C_{\mathrm{ts}}) - \Omega^2 M]^{-1}$ $[\Gamma_z\Omega^2 - i\Omega C_{\mathrm{ts}}l_z - K_{\mathrm{ts}}l_z]$	$H_\theta^{\mathrm{ts}}(\Omega) = \dfrac{U}{\theta_g}$ $= [(K + K_{\mathrm{ts}}) + i\Omega(C + C_{\mathrm{ts}}) - \Omega^2 M]^{-1}$ $[\Gamma_\theta\Omega^2 - i\Omega C_{\mathrm{ts}}l_\theta - K_{\mathrm{ts}}l_\theta]$	$H_{\mathrm{lat}}(\Omega) = \dfrac{U}{a_g}$ $= [(K + K_{\mathrm{ts}}) + i\Omega(C + C_{\mathrm{ts}}) - \Omega^2 M]^{-1}\Gamma_{\mathrm{lat}}$

The corresponding element nodal force vector, consisting of three shear forces and three moments at each node, is

$$f^e = \{F_{x1}, F_{y1}, F_{z1}, M_{x1}, M_{y1}, M_{z1}, F_{x2}, F_{y2}, F_{z2}, M_{x2}, M_{y2}, M_{z2}\}^T \quad (7.86)$$

The element stiffness matrix expresses the relation of element nodal force vector f^e with element nodal displacement vector d^e. For a 3D beam element with a length of L_e, the element stiffness and mass matrices k^e and m^e are expressed in Appendix A.1. They are functions of geometry and material properties of the beam element. If the mass and moments of inertia of the tip are not negligible, the tip will contribute to the system mass matrix through node C. The mass matrix of the tip m_t is given in Appendix A.2 [55].

The system mass and stiffness matrices M and K are obtained by assembling the contributions from all the beam elements and the tip

$$M = \sum_e^{\text{cantilever}} m^e + G_t^T m_t G_t, \quad K = \sum_e^{\text{cantilever}} k^e \quad (7.87)$$

in which $\sum_e^{\text{cantilever}}$ represents the assembly of the element mass or stiffness matrices of the cantilever, G_t is the Kronecker matrix reflecting the position information of the nodal displacements at node C in the global displacement vector. The determination of the system damping matrix C is usually a difficult task. The proportional damping matrix may be employed as

$$C = \Phi^{-T} C_\Lambda \Phi^{-1} \quad (7.88)$$
$$C_\Lambda = \text{diag}(2\varsigma_1\omega_1, 2\varsigma_2\omega_2, \ldots, 2\varsigma_n\omega_n) \quad (7.89)$$
$$\Phi = [\varphi_1, \varphi_2, \ldots, \varphi_n] \quad (7.90)$$

where n is the total number of DOFs of the system, ω_i, φ_i and ς_i ($i = 1, 2, \ldots, n$) are the ith circular natural frequency, eigenmode vector and damping ratio of the system. Natural frequencies and eigenmodes of the system are obtained by solving a generalized eigenvalue problem

$$K\Phi = M\Phi\Lambda^2 \quad (7.91)$$
$$\Phi^T M\Phi = I \quad (7.92)$$
$$\Lambda^2 = \text{diag}\left[\omega_1^2, \omega_2^2, \ldots, \omega_n^2\right] \quad (7.93)$$

where I is the identity matrix.

The FE motion equation for the tip-cantilever system is

$$M\ddot{d} + C\dot{d} + Kd = F_{\text{ext}} + F_{\text{ts}} \quad (7.94)$$

Here, d, $\dot{d}$ and $\ddot{d}$ are the system displacement, velocity and acceleration vectors, respectively, F_{ext} is the force vector due to the external forces except that from the tip–sample interaction, F_{ts} is the force vector due to the tip–sample interaction. F_{ts} is solely contributed by the forces and moments at node C as

$$F_{\text{ts}} = G_t^T f_{\text{ts}}^C \quad (7.95)$$

As shown in Fig. 7.22, the force vector at node C, $f_{ts}^C = \{f_x^C, f_y^C, f_z^C, M_x^C, M_y^C, M_z^C\}^T$, is caused by the interaction force vector at the tip end (point P), $f_{ts}^P = \{f_x, f_y, f_z\}^T$. The vector $f_{ts}^\alpha = \{f_t, f_{lat}, f_n\}^T$ contains the interaction forces on the sample surface. Subscripts t, lat and n represent the tangential, lateral and normal directions of the sample surface. The transformation relation between f_{ts}^C and f_{ts}^α are given by

$$f_{ts}^C = A_{C\alpha} f_{ts}^\alpha \tag{7.96}$$

where

$$A_{C\alpha} = \begin{bmatrix} \cos\alpha & 0 & -\sin\alpha \\ 0 & 1 & 0 \\ \sin\alpha & 0 & \cos\alpha \\ e_t\sin\alpha & l & e_t\cos\alpha \\ -l\cos\alpha & 0 & l\sin\alpha \\ -e_t\cos\alpha & 0 & e_t\sin\alpha \end{bmatrix} \tag{7.97}$$

The interaction force vector f_{ts}^α is usually a function of the relative displacement and velocity between the sample and the point P as $f_{ts}^\alpha = f_{ts}^\alpha(d^\alpha - a^\alpha, \dot{d}^\alpha - \dot{a}^\alpha)$, in which $d^\alpha = \{d_t^P, d_{lat}^P, d_n^P\}^T$ is the displacement vector at the point P, $a^\alpha = \{a_t, a_{lat}, a_n\}^T$ is the displacement vector of the sample. The vector d^α can be calculated from the displacement vector at node C as

$$d^\alpha = A_{C\alpha}^T d^C \tag{7.98}$$

Here, vector $d^C = G_t d$ is part of the solution of the system motion equation. Assuming that a^α is known, the interaction force vector can be expressed as $F_{ts} = F_{ts}(d, \dot{d})$. Equation (7.94) can be rewritten as a typical non-linear motion equation

$$M\ddot{d} + C\dot{d} + Kd = F_{ext} + F_{ts}(d, \dot{d}) \tag{7.99}$$

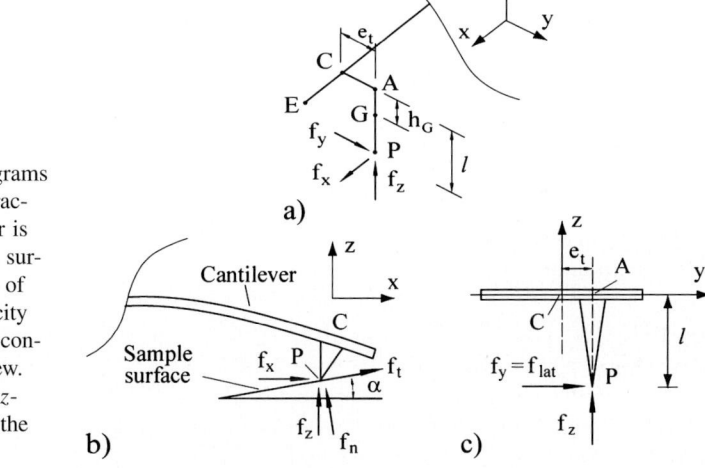

Fig. 7.22. Force diagrams of tip–sample interaction. The cantilever is tilted to the sample surface with an angle of α and the eccentricity of the cantilever is considered. (**a**) 3D view. (**b**) View in the x-z-plane. (**c**) View in the y-z-plane

The tip–sample interaction is discussed in Sect. 7.2.1. If the tip–sample interaction is linearized, the interaction force vector f_{ts}^{α} can be expressed as

$$f_{ts}^{\alpha} = -k_{ts}^{\alpha}(d^{\alpha} - a^{\alpha}) - c_{ts}^{\alpha}(\dot{d}^{\alpha} - \dot{a}^{\alpha}) \tag{7.100}$$

$$k_{ts}^{\alpha} = \begin{bmatrix} k_t & 0 & 0 \\ 0 & k_{lat} & 0 \\ 0 & 0 & k_n \end{bmatrix}, \quad c_{ts}^{\alpha} = \begin{bmatrix} c_t & 0 & 0 \\ 0 & c_{lat} & 0 \\ 0 & 0 & c_n \end{bmatrix} \tag{7.101}$$

The nodal force vector at node C can be expressed as

$$
\begin{aligned}
f_{ts}^{C} &= A_{C\alpha} f_{ts}^{\alpha} = A_{C\alpha} \left(-k_{ts}^{\alpha} d^{\alpha} - c_{ts}^{\alpha} \dot{d}^{\alpha} \right) + A_{C\alpha} \left(k_{ts}^{\alpha} a^{\alpha} + c_{ts}^{\alpha} \dot{a}^{\alpha} \right) \\
&= -A_{C\alpha} k_{ts}^{\alpha} A_{C\alpha}^{T} d^{C} - A_{C\alpha} c_{ts}^{\alpha} A_{C\alpha}^{T} \dot{d}^{C} + A_{C\alpha} \left(k_{ts}^{\alpha} a^{\alpha} + c_{ts}^{\alpha} \dot{a}^{\alpha} \right) \\
&= -k_{ts} d^{C} - c_{ts} \dot{d}^{C} + f_{ts}^{a}
\end{aligned} \tag{7.102}
$$

where $f_{ts}^{a} = A_{C\alpha}(k_{ts}^{\alpha} a^{\alpha} + c_{ts}^{\alpha} \dot{a}^{\alpha})$ is the force vector due to the motion of the sample surface. The terms $k_{ts} = A_{C\alpha} k_{ts}^{\alpha} A_{C\alpha}^{T}$ and $c_{ts} = A_{C\alpha} c_{ts}^{\alpha} A_{C\alpha}^{T}$ are the equivalent stiffness and damping contributions due to the linearized tip–sample interaction. Substituting (7.102) into (7.95), and using the relation $d^{C} = G_t d$, one can rewrite the system motion equation (7.94) as

$$M\ddot{d} + (C + C_{ts})\dot{d} + (K + K_{ts})d = F_{ext} + G_t^{T} f_{ts}^{a} \tag{7.103}$$

$$C_{ts} = G_t^{T} c_{ts} G_t, \quad K_{ts} = G_t^{T} k_{ts} G_t \tag{7.104}$$

Equation (7.103) demonstrates that the linear viscoelastic tip–sample interaction is equivalent to adding additional damping and stiffness matrices to the original system. Appendix A.3 gives the detailed forms of k_{ts} and c_{ts}, from which it is seen that under tip–sample interaction, extension of the cantilever is coupled with vertical bending, and torsion is coupled with lateral bending [55]. If $e_t \neq 0$, all the four deformations, vertical bending, lateral bending, torsion and extension become coupled.

The natural frequencies and modal shapes of the tip-cantilever system without tip–sample interaction are obtained by solving the generalized eigenvalue problems in (7.91). The modal characteristics for a system under linear viscoelastic tip–sample interaction can be determined by solving a similar eigenvalue problem as

$$(K + K_{ts})\Phi_c = M\Phi_c \Lambda_c^2 \tag{7.105}$$

$$\Phi_c^{T} M\Phi_c = I, \quad \Phi_c = [\varphi_{c1} \varphi_{c2} \dots \varphi_{cn}] \tag{7.106}$$

$$\Lambda_c^2 = \text{diag}\left[\omega_{c1}^2 \omega_{c2}^2 \dots \omega_{cn}^2\right] \tag{7.107}$$

where ω_{ci} and φ_{ci} ($i = 1, 2, \dots n$) are the ith circular frequency and eigenvector of the system under tip–sample interaction.

7.3.2
Modeling of TappingMode

In TM, the cantilever is driven to vibrate by the vertical harmonic motion of its holder $g_z(t) = h_g e^{i\Omega t}$, as shown in Fig. 7.23. Except the interaction force, no

Fig. 7.23. Schematic diagram of a cantilever in TM. The cantilever is driven to vibrate by the harmonic motion of its holder along the z-axis

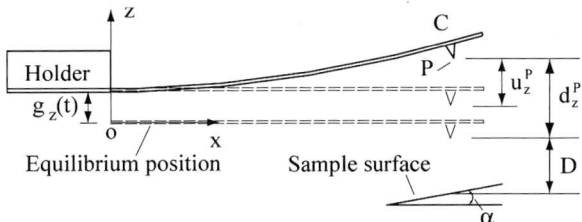

external force exists ($F_{\text{ext}} = \mathbf{0}$). The sample surface is fixed ($a^\alpha = \mathbf{0}$). The total dynamic displacement vector of the cantilever can be expressed as

$$\boldsymbol{d}(t) = \boldsymbol{u}(t) + \boldsymbol{l}_z g_z(t) \tag{7.108}$$

where $\boldsymbol{u}(t)$ is the cantilever deflection relative to the moving holder. The ith element of the positioning vector $\boldsymbol{l}_z$ equals 1 if the ith DOF of $\boldsymbol{d}(t)$ corresponds to the translational displacement in the z-axis; otherwise, it equals zero. The motion equation (7.94) can be rewritten as

$$\boldsymbol{M}\ddot{\boldsymbol{u}} + \boldsymbol{C}\dot{\boldsymbol{u}} + \boldsymbol{K}\boldsymbol{u} = -\Gamma_z \ddot{g}_z(t) + \boldsymbol{F}_{\text{ts}}(\boldsymbol{u}, \dot{\boldsymbol{u}}) \tag{7.109}$$

$$\Gamma_z = \boldsymbol{M}\boldsymbol{l}_z \tag{7.110}$$

When the cantilever is vibrating far away from the surface ($\boldsymbol{F}_{\text{ts}} = \mathbf{0}$), the vibration amplitude due to the excitation from the holder can be determined from the FRF of the system. Assuming $\boldsymbol{u} = \boldsymbol{U}e^{\text{i}\Omega t}$, one can obtain the FRF vector of the cantilever due to the z-direction motion of the holder from (7.109)

$$\boldsymbol{H}_z(\Omega) = \frac{\boldsymbol{U}}{h_g} = [\boldsymbol{K} + \text{i}\Omega\boldsymbol{C} - \Omega^2\boldsymbol{M}]^{-1}\Gamma_z\Omega^2 \tag{7.111}$$

The ith element of $\boldsymbol{H}_z(\Omega)$, $H_{zi}(\Omega)(i = 1, 2, \ldots n)$, represents the response of the ith DOF of the system when the holder is moving harmonically with a unit amplitude at circular frequency Ω. If the total vertical displacement (d_z^C) and rotation about the y-axis (θ_y^C) at node C are the pth and qth DOFs of the system, the free vibration amplitudes and phases of d_z^C and θ_y^C under the excitation $h_g e^{\text{i}\Omega t}$ are the amplitudes and arguments of the complex values of $[H_{zp}(\Omega) + 1]h_g$ and $H_{zq}(\Omega)h_g$, respectively. Note that the readout of the measuring system of AFM is the signal about θ_y^C.

Under linear tip–sample interaction, the dynamic response of the tip-cantilever system can be obtained in the same way as that used in the free vibration analysis except that the system damping and stiffness matrices become $\boldsymbol{C} + \boldsymbol{C}_{\text{ts}}$ and $\boldsymbol{K} + \boldsymbol{K}_{\text{ts}}$. The surface-coupled FRF vector due to the z-direction motion of the holder can be expressed as [55]

$$\boldsymbol{H}_z^{\text{ts}}(\Omega) = \frac{\boldsymbol{U}}{h_g}$$

$$= [(\boldsymbol{K} + \boldsymbol{K}_{\text{ts}}) + \text{i}\Omega(\boldsymbol{C} + \boldsymbol{C}_{\text{ts}}) - \Omega^2\boldsymbol{M}]^{-1}[\Gamma_z\Omega^2 - \text{i}\Omega\boldsymbol{C}_{\text{ts}}\boldsymbol{l}_z - \boldsymbol{K}_{\text{ts}}\boldsymbol{l}_z] \tag{7.112}$$

With non-linear tip–sample interaction, the motion equation (7.109) governs the non-linear dynamic response of the cantilever. The temporal response of the cantilever is solved using the Runge–Kutta algorithm. There might be couplings among the vertical bending, lateral bending, extension, and torsion of the cantilever. Equation (7.109) takes all the couplings, if any, into consideration. Numerical simulations have shown that in TM, compared with vertical displacement and rotation about the y-axis, the displacements/rotations related to the extension, torsion and lateral bending of the cantilever are all negligible. The reasons are that for a typical cantilever in AFM, the stiffnesses of extension, torsion and lateral bending are much higher than that of vertical bending, and in TM, the contact time of the tip on the sample surface is only a small fraction of its vibration period. For a rectangular cantilever, the torsional and lateral bending stiffness are two orders, and the extension stiffness is four to five orders, of magnitude higher than that of vertical bending. Practically all displacements/rotations related to the extension, torsion and lateral bending of the cantilever can be neglected (constrained as zero). Equations (7.96) and (7.98) are simplified as

$$f_z^C = f_t \sin\alpha + f_n \cos\alpha \,, \quad M_y^C = l(-f_t \cos\alpha + f_n \sin\alpha) \tag{7.113}$$

$$d_t^P = d_z^C \sin\alpha - l\theta_y^C \cos\alpha \,, \quad d_n^P = d_z^C \cos\alpha + l\theta_y^C \sin\alpha \tag{7.114}$$

Equation (7.113) describes the relations between the interaction forces and forces/moments at node C. Equation (7.114) describes the relations between the displacements of the tip end (point P) and displacements/rotations at node C. For a tilted cantilever ($\alpha \neq 0$), (7.113) demonstrates that a fraction of the tangential interaction force f_t will also contribute to the vertical bending of the cantilever.

A TappingMode etched silicon cantilever MPP-11100 (Veeco Probes, CA) is studied. The dimensional and material parameters of the cantilever are: length $L = 125\,\mu m$, width $b = 35\,\mu m$, thickness $h = 4\,\mu m$, mass density $\rho = 2330\,kg/m^3$, Young's modulus $E = 1.3 \times 10^{11}\,Pa$, Poisson's ratio $\nu = 0.28$, tip radius $R = 10\,nm$, quality factor $Q = 100$. A proportional damping is assumed with a uniform

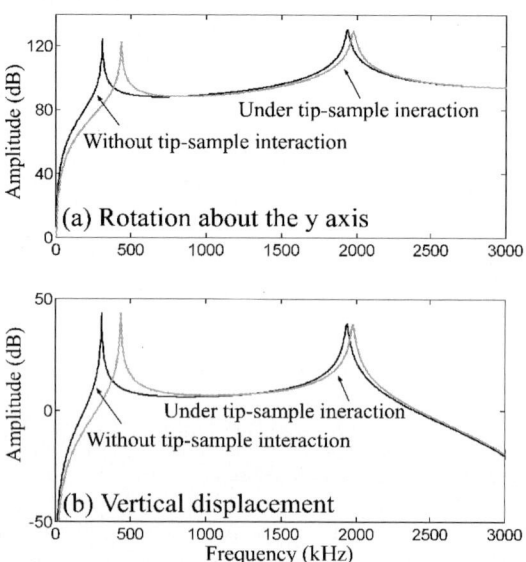

Fig. 7.24. Amplitudes of FRFs of θ_y^C (a) and d_z^C (b). The linear tip–sample interaction is in repulsive regime, $D = 0.08\,nm$, $k_n = 35\,N/m$, $k_t = k_{lat} = 58\,N/m$

modal damping ratio $\varsigma_i = 1/(2Q)$. The cantilever is discretized by ten 3D beam elements with the same length. It is set that $\alpha = 15°$, $\lambda = 1.0$, $l = 12.5\,\mu m$, $e_t = 0$ or $0.1b$. The interaction-related parameters of the cantilever and a HOPG sample are: $H = 2.96 \times 10^{-19}\,J$, $E^* = 10.2\,GPa$, $G^* = 4.2\,GPa$, $a_0 = 0.38\,nm$. No energy dissipation due to the tip–sample interaction is considered.

The cantilever responses for free vibration and under linear tip–sample interaction can be conveniently determined from the system FRF. Figure 7.24 gives the amplitudes of the FRFs of θ_y^C and d_z^C [55]. The resonance shift due to the repulsive interaction is demonstrated clearly from the figure. For TM, non-linear temporal analysis is needed. Figure 7.25 shows the steady state responses of θ_y^C, d_z^C, and interaction force f_n, when the cantilever is driven at its fundamental frequency of vertical bending [55]. To demonstrate the contribution of in-plane mechanical properties to cantilever vertical bending, two samples with different effective shear moduli (all the other parameters are the same) are considered. In both simulations, the response of θ_y^C contains higher harmonics. In TM, the readout of the measuring system is about θ_y^C. The term d_z^C is calculated from θ_y^C by assuming a one-to-one relation

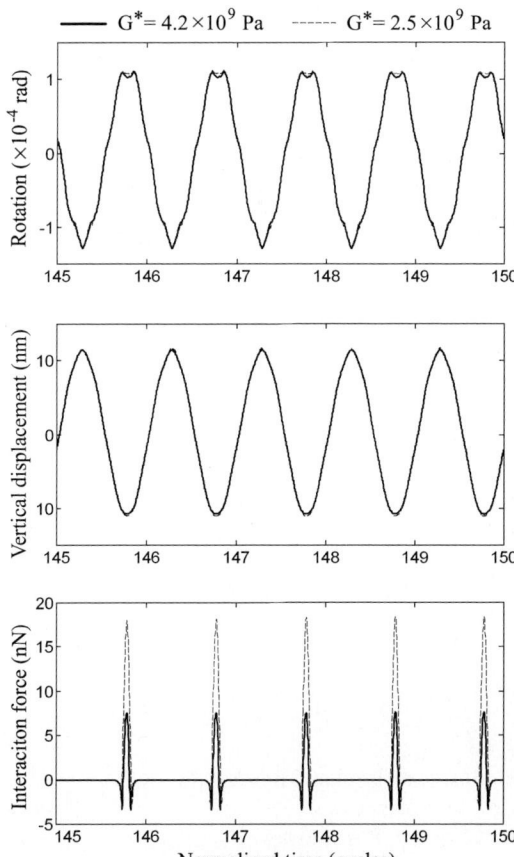

Fig. 7.25. Steady state responses of θ_y^C (**a**), d_z^C (**b**), and interaction force f_n (**c**). The cantilever is excited at the first resonance frequency of vertical bending (308.9 kHz). The excitation amplitude of the holder is $h_g = 0.32\,nm$, which corresponds to a free vibration amplitude of 50 nm at the cantilever tip. The equilibrium tip–sample distance is $D = 10\,nm$

between them. The feedback system uses the obtained d_z^C as an input for amplitude modulation. The involvement of higher frequency component in θ_y^C invalidates the one-to-one assumption and could give rise to inaccuracy in the feedback system. Figure 7.25a,b shows that the change of in-plane mechanical property has little affect on the amplitude of θ_y^C and d_z^C but significant effects on interaction forces. In TM, the interaction force is much more sensitive to the changes of in-plane mechanical properties than cantilever response amplitudes.

In commercial AFM, phase shift is measured using a lock-in amplifier. For a non-harmonic signal, the lock-in amplifier actually measures the phase shift between the driving signal and the first component of the Fourier's series of θ_y^C. Figure 7.26 gives the detailed views of θ_y^C in a short period of time [55]. The *true* phase shifts between the two responses for different G^* can be calculated by comparing the time differences at the cross zero points of the two signals. Due to the presence of higher frequency modes, the phase shifts are different when the signal crosses zero from above or from below. A careful calculation shows that a 40% change of G^* could lead to a phase difference of $\pm 2°$, which is big enough for imaging purposes.

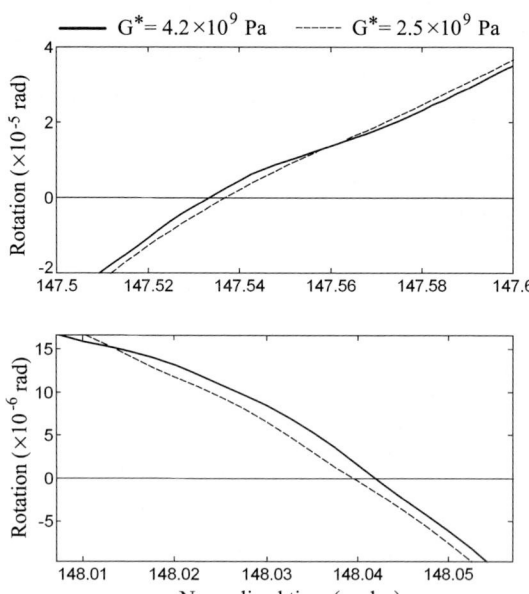

Fig. 7.26. Detailed views of θ_y^C in Fig. 7.25a. The phase differences between the two signals are different when the signals cross zero from above (**a**) and from below (**b**)

7.3.3
Modeling of Torsional Resonance Mode

In TRmode, the cantilever is excited by the torsional vibration of its holder $g_\theta(t) = \theta_g e^{i\Omega t}$. Except for the interaction force, no external force exists ($F_{ext} = 0$). The sample surface is fixed ($a^\alpha = 0$). The total dynamic displacement vector of the cantilever can be expressed as

$$d(t) = u(t) + l_\theta g_\theta(t) \tag{7.115}$$

where $u(t)$ is the cantilever deflection relative to the rotating holder, $g_\theta(t)$ is the torsional motion of the holder. The ith element of the positioning vector l_θ equals 1 if the ith DOF of $d(t)$ corresponds to the torsion about the x-axis; otherwise, it equals zero. The motion equation is

$$M\ddot{u} + C\dot{u} + Ku = -\Gamma_\theta \ddot{g}_\theta(t) + F_{ts}(u, \dot{u}) \tag{7.116}$$

$$\Gamma_\theta = Ml_\theta \tag{7.117}$$

In free vibration ($F_{ts} = 0$), the FRF vector due to the torsion of the holder is [55]

$$H_\theta(\Omega) = \frac{U}{\theta_g} = [K + i\Omega C - \Omega^2 M]^{-1} \Gamma_\theta \Omega^2 \tag{7.118}$$

If θ_x^C is the mth DOF of the system, the total free vibration amplitude and phase of θ_x^C under a harmonic motion of the holder $g_\theta(t) = \theta_g e^{i\Omega t}$ are the amplitude and argument of the complex value of $[H_{\theta m}(\Omega) + 1]\theta_g$, respectively.

If the tip–sample separation (normal pressure) remains almost constant during measurement, the tip–sample interaction can be modeled using a linear viscoelastic model. The FRF vector due to the torsion of the holder under tip–sample interaction is [55]

$$H_\theta^{ts}(\Omega) = \frac{U}{\theta_g}$$
$$= [(K + K_{ts}) + i\Omega(C + C_{ts}) - \Omega^2 M]^{-1}[\Gamma_\theta \Omega^2 - i\Omega C_{ts} l_\theta - K_{ts} l_\theta] \tag{7.119}$$

The temporal response of the cantilever can be obtained by solving the motion equation (7.116) using the Runge–Kutta algorithm. The deflection of the cantilever in TRmode is a combination of torsion and lateral bending. If the tip eccentricity exists ($e_t \neq 0$), the twist of the cantilever will change the tip–sample distance and so the normal interaction force. In that case, the vertical bending and extension will be coupled with the torsion/lateral bending of the cantilever. However, in TRmode, the deflections of the cantilever corresponding to the vertical bending and extension are small and can be ignored. Equations (7.96) and (7.98) are simplified as

$$f_y^C = f_{lat}, \quad M_x^C = e_t(f_t \sin\alpha + f_n \cos\alpha) + l f_{lat},$$
$$M_z^C = e_t(-f_t \cos\alpha + f_n \sin\alpha) \tag{7.120}$$
$$d_t^P = e_t\left(\theta_x^C \sin\alpha - \theta_z^C \cos\alpha\right), \quad d_{lat}^P = d_y^C + l\theta_x^C,$$
$$d_n^P = e_t\left(\theta_x^C \cos\alpha + \theta_z^C \sin\alpha\right) \tag{7.121}$$

TRmode simulations are performed in frequency and time domains for a silicon cantilever with the following dimensional and material parameters: $L = 252\,\mu m$,

$b = 35\,\mu m$, $h = 2.3\,\mu m$, $\rho = 2330\,kg/m^3$, $E = 1.3 \times 10^{11}\,Pa$, $\nu = 0.28$, $R = 10\,nm$, $Q = 33.3$. A proportional damping is assumed with a uniform modal damping ratio $\varsigma_i = 1/(2Q)$. In these analyses, the cantilever is discretized by ten 3D beam elements with the same length and the mass and moments of inertia of the tip are ignored. Under tip–sample interaction, we set $\alpha = 15°$, $\lambda = 1.0$, $l = 12.5\,\mu m$, $e_t = 0$ or $0.1b$. The interaction-related parameters are: $H = 2.96 \times 10^{-19}\,J$, $E^* = 10.2\,GPa$, $G^* = 4.2\,GPa$, $a_0 = 0.38\,nm$. No energy dissipation due to the tip–sample interaction is considered.

Figure 7.27 gives the amplitudes of FRFs of θ_x^C in free vibration and under linear tip–sample interaction [55]. It is clear that pure torsional analysis leads to different FRF from that with both torsion and lateral bending taken into consider-

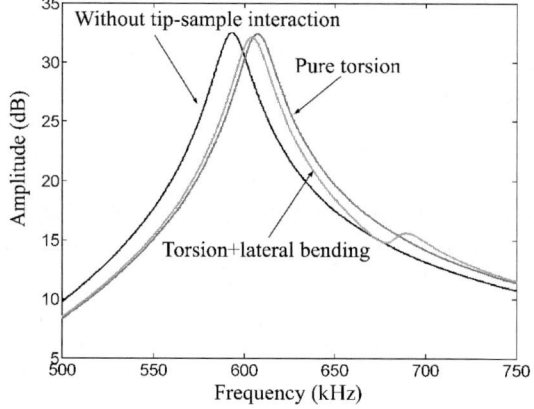

Fig. 7.27. Amplitudes of FRFs of θ_x^C in TRmode. The linear tip–sample interaction is in repulsive regime, $D = 0.37\,nm$, $k_n = 6.5\,N/m$, $k_t = k_{lat} = 10.6\,N/m$

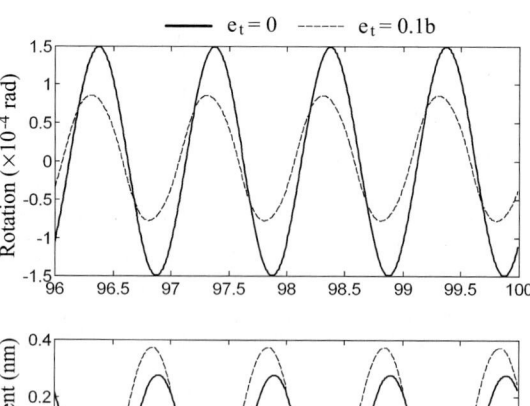

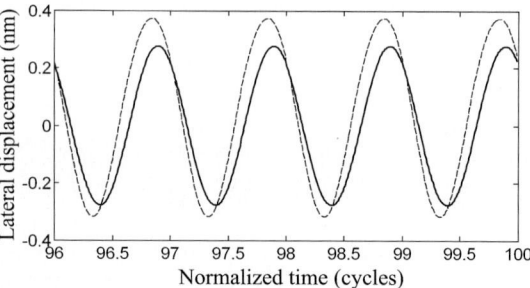

Fig. 7.28. Steady state responses of θ_x^C and d_y^C in TRmode, with and without tip eccentricity. The cantilever is driving at its first torsional resonance frequency and the excitation amplitude of the holder is $\theta_g = 5 \times 10^{-6}$ (rad). The equilibrium position is $D = 0.375\,nm$

ation. On the curve considering both torsion and lateral bending, there is a small peak around 690 kHz, which corresponds to the shifted resonance of lateral bending. The steady state responses of θ_x^C and d_y^C, for $e_t = 0$ and $e_t = 0.1b$, are given in Fig. 7.28 [55]. The results demonstrate that tip eccentricity has significant effects on both the amplitude and phase of cantilever responses and therefore cannot be ignored in TRmode. For the responses of $e_t = 0.1b$, due to the existence of tip eccentricity, the response histories become asymmetric about the time coordinate. This is due to the fact that for a positive (negative) e_t, a positive twist of the cantilever lifts up (presses down) the tip and therefore decreases (increases) the lateral contact force, while a negative twist of the cantilever presses down (lifts up) on the tip and increases (decreases) the lateral contact force. An asymmetric twist response in TRmode could be used as the indication of the existence of tip eccentricity.

7.3.4
Modeling of Lateral Excitation Mode

In LE mode, the cantilever is excited and vibrates by the harmonic motion of the sample surface ($a^\alpha \neq 0$) through the lateral tip–sample interaction. The driving frequency could be a value in a wide range around the resonance frequencies of torsion or lateral bending. Except the interaction force, no external force exists ($F_{ext} = 0$). The system motion equation of LE mode is

$$M\ddot{d} + C\dot{d} + Kd = F_{ts}(d, \dot{d}) \tag{7.122}$$

The motion of the sample surface is

$$a^\alpha = \{a_t, a_{lat}, a_n\}^T = \{0, a_g e^{i\Omega t}, 0\}^T \tag{7.123}$$

If the tip–sample separation (normal pressure) remains almost constant during the measurement, tip–sample interaction can be modeled using a linear viscoelastic model. The motion equation of (7.122) can then be expressed as

$$M\ddot{d} + (C + C_{ts})\dot{d} + (K + K_{ts})d = G_t^T f_{ts}^a \tag{7.124}$$

$$f_{ts}^a = A_{C\alpha}\left(k_{ts}^\alpha a^\alpha + c_{ts}^\alpha \dot{a}^\alpha\right) \tag{7.125}$$

By substituting (7.123) and (7.125) into (7.124) and assuming $d = Ue^{i\Omega t}$, one can rewrite the motion equation as

$$[-\Omega^2 M + i\Omega(C + C_{ts}) + (K + K_{ts})]U = \Gamma_{lat} a_g \tag{7.126}$$

$$\Gamma_{lat} = G_t^T A_{C\alpha}\left(k_{ts}^\alpha \begin{Bmatrix} 0 \\ 1 \\ 0 \end{Bmatrix} + c_{ts}^\alpha \begin{Bmatrix} 0 \\ i\Omega \\ 0 \end{Bmatrix}\right) = G_t^T \begin{Bmatrix} 0 \\ k_{lat} + i\Omega c_{lat} \\ 0 \\ l(k_{lat} + i\Omega c_{lat}) \\ 0 \\ 0 \end{Bmatrix} \tag{7.127}$$

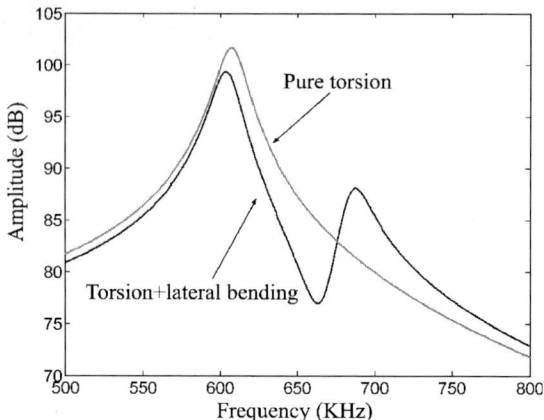

Fig. 7.29. Amplitudes of FRFs of θ_x^C in LE mode. The tip–sample interaction is in repulsive regime, $D = 0.37$ nm, $k_n = 6.5$ N/m, $k_t = k_{lat} = 10.6$ N/m [55]

The FRF vector due to the motion of the sample in LE mode is [55]

$$H_{lat}(\Omega) = \frac{U}{a_g} = [(K + K_{ts}) + i\Omega(C + C_{ts}) - \Omega^2 M]^{-1} \Gamma_{lat} \qquad (7.128)$$

The readout of the measuring system is the signal about θ_x^C. If θ_x^C is the mth DOF of the system, the total free vibration amplitude and phase of θ_x^C under a harmonic motion of the holder $a_{lat}(t) = a_g e^{i\Omega t}$ are the amplitude and argument of the complex value of $H_{lat\,m}(\Omega)a_g$, respectively.

The coupling relations in LE mode are the same as these in TRmode. The response of the cantilever can be solved (7.124) using the Runge–Kutta algorithm. For the same cantilever and sample as those in Sect. 7.3.3. Figure 7.29 gives the amplitudes of FRFs of θ_x^C [55]. On the curve considering both torsion and lateral bending, the first and second peaks correspond to the shifted resonances of torsion and lateral bending, respectively. Figure 7.30 gives the steady state responses of θ_x^C and d_y^C, with $e_t = 0$ and $e_t = 0.1b$, for three different driving frequencies [55]. Similar to the analysis in TRmode, the tip eccentricity has a significant effect on both of the amplitude and phase of cantilever responses. Due to the existence of the eccentricity, the response histories become asymmetric about the time coordinate. Tip eccentricity, if any, has to be considered in response analysis of LE mode and the asymmetric phenomenon in twist response of cantilever could be used as the indication of the existence of tip eccentricity.

7.4
Atomic-Scale Topographic and Friction Force Imaging in FFM

FFM is usually categorized as one of the static AFM modes because during measurement, the cantilever is not excited to vibrate and under certain conditions, the cantilever is indeed in the quasistatic motion. Whereas, stick-slip tip motion is often observed in friction measurement, as shown in Fig. 7.31 [48]. In that case, cantilever

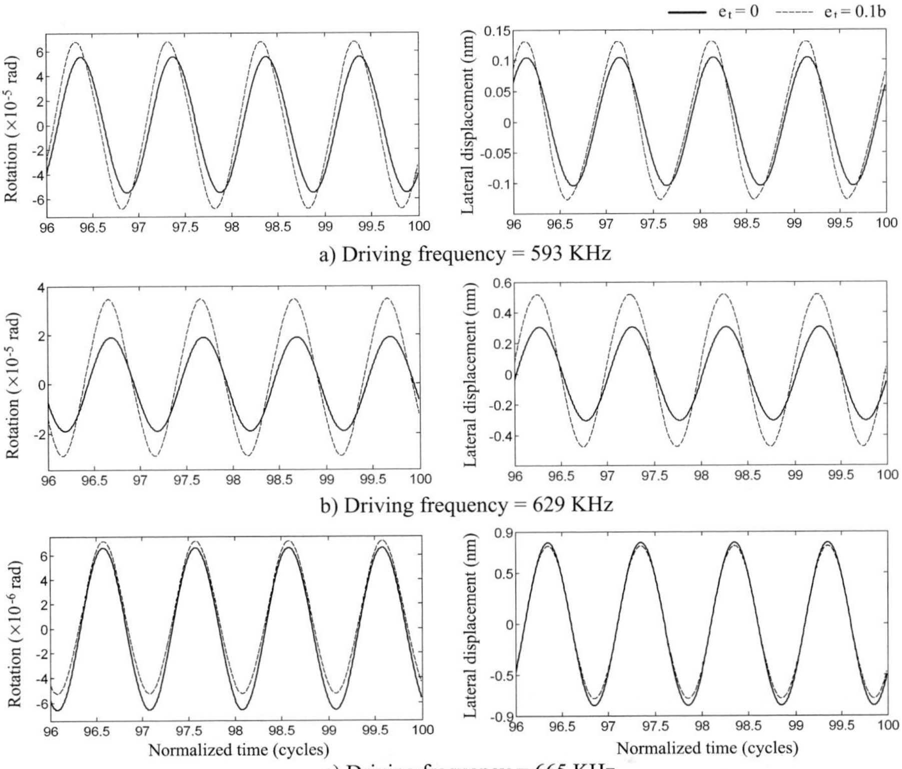

$$ \rule{0pt}{0pt} \quad\quad \text{—— } e_t = 0 \quad \text{----- } e_t = 0.1b $$

a) Driving frequency = 593 KHz

b) Driving frequency = 629 KHz

c) Driving frequency = 665 KHz

Fig. 7.30. Steady state responses of θ_x^C and d_y^C in LE mode, with or without tip eccentricity. The cantilever is driving at its first torsional resonance frequency (**a**), at its first lateral bending resonance frequency (**c**), and at a frequency in the middle of them (**b**). The excitation amplitude of the sample surface is $a_g = 1$ nm. The equilibrium position is $D = 0.375$ nm [55]

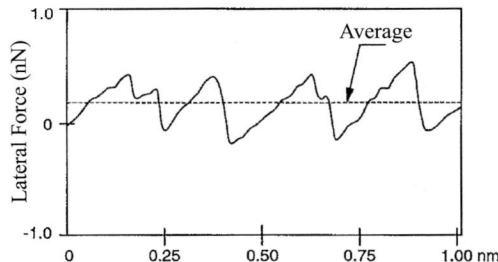

Fig. 7.31. Stick-slip tip motion in friction measurement [48]

dynamics is closely correlated to the measured topographic and friction images and has to be considered. Using the 3D FE beam model described in Sect. 7.3, this section simulates the friction profiling process in FFM and studies how the cantilever dynamics affects the measured atomic-scale topographic and friction maps of graphite surface [57].

7.4.1
FFM Images of Graphite Surface

The working mechanism of an optical beam-deflection FFM can be explained in
Fig. 7.1. FFM allows simultaneous measurements of surface topography, normal
and lateral forces. The four-segment photodiodes are used to measure the cantilever
flexural angle θ_y^C and twist angle θ_x^C. With the tilting of the cantilever to the sample
surface being neglected, the vertical bending is caused by the normal force f_z and
the moment $M_y^C = -f_x l$ resulting from the lateral force f_x. The lateral force f_y
causes the lateral bending, and its resulting torque $f_y l$ twists the cantilever. The term

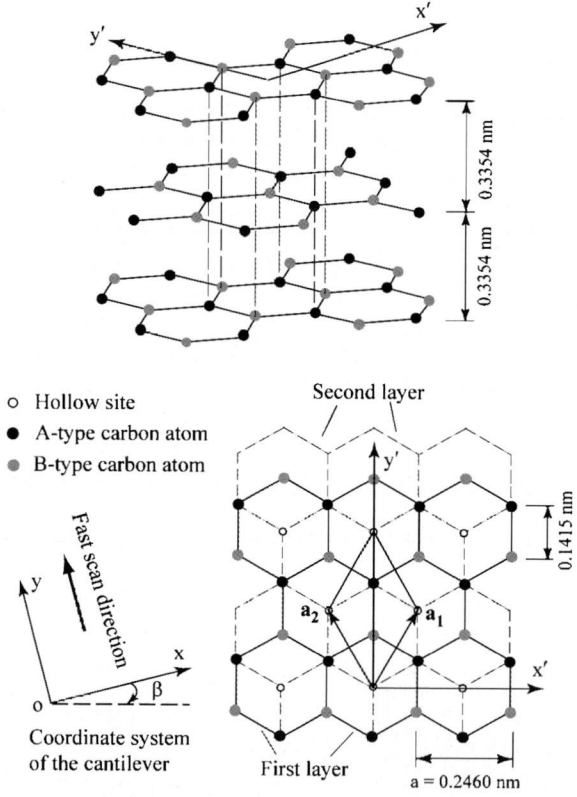

Fig. 7.32. Schematic view of the periodic hexagonal lattice structure of (0001) graphite surface.
Layers of the hexagonal structures are staggered with a distance of 0.3354 nm. Two types of
atoms are distinguished. Hollow sites are the centers of each hexagon. In the coordinate system
for the cantilever, the x-axis is along the cantilever longitudinal direction. The cantilever is
always scanned along the y-axis in constant-force mode of FFM. Another coordinate system
(x', y') is introduced for convenient determination of atom locations on the graphite surface.
The shown parallelogram is a primitive unit cell which includes two carbon atoms. Vectors
a_1 and a_2 are the unit lattice vectors for the primitive unit cell. The angle β is defined to
represent the relation between the coordinate systems for the cantilever and graphite lattice
structure

θ_y^C is the flexural angle of the cantilever due to vertical bending. It is related to the normal load f_z and lateral force f_x. The contribution from f_x to θ_y^C is much smaller than that from f_z (less than 5%). Approximately, the normal load f_z can be viewed as the sole cause of θ_y^C. Therefore, normal force f_z can be obtained by measuring θ_y^C. With the help of a feedback loop in FFM, surface topography can be obtained by keeping a constant θ_y^C through the z-direction motion of the piezotube when the cantilever tip is scanned over the sample surface. Since the lateral force f_y is the only interaction force that is responsible for the cantilever torsion, the twist angle θ_x^C is a good measurement for f_y. In constant-force mode of FFM, a constant normal load (or, equivalently, a constant θ_y^C) is maintained to obtain the surface topography and make the friction measurement meaningful. During the measurement, the scan direction is perpendicular to the longitudinal direction of the cantilever (along the y-axis) and the lateral force f_y is obtained by measuring the twist angle θ_x^C.

Atomic-scale topographic imaging has been carried out by researchers on HOPG and other samples. Figure 7.32 shows the hexagonal structure of (0001) graphite surface. Layers of the hexagonal structures are staggered with a distance of 0.3354 nm. Two types of carbon atoms, A-type and B-type, exist due to the way the layers staggered. A-type atoms have a direct neighbor in the adjacent layers while B-type atoms do not. Within one layer, the distance between the adjacent A-type or B-type atoms is 0.2460 nm and it is 0.1415 nm between any two adjacent carbon atoms. Binnig et al. [9] successfully observed the hexagonal structure of the graphite surface (see Fig. 7.33a). In their experiment, a cantilever without a tip was scanned over the graphite surface with its corner touching the surface to obtain the topographic images. Marti et al. [35] also obtained the full hexagonal topographic image of graphite surface covered with paraffin oil using a diamond epoxied to one of the four cross points of four platinum wires. However, many others [2, 48] could only show the trigonal lattice of three peaks with a distance of about 0.246 nm in their experimentally-obtained topographic images, i.e., so-called atomic resolution of every other atom (see Fig. 7.33b).

FFM topographic and friction images are closely correlated to the cantilever dynamics. The friction force map measured in FFM is actually the map about the cantilever torsional angle θ_x^C and the topographic map is about the z-direction holder

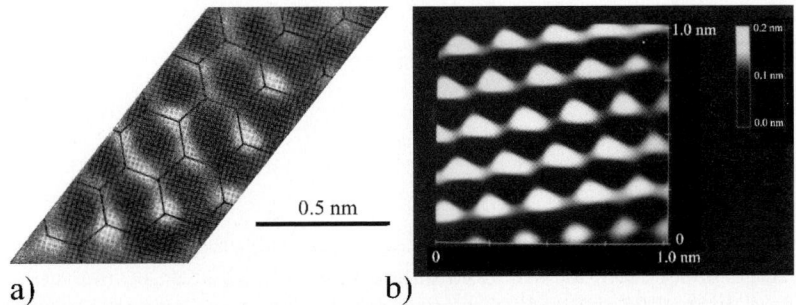

a)   b)

Fig. 7.33. AFM Topography images of graphite surface. (**a**) Topography image of graphite obtained with a cantilever without a tip [9]. Hexagonal structure of graphite surface is shown. (**b**) Topography map of HOPG shown atomic resolution of every other atoms [48]

motion z_h. In the measured maps of FFM, the topography and lateral forces at the locations of the tip $(x_t(t), y_t(t))$ are plotted against the nominal coordinates of the tip, i.e., the tip position at time t if the cantilever is rigid (no deflection). The nominal tip coordinates are determined from the coordinate of the cantilever holder. Here we represent them as $(x_h(t) + L_h, y_h(t))$, where $(x_h(t), y_h(t))$ are coordinates of the holder and L_h is a constant in each experiment. Due to the cantilever deflection during measurements, usually one will find that the nominal tip coordinates $(x_h(t) + L_h, y_h(t))$ are different from the real tip coordinates $(x_t(t), y_t(t))$. This means that cantilever dynamics have to be considered in order to explain the afore-mentioned different experimentally-obtained topography of graphite.

7.4.2
Interatomic Forces Between Tip and Surface

To be able to simulate the cantilever behavior in the profiling process of FFM, the interatomic forces between cantilever tip and surface has to be determined first. The interaction forces between the scanning cantilever tip and the graphite surface can be calculated from the spatial derivations of an interaction potential between the atoms of the tip and graphite surface. In simplified interaction models, both the tip and sample are assumed to be rigid. The tip is represented as either one atom or multiatoms, and only one layer of the lattice is involved. Another assumption is that the interaction system is always in an equilibrium state. These simplifications are justified considering that the interaction potential drops very rapidly as the atom distance increases, and that the scan speed of the cantilever tip in FFM is much lower than the characteristic velocity of the lattice vibration.

The periodic structure of the graphite surface induces a periodic interaction potential

$$V_{ts}(r_t) = V_{ts}(r_t + i_1 a_1 + i_2 a_2) \tag{7.129}$$

where r_t is the tip position vector, a_1 and a_2 are the unit lattice vectors in the surface plane (refer to Fig. 7.32), and i_1 and i_2 are arbitrary integers. It is assumed that the interaction potential between the tip and graphite surface equals the sum of the interaction potential between individual tip atoms and carbon atoms of graphite surface. Approximating the pairwise interaction potential between two atoms by the Lennard–Jones potential, one has

$$V_{ts}(r_t) = \sum_k \sum_i 4\varepsilon_{ts} \left[\left(\frac{\sigma}{r_{ki}}\right)^{12} - \left(\frac{\sigma}{r_{ki}}\right)^6 \right] \tag{7.130}$$

where the parameters can be chosen as $\varepsilon_{ts} = 0.87 \times 10^{-2}$ eV and $\sigma = 0.249$ nm [50], r_{ki} is the distance from the kth atom on the tip to the ith atom on the surface. In one-atom tip models, the tip location coincides to the atom location. In multiatom tip models, the coordinates of the tip atoms can be determined from the tip position vector r_t given that the relative positions of the atoms to the tip position are known. In the (x, y, z) coordinate system of the cantilever, the tip position vector is denoted as $r_t = (x_t, y_t, z_t)$. The interaction forces between the tip and the graphite surface are given by

$$f_x(x_t, y_t, z_t) = -\frac{\partial V_{ts}(x_t, y_t, z_t)}{\partial x_t}, \quad f_y(x_t, y_t, z_t) = -\frac{\partial V_{ts}(x_t, y_t, z_t)}{\partial y_t},$$

$$f_z(x_t, y_t, z_t) = -\frac{\partial V_{ts}(x_t, y_t, z_t)}{\partial z_t}$$

$$(7.131)$$

The angle β is defined as the angle from the x-axis of the cantilever to the x'-axis of the graphite surface. Using a one-atom tip model, interaction force maps are calculated for a constant tip-surface distance $z_t = 0.17\,\mathrm{nm}$ when $\beta = 0°$, as shown in the left-hand column of Fig. 7.34a [57]. If the normal force keeps constant, z_t has to change at different locations on the graphite surface. For a constant normal load $W = 25\,\mathrm{nN}$, the proper z_t map is obtained by solving the non-linear equation

$$f_z(x_t, y_t, z_t) = W \qquad (7.132)$$

The corresponding lateral force maps are then calculated. The maps in the right-hand column of Fig. 7.34a give the tip-surface distance and lateral forces exerted on the cantilever tip when the tip is at different locations of the graphite surface. The topography of the graphite surface can be obtained from the z_t map by subtracting the $\min(z_t)$ from z_t.

Figure 7.34b shows the relative locations of the carbon atoms of graphite surface and the maxima of the lateral forces under the constant normal load $W = 25\,\mathrm{nN}$ [57]. Without considering any cantilever dynamics, the topographic map (or constant normal force profile) does reflect the surface atomic structure. The locations of the maxima of the normal force coincide to the atom locations. Whereas the lateral force maps are not straightforward tools for observation of atomic structures since the maxima of the lateral forces are not located at the atom locations.

7.4.3
Modeling of FFM Profiling Process

Using the 3D finite element beam model described in Sect. 7.3, the motion equation governing the cantilever response is expressed as

$$\boldsymbol{M\ddot{u}} + \boldsymbol{C\dot{u}} + \boldsymbol{Ku} = \boldsymbol{F}_{ts} - \boldsymbol{M\Gamma\ddot{g}}_h \qquad (7.133)$$

Here, $\boldsymbol{u}, \dot{\boldsymbol{u}}$ and $\ddot{\boldsymbol{u}}$ are displacement, velocity and acceleration vectors of the cantilever relative to its holder, $\boldsymbol{g}_h = \{x_h, y_h, z_h\}^T$ is the vector of holder motion, Γ is the position matrix describing the relation between the total displacement vector $\boldsymbol{u}_{tol}$ and $\boldsymbol{u}$. The total displacement vector of the cantilever in the (x, y, z) coordinate system is expressed as

$$\boldsymbol{u}_{tol} = \boldsymbol{u} + \Gamma\boldsymbol{g}_h \qquad (7.134)$$

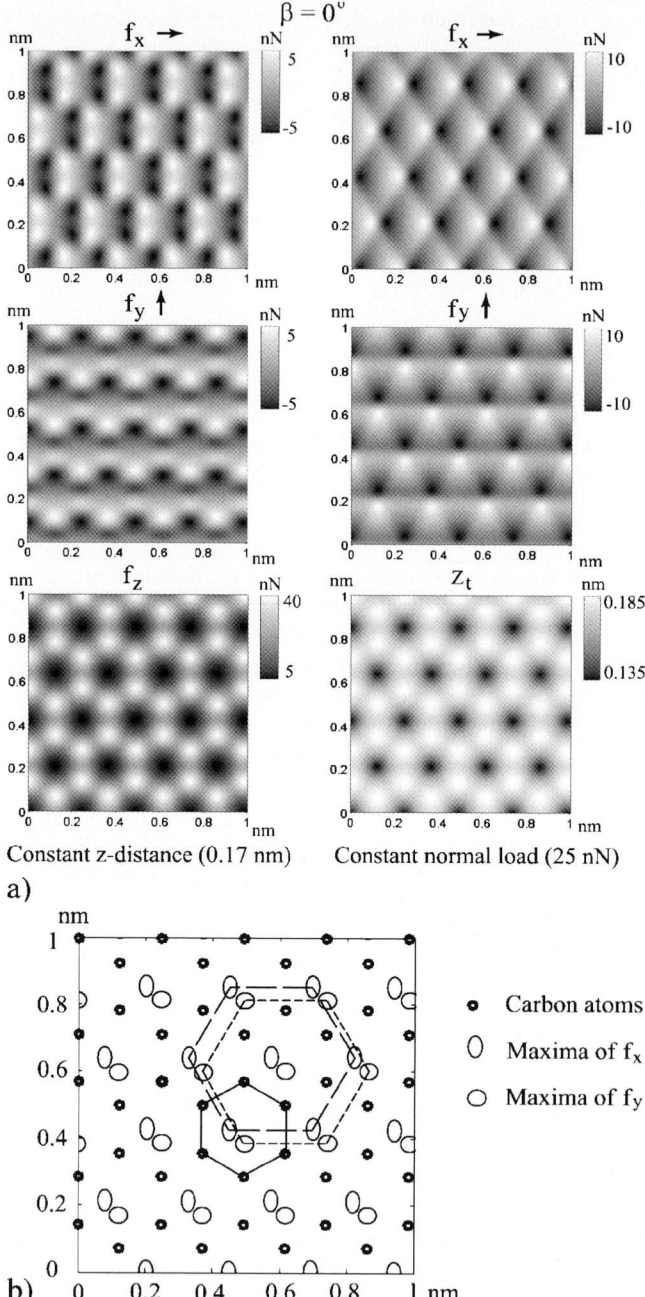

Fig. 7.34. (**a**) 3D interatomic force maps with a constant tip-surface distance $z_t = 0.17$ nm (*left-hand column*). Lateral interatomic force and tip-surface distance maps with a constant normal load $f_z = 25$ nN (*right-hand column*). (**b**) The relative positions of the carbon atoms and the maxima of lateral forces under the constant normal load $f_z = 25$ nN [57]

For easy illustration, it is assumed that the tip is at the end of the cantilever and the cantilever is represented by only one-beam element. There are six DOFs at the end of the cantilever: d_x^C, d_y^C, d_z^C, θ_x^C, θ_y^C, and θ_z^C. The extension of the cantilever can be neglected, i.e., $d_x^C = 0$. In constant-force mode of FFM, the tip is always in a stable equilibrium state in the vertical direction if the holder is moving with a moderate scan speed. In that case, the vertical displacement and flexural angle can be obtained by

$$d_z^C = \text{constant} = \frac{L^3}{3EI_y} f_z + \frac{L^2 l}{2EI_y} f_x , \quad \theta_y^C = \text{constant} = -\frac{L^2}{2EI_y} f_z - \frac{Ll}{EI_y} f_x$$

(7.135)

For a typical cantilever, the tip length l is only about $1/10 \sim 1/30$ of the cantilever length L. The lateral force f_x could be one order of amplitude smaller than the normal force f_z. From (7.135), the contribution of f_x to d_z^C and θ_y^C is much smaller than that from f_z. With the contribution of f_x being neglected, (7.134) is rewritten as

$$d_z^C \approx \frac{L^3}{3EI_y} f_z , \quad \theta_y^C \approx -\frac{L^2}{2EI_y} f_z$$

(7.136)

Therefore, the topography map $z_t(x_t, y_t)$ can be obtained by solving the non-linear equation (7.132).

At point C, the cantilever deflections corresponding to lateral bending (d_y^C and θ_z^C) and torsion (θ_x^C) are governed by (7.133). The cantilever is scanned with a constant velocity, i.e., $\ddot{g}_h = \{0, 0, 0\}^T$. Equation (7.133) is expressed as

$$\rho AL \begin{bmatrix} 39/105 & 0 & -11L/210 \\ 0 & I_p/3A & 0 \\ -11L/210 & 0 & L^2/105 \end{bmatrix} \begin{Bmatrix} \ddot{d}_y^C \\ \ddot{\theta}_x^C \\ \ddot{\theta}_z^C \end{Bmatrix} + \begin{bmatrix} c_{11} & c_{12} & c_{13} \\ c_{21} & c_{22} & c_{23} \\ c_{31} & c_{32} & c_{33} \end{bmatrix} \begin{Bmatrix} \dot{d}_y^C \\ \dot{\theta}_x^C \\ \dot{\theta}_z^C \end{Bmatrix}$$

$$+ \begin{bmatrix} 12EI_z/L^3 & 0 & -6EI_z/L^2 \\ 0 & GJ/L & 0 \\ -6EI_z/L^2 & 0 & 4EI_z/L \end{bmatrix} \begin{Bmatrix} d_y^C \\ \theta_x^C \\ \theta_z^C \end{Bmatrix} = \begin{Bmatrix} f_y^C \\ M_x^C \\ M_z^C \end{Bmatrix}$$

(7.137)

The force and moment at point C are related with f_y as

$$\begin{Bmatrix} f_y^C \\ M_x^C \\ M_z^C \end{Bmatrix} = \begin{Bmatrix} 1 \\ l \\ 0 \end{Bmatrix} f_y(x_t, y_t, z_t)$$

(7.138)

The lateral force f_y is a function of the tip location, which can be determined from the constant force condition in (7.132) and the cantilever deflection as follows

$$x_t = x_h + L - l\theta_y^C , \quad y_t = y_h + d_y^C + l\theta_x^C ,$$

(7.139)

In (7.139), $-l\theta_y^C \approx \frac{L^2 l}{2EI_y} f_z = \text{constant}$, i.e., along each scan line, $x_t = \text{constant}$. Equation (7.137) is non-linear and needs to be solved numerically with (7.132) to simulate the cantilever response in FFM. The resulting maps of z_t and θ_x^C are the FFM imaging results of topographic and friction force, respectively.

The damping matrix in (7.137) addresses the energy dissipation mechanism in the tip-cantilever-surface system. It includes both the material damping in the cantilever and the energy dissipation induced from the tip–sample interaction, such as phonon generation. It is demonstrated in Sect. 7.3 that the damping effects due to the tip–sample interaction can be equivalently addressed as an additional damping term to the material damping matrix of the cantilever. Here, the damping matrix is calculated as

$$C = [\varphi_1, \varphi_2, \varphi_3]^{-T} \text{diag}(2\varsigma_1\omega_1, 2\varsigma_2\omega_2, 2\varsigma_3\omega_3)[\varphi_1, \varphi_2, \varphi_3]^{-1} \qquad (7.140)$$

where ω_i, φ_i and $\varsigma_i (i = 1, 2, 3)$ are the ith circular natural frequency, normalized eigenmode vector and damping ratio of the system. Large damping ratios (close to 1.0) are adopted to simulate the damping effects in the profiling process of FFM.

7.4.4
Simulations on Graphite Surface

The profiling process of FFM is simulated to obtain the topographic and lateral force maps of graphite surface for different combinations of normal loads, tip lengths, and scan directions. The rectangular silicon cantilever considered here has the following dimensional and material parameters: $L = 252\,\mu\text{m}$, $b = 35\,\mu\text{m}$, $h = 2.3\,\mu\text{m}$, $\rho = 2330\,\text{kg/m}^3$, $E = 1.3 \times 10^{11}$ Pa, $\upsilon = 0.28$. Two scan directions are considered. In one case, $\beta = 0°$, i.e., the scan is carried out along the y'-axis, and in the other case $\alpha = 30°$, which is equivalent to $\beta = 90°$ due to the hexagonal structure of graphite, i.e., the scan direction is along the x'-direction. The scan size is 1 nm × 1 nm. The scan velocity in the fast scan direction is 200 nm/s. Figure 7.35 shows the simulated maps of the cantilever twist angle $-\theta_x^C$, the tip-distance map z_t, and the paths of cantilever tip [57]. The maps of $-\theta_x^C$ can be viewed as a measurement of the lateral force that resists the movement of the cantilever, i.e., the lateral force whose direction is opposite to the scan direction.

7.4.4.1
Slow-Fast Motion Pattern

In Fig. 7.35a, the results are for the normal load $W = 10$ nN, tip length $l = 12.5\,\mu\text{m}$, and $\beta = 0°$. The full hexagonal structure of the graphite surface can be seen in the

Fig. 7.35. The simulated maps of cantilever twist angle $-\theta_x^C$ (*left-hand column*), tip-surface distance (*middle column*) z_t, and 13 paths of the cantilever tip (*right-hand column*) for different combinations of the normal loads, tip lengths, and scan directions. The maps of cantilever twist angle are equivalent to the lateral force maps. The tip-surface distance maps are equivalent to the topographic maps. The data on the maps are unit-cell averaged. The paths of the tip is time-resolved, i.e., the paths are plotted by dots separated by equal time interval $\Delta t = 0.05$ ms

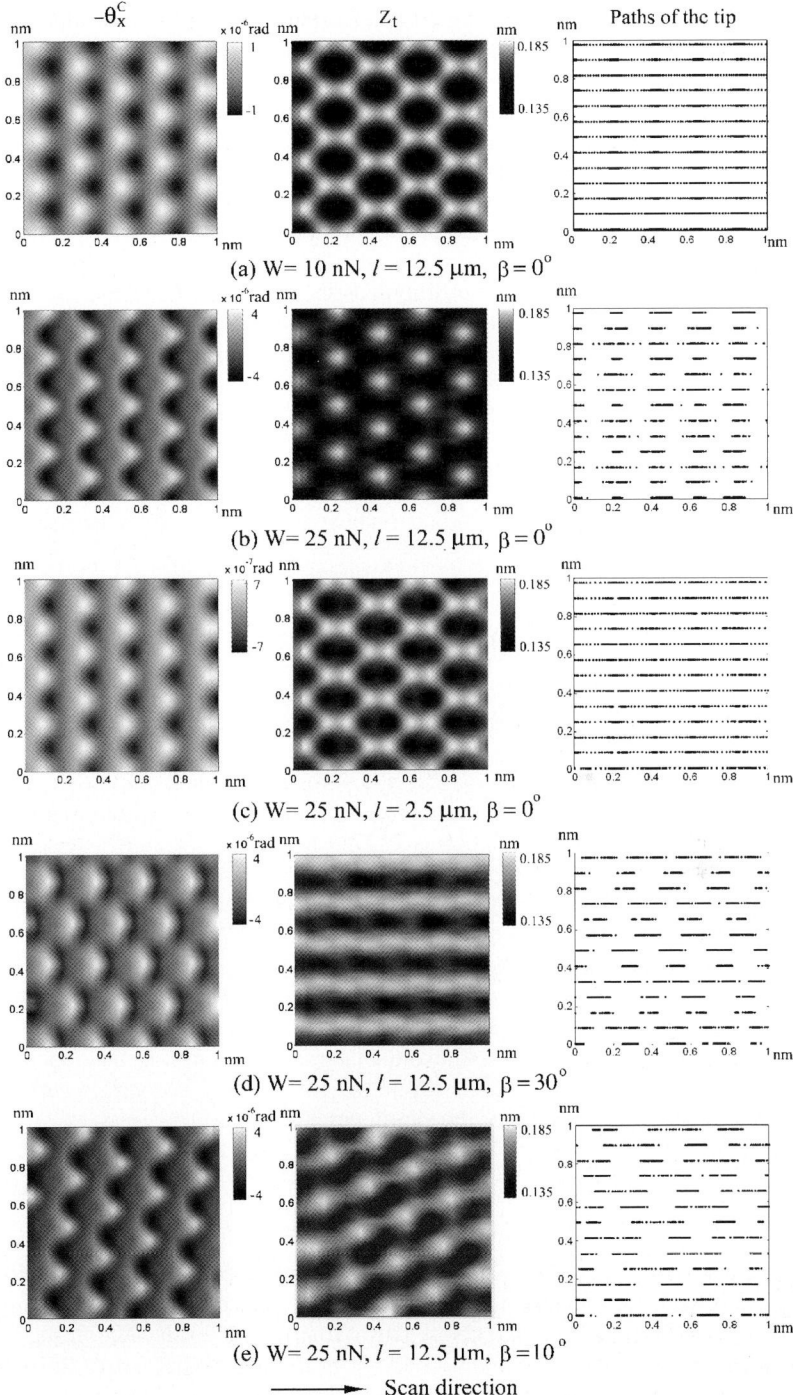

(a) W= 10 nN, l = 12.5 μm, β = 0°

(b) W= 25 nN, l = 12.5 μm, β = 0°

(c) W= 25 nN, l = 2.5 μm, β = 0°

(d) W= 25 nN, l = 12.5 μm, β = 30°

(e) W= 25 nN, l = 12.5 μm, β = 10°

⟶ Scan direction

topographic map. The paths of the cantilever tip are plotted by dots separated by equal time interval. Although the cantilever holder moves in a constant velocity, the cantilever tip does not slide over the surface smoothly. The dense part on the tip paths indicates that the tip is moving slowly over the surface while the sparse part represents a faster motion. This "slow-fast" motion of the cantilever tip is slightly different from what is well-known as the "stick-slip" behavior since "stick" means that there is absolutely no relative motion between the cantilever tip and sample surface. However, as the dense and sparse parts become remarkable, we refer the slow-fast motion pattern as the stick-slip, as most of researchers do. Figure 7.35b gives the results under the same conditions as those in Fig. 7.35a, except that the normal load is increased from 10 nN to 25 nN. The slow-fast tip motion in Fig. 7.35b is more remarkable than that in Fig. 7.35a due to the increased normal load, as shown clearly in the tip paths in Fig. 7.35a,b. Consequently, the hexagonal structure in topography cannot be observed in Fig. 7.35b and only the "resolution of every other atom" is shown. Also, dramatic difference can be seen in the lateral force maps in Fig. 7.35a,b.

7.4.4.2
Conditions for Stick-Slip Occurrence

One would ask under what circumstances the stick-slip (or remarkable slow-fast) tip motion will not occur so that the detection of the full atomic structure becomes possible. It is generally recognized that stick-slip will occur if a soft cantilever is scanned over a surface with large lateral forces [23, 36, 50]. Experimentally, it is also observed that the sticking domain decreased with the increased normal load, and under lower normal load the tip shows smoother motion [17]. Johnson and Woodhouse [29] assumed a sinusoidal lateral force and gave the analytical condition under which the tip motion is steady and no stick-slip occurs

$$T^* < T_c^* = \lambda_1 c_e / 2\pi \tag{7.141}$$

where T^* is the magnitude of the sinusoidal lateral force, T_c^* is the critical lateral force magnitude at which stick-slip will occur, λ_1 is the periodic lattice spacing of the sample surface, and c_e is the effective lateral stiffness. Although (7.141) is obtained under the assumption of a sinusoidal lateral force, it may be used for a rough estimation of the occurrence of stick-slip in the simulations. For a rigid surface, c_e equals the static lateral stiffness of the cantilever, i.e.,

$$c_e = \left(\frac{L^3}{3EI_z} + \frac{Ll^2}{GJ} \right)^{-1} \tag{7.142}$$

For the cantilever in Fig. 7.35a,b, $c_e = 94.4\,\text{N/m}$. If it is chosen as $\lambda_l = 0.426\,\text{nm}$, the critical lateral force magnitude is $T_c^* = 6.4\,\text{nN}$. The maximum lateral force is about 3 nN for $W = 10.0\,\text{nN}$ and about 10 nN for $W = 25.0\,\text{nN}$. According to the condition in (7.141), stick-slip should happen in Fig. 7.35b but not in Fig. 7.35a, which is consistent with the simulated results.

7.4.4.3
Methods for Stick-Slip Prevention

To observe the full atomic structure of the surface, the velocity of the cantilever tip should not oscillate too much during the scanning process. As pointed out earlier, this usually requires a relatively small lateral force and large lateral stiffness of the cantilever. A small lateral force can be achieved simply by applying a small normal load during measurement. However, for a typical cantilever, a small normal load means a small lateral force and therefore small cantilever flexural and twist angles (θ_y^C and θ_x^C) that may not be measured effectively. Another way would be to increase the lateral stiffness of the cantilever. For a rectangular cantilever, considering $G \approx E/2$ and $J \approx \frac{1}{3}bh^3$, one has

$$s_{\text{lat}} \approx \frac{Ebh}{L}\left(\frac{4L^2}{b^2} + \frac{6l^2}{h^2}\right)^{-1}, \quad s_n = \frac{Ebh^3}{4L^3}, \quad s_\theta = \frac{Ebh^3}{6Ll} \tag{7.143}$$

where s_{lat} is the lateral cantilever stiffness considering both cantilever torsion and lateral bending, s_n is the vertical spring constant of the cantilever, and s_θ is the twist stiffness ($\theta_x^C = s_\theta f_y$). The term s_{lat} can be increased by adopting the following three methods, separately or in combination: (1) increasing the cantilever width b and/or the thickness h; (2) decreasing the cantilever length L; (3) decreasing the tip length l. All the three methods may suffer some limitations to applications. Methods (1) and (2) will increase s_n and s_θ simultaneously. Too big s_n and s_θ will lead to small θ_y^C and θ_x^C, just like the effects resulting from applying a small normal load. Although method (3) has no effect on s_n, it results in an increase of s_θ. Therefore, care has to be taken in choosing the cantilever geometry, tip length, and applied load in order that the remarkable unsmooth tip motion does not occur, and at the same time, the signal is big enough to be measured.

With the same normal load as that in Fig. 7.35b, Fig. 7.35c show the results with a smaller tip length $l = 2.5\,\mu\text{m}$. As expected, the slow-fast tip motion in Fig. 7.35c becomes less remarkable than that in Fig. 7.35b and the full hexagonal lattice structure is shown in a topographic map, although the normal load is the same as that in Fig. 7.35b. With a tip length of 2.5 μm, $c_e = 194.1\,\text{N/m}$ and the critical lateral force magnitude $T_c^* = 13.2\,\text{nN}$. According to (7.141), the stick-slip motion should not occur. Actually, this was exactly what Binning et al. [9] did in their experiment where the full atomic structure was obtained successfully. In their experiment, a cantilever without a tip was used. During measurement, the cantilever corner touched the sample surface for imaging. In the other experiment by Marti et al. [35] in which the full hexagonal structure of graphite surface was

Table 7.6. Conditions for stick-slip occurrence

	Applied load	Cantilever lateral stiffness	Cantilever geometry
Stick-slip	Large	Small	Smaller width, smaller thickness, larger length, larger tip length
No stick-slip	Small	Large	Larger width, larger thickness, smaller length, smaller tip length

observed, a totally different detecting-sensing design was employed. The lateral stiffness of the wires they used was about $8 \times 10^4 \, N/m$, which is much stiffer than the commercially-available cantilevers whose lateral stiffness is typically 10–500 N/m. Table 7.6 summarizes the conditions for stick-slip occurrence.

7.4.4.4
Image Patterns Due to Different Scan Directions

In Fig. 7.35d, the results are for the normal load $W = 25$ nN, tip length $l = 12.5 \, \mu m$, and $\beta = 30°$. Compared with Fig. 7.35b, the effects of scan direction on topographic and lateral force maps can be seen. Different pattern of tip paths are also shown. The differences of Fig. 7.35b,d can be explained by the different distributions of atom locations. As shown in Fig. 7.36, the peaks of topography for $\beta = 0°$ appear at the place where the tip is scanned between the two closely-place atoms [57]. When $\beta = 30°$, two different areas can be distinguished. In area A, there is no carbon atom in the way of the scanned tip, resulting in a stripe-like dark area. Area B is the narrow stripe where the atoms are located. Due to the unsmooth motion of the tip, the topography in this area is "averaged" and thus a stripe-like bright area is shown.

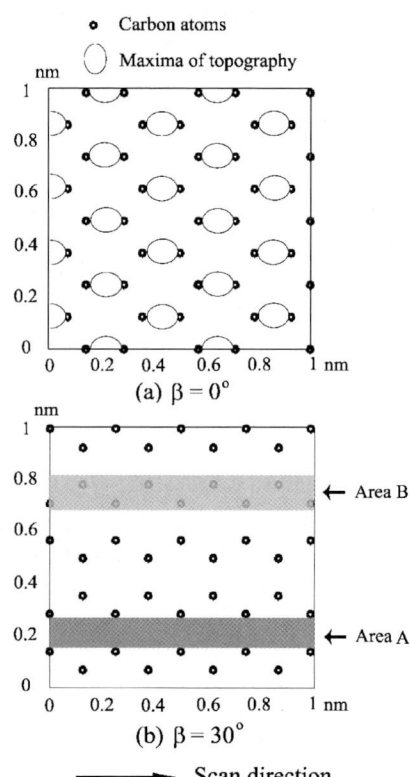

Fig. 7.36. (a) The relative positions of the carbon atoms in graphite surface and the maxima of topography in Fig. 7.34b, when $\beta = 0°$. (b) When $\beta = 30°$, two different areas can be distinguished. In area A, there is no carbon atom in the way of the scanned tip. Area B is the narrow stripe consisting of the carbon atoms

Table 7.7. Relations of scan direction and characteristics of topographic maps of graphite, when stick-slip tip motion occurs or does not occur

	Scan angle β		
	Close to 0° (or 60°)	Close to 30°	Between 0° and 30°, or between 30° and 60°
Stick-slip	Trigonal structure	Stripe-like structure	Between trigonal and stripe-like structure
No stick-slip	Hexagonal structure	Hexagonal structure	Hexagonal structure

Figure 7.35e shows the results for the normal load $W = 25$ nN, tip length $l = 12.5$ μm, and $\beta = 10°$. The simulated topographic map is very similar to some of the scanning tunneling microscopy (STM) experimental results [38]. Compared with Fig. 7.35b,d (trigonal topography and stripe-like topography, respectively), Fig. 7.35e can be viewed as something between them. The scan directions of $\beta = 0°$ and 30° are two extreme cases regarding the atom location distribution. Any scan direction in the range of (0°, 30°) and (30°, 60°) should result in lateral force and topography images that are something between the results for $\beta = 0°$ and 30°. The closer β is to 0° or 60°, the more the images are similar to those in Fig. 7.35b. While β is close to 30°, the resulting images should look more like those shown in Fig. 7.35d. Table 7.7 summarizes the relations of scan direction and characteristics of topographic maps of graphite, when stick-slip tip motion occurs or does not occur.

7.5
Quantitative Evaluation of the Sample's Mechanical Properties

The contact stiffnesses in the normal and lateral directions k_n and k_{lat} are given in (7.5) and (7.6). If one can measure k_n and k_{lat}, E^* and G^* can be obtained. The material parameters of the tip (E_t, G_t, and v_t) are known. The elastic and shear moduli, and Poisson's ratio of the sample can then be calculated utilizing the following relations

$$E^* = \left[\left(1 - v_t^2\right)/E_t + \left(1 - v_s^2\right)/E_s \right]^{-1} \tag{7.144}$$

$$G^* = \left[(2 - v_t)/G_t + (2 - v_s)/G_s \right]^{-1} \tag{7.145}$$

$$G_s = \frac{E_s}{2(1 + v_s)} \tag{7.146}$$

In AFAM, the characteristic equation to calculate the contact resonance frequency of vertical bending (ω_c) can be obtained following the method described in Sect. 7.2.3. With the normal contact viscosity being neglected ($\eta_n = 0$) and assuming that the tip is located at the end of the cantilever ($\lambda = 1$), the characteristic equation is

$$(\beta_w L)^3 [1 + \cos(\beta_w L)\cosh(\beta_w L)] + \frac{k_n}{EI_y/L^3}[\cosh(\beta_w L)\sin(\beta_w L)$$
$$- \cos(\beta_w L)\sinh(\beta_w L)] = 0 \tag{7.147}$$

$$\beta_w^4 = \frac{\rho A}{EI_y}\omega_c^2 \tag{7.148}$$

Once ω_c is measured, from (7.147), the normal contact stiffness can be obtained by [41, 66]

$$k_n = -\frac{EI_y}{L^3} \frac{(\beta_w L)^3 [1 + \cos(\beta_w L) \cosh(\beta_w L)]}{[\cosh(\beta_w L) \sin(\beta_w L) - \cos(\beta_w L) \sinh(\beta_w L)]} \tag{7.149}$$

In TR and LE mode, under a certain normal load f_c, the lateral contact stiffness can be calculated from the characteristic equation (7.51) [54]

$$k_{lat} = \frac{GJ\beta_\theta}{l^2} \{ [\tan[\beta_\theta(1-\lambda)L] - 1/\tan(\beta_\theta \lambda L)] \} \tag{7.150}$$

$$\beta_\theta^2 = \frac{\rho I_P}{GJ} \omega_c^2 \tag{7.151}$$

Here ω_c represents the torsional contact resonance frequency. Since (7.51) is obtained from the pure torsional analysis, (7.150) can be used only if the pure torsional approximation is valid.

The lateral viscosity η_{lat} is related to the energy dissipated due to the tip–sample interaction [56]. For TRmode

$$E_{dis} = \pi \eta_{lat} \Omega [l^2 |\Theta_c|^2 + |\phi_v(x=L)|^2 + 2l |\Theta_c| |\phi_v(x=L)| \cos(\varphi_1 - \varphi_2)] \tag{7.152}$$

and for LE mode

$$E_{dis} = \pi \eta_{lat} \Omega \big[g_0^2 + l^2 |\Theta_c|^2 + |\phi_v(x=L)|^2 + 2l |\Theta_c| |\phi_v(x=L)| \\ \cdot \cos(\varphi_1 - \varphi_2) \big] \tag{7.153}$$

Here, φ_1 and φ_2 are the phase angles of cantilever responses Θ_c and $\phi_v(x = L)$, respectively.

In TRmode, if k_{lat} is known, η_{lat} can be obtained from (7.61) using the measured phase shift [53]

$$\eta_{lat} = \frac{I_a k_{lat} l^2 [e^{2R_a L} s_1 - s_2] + (k_{lat} l^2 R_a - G J r_1) s_4 - (k_{lat} l^2 R_a + G J r_1) e^{2R_a L} s_3}{R_a l^2 \Omega [e^{2R_a L} s_1 - s_2] + I_a l^2 \Omega [e^{2R_a L} s_3 - s_4]} \tag{7.154}$$

Here, it is assumed $\lambda = 1$, $s_1 = \cos(\phi + I_a L)$, $s_2 = \cos(\phi - I_a L)$, $s_3 = \sin(\phi + I_a L)$, $s_4 = \sin(\phi - I_a L)$, R_a and I_a are the real and imaginary parts of the complex number a_{tr}. Again, (7.154) can only be applied if the pure torsional approximation is valid.

Unlike the pure torsional analysis, the coupled torsional-bending analysis of TR and LE mode could not lead to the explicit relations between the lateral contact stiffness/viscosity and cantilever torsional amplitude/phase shift, due to the mathematical complexity. In the case where pure torsional analysis cannot provide a good approximation for the cantilever torsional response, parameter identification methods, e.g., curve fitting, are needed for the extraction of the sample's mechanical properties.

Geometry and material properties of the cantilever and tip are required to calculate the contact stiffness, viscosity and sample's elastic parameters using (7.5), (7.6), (7.149), (7.150) and (7.154). In experiments, there are practical problems that make the reliable quantitative measurement difficult. AFM cantilevers may not behave perfectly like a clamped beam. It is difficult to obtain precisely the geometric dimensions and elastic constants of the cantilever and tip. Figure 7.37 show an example of contact resonances of a cantilever measured in AFAM experiments [3]. After two series of measurements, different contact spectra were detected. It is believed that the silicon tip radius was changed after successive experiments, as shown in Fig. 7.38 [3]. The increase of the tip radius is caused by tip wear. Consequently, the measured contact stiffness and contact resonance frequency increase.

In the cases where the geometry and material parameters of the cantilever are not known precisely, reference measurements on samples with known elastic constant may be used to derive the contact stiffness of interest without needing any cantilever information except the tip radius. Using (7.149), the normal contact stiffness of the sample of interest can be calculated by

$$k_n = k_n^{\text{ref}} \frac{\Phi(\beta_w L)}{\Phi(\beta_w^{\text{ref}} L)} \tag{7.155}$$

$$\Phi(x) = \frac{x^3[1 + \cos x \cosh x]}{[\cosh x \sin x - \cos x \sinh x]} \tag{7.156}$$

where k_n^{ref} can be obtained from (7.5) given the normal load and tip radius. Using (7.148), $\beta_w L$ can be determined from

$$(\beta_w L)^4 = \frac{\omega_c^2 (\beta_w^{\text{ref}} L)^4}{\omega_c^{\text{ref2}}} = \frac{\omega_c^2 (\beta_w^{\text{free}} L)^4}{\omega_{\text{free}}^2} \tag{7.157}$$

where ω_{free} is the measured resonance frequency of the cantilever in free vibration, and $\beta_w^{\text{free}} L$ is mode constant of the clamped-free beam, which is determined from the characteristic equation.

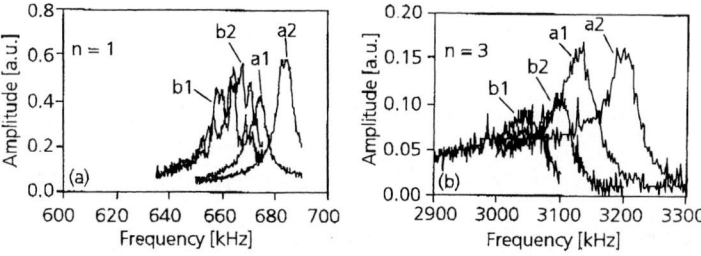

Fig. 7.37. Experimental contact resonance spectra of a cantilever with a silicon tip. The sample is Si (100) with RF-sputtered coating of thickness 5 nm. The first (**a**) and third (**b**) flexural resonances were detected. The contact resonance spectra b1 and b2 were measured under two static normal loads 410 nN and 820 nN, respectively. The contact resonance spectra a1 and a2 obtained after two series of AFAM measurements under two static normal loads 410 nN and 820 nN, respectively [3]

Fig. 7.38. Increase of the tip radius after successive
AFAM measurements on a sample of Si (100) with
RF-sputtered coating of thickness 5 nm [3]

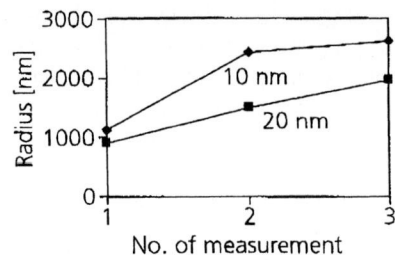

7.6
Closure

In dynamic AFM modes, resonance frequency, vibration amplitude and phase angle information of AFM cantilevers are exploited for measurement of surface topography and material mechanical properties. In this chapter, analytical and numerical models are described for dynamic simulation of cantilever responses in various dynamic modes. The relationships between cantilever responses and material mechanical properties established through these models are used to explain measured results and evaluate material properties.

Cantilever vertical bending is modeled by point-mass models and 1D beam models. In TM and NC-AFM, the vibration amplitude and frequency shift are dependent on the average work done by the interaction force per period. The phase angle is related to the energy dissipation due to the tip–sample interaction. Direct relation between the material properties and vibrational information cannot be established. Linearized tip–sample interaction in AFAM enables the obtainment of relations between the contact resonance frequencies corresponding to cantilever vertical bending and contact stiffnesses.

In TR and LE modes, cantilever deflection is a combination of torsion and lateral bending. Lateral tip–sample interaction can be represented by linear spring and dashpot. For TRmode, pure torsional analysis leads to the relations between torsional resonace frequency/amplitude/phase and the lateral contact stiffness/viscosity. Considering both the torsional and lateral bending, coupled analysis describes the cantilever responses in TR and LE modes. Parametric analyses demonstrate that the pure torsional analysis can provide a good approximation to cantilever responses only if the lateral tip–sample interaction is relatively weak compared to the torsional or lateral bending stiffness of the cantilever.

The 3D FE beam model is capable of accurately simulating the cantilever dynamics under complex tip–sample interaction for all dynamic AFM modes. The exact excitation mechanisms, tip geometry/location, tilting of the cantilever to the sample surface, and all the possible couplings among the different deflections of AFM cantilevers, can be addressed using this model. The effects of higher-order modes of vertical bending in TM and tip eccentricity in TR and LE modes on cantilever responses are studied. Using the FE model, friction profiling process in FFM is simulated. The topographic and friction force maps obtained in FFM experiments are the combined results of the real spatial distributions of 3D tip–sample interaction forces and cantilever dynamics. The experimentally-obtained hexagonal (full atomic structure) and trigonal (atomic resolution of every other atom) topographic images of graphite surfaces can be achieved in simulations with different combinations of cantilever geometry, applied normal load, and scan directions.

Mechanical properties of the sample are evaluated from the measured contact stiffness/viscosity between the tip and sample surface in normal and lateral directions. The normal contact stiffness is obtained in AFAM from the flexural contact resonance frequencies. If pure torsional analysis is valid, lateral stiffness and viscosity can be expressed explicitly as the functions of torsional contact resonance frequency and phase shift. In the cases where pure torsional analysis cannot provide a good approximation for the cantilever torsional response, parameter identification methods, e.g., curve fitting, are needed for the extraction of the sample's mechanical properties.

Acknowledgements. The authors would like to thank Dr. Ute Rabe and Dr. Walter Arnold of Fraunhofer Institute for Nondestructive Testing (IZFP), Saarbruecken, Germany, and Prof. Joseph Turner of University of Nebraska for critically reading our manuscript.

A
Appendices

A.1
Stiffness and Mass Matrices of 3D Beam Element

For a 3D beam element with a length of L_e, the element stiffness matrix is

$$k^e = \begin{bmatrix} k_{11} & k_{12} \\ k_{21} & k_{22} \end{bmatrix} \tag{A.1}$$

in which

$$k_{11} = \begin{bmatrix} \frac{EA}{L_e} & 0 & 0 & 0 & 0 & 0 \\ 0 & \frac{12EI_z}{L_e^3} & 0 & 0 & 0 & \frac{6EI_z}{L_e^2} \\ 0 & 0 & \frac{12EI_y}{L_e^3} & 0 & \frac{-6EI_y}{L_e^2} & 0 \\ 0 & 0 & 0 & \frac{GJ}{L_e} & 0 & 0 \\ 0 & 0 & \frac{-6EI_y}{L_e^2} & 0 & \frac{4EI_y}{L_e} & 0 \\ 0 & \frac{6EI_z}{L_e^2} & 0 & 0 & 0 & \frac{4EI_z}{L_e} \end{bmatrix} \tag{A.2}$$

$$k_{21} = k_{12}^T = \begin{bmatrix} \frac{-EA}{L_e} & 0 & 0 & 0 & 0 & 0 \\ 0 & \frac{-12EI_z}{L_e^3} & 0 & 0 & 0 & \frac{-6EI_z}{L_e^2} \\ 0 & 0 & \frac{-12EI_y}{L_e^3} & 0 & \frac{6EI_y}{L_e^2} & 0 \\ 0 & 0 & 0 & \frac{-GJ}{L_e} & 0 & 0 \\ 0 & 0 & \frac{-6EI_y}{L_e^2} & 0 & \frac{2EI_y}{L_e} & 0 \\ 0 & \frac{6EI_z}{L_e^2} & 0 & 0 & 0 & \frac{2EI_z}{L_e} \end{bmatrix} \tag{A.3}$$

$$
k_{22} = \begin{bmatrix}
\frac{EA}{L_e} & 0 & 0 & 0 & 0 & 0 \\
0 & \frac{12EI_z}{L_e^3} & 0 & 0 & 0 & \frac{-6EI_z}{L_e^2} \\
0 & 0 & \frac{12EI_y}{L_e^3} & 0 & \frac{6EI_y}{L_e^2} & 0 \\
0 & 0 & 0 & \frac{GJ}{L_e} & 0 & 0 \\
0 & 0 & \frac{6EI_y}{L_e^2} & 0 & \frac{4EI_y}{L_e} & 0 \\
0 & \frac{-6EI_z}{L_e^2} & 0 & 0 & 0 & \frac{4EI_z}{L_e}
\end{bmatrix}
\tag{A.4}
$$

The corresponding element mass matrix is

$$
m^e = \begin{bmatrix} m_{11} & m_{12} \\ m_{21} & m_{22} \end{bmatrix}
\tag{A.5}
$$

in which

$$
m_{11} = \rho A L_e \begin{bmatrix}
\frac{1}{3} & 0 & 0 & 0 & 0 & 0 \\
0 & \frac{156}{420} & 0 & 0 & 0 & \frac{22}{420}L_e \\
0 & 0 & \frac{156}{420} & 0 & \frac{-22}{420}L_e & 0 \\
0 & 0 & 0 & \frac{I_p}{3A} & 0 & 0 \\
0 & 0 & \frac{-22}{420}L_e & 0 & \frac{4}{420}L_e^2 & 0 \\
0 & \frac{22}{420}L_e & 0 & 0 & 0 & \frac{4}{420}L_e^2
\end{bmatrix}
\tag{A.6}
$$

$$
m_{21} = m_{12}^T = \rho A L_e \begin{bmatrix}
\frac{1}{6} & 0 & 0 & 0 & 0 & 0 \\
0 & \frac{54}{420} & 0 & 0 & 0 & \frac{13}{420}L_e \\
0 & 0 & \frac{54}{420} & 0 & \frac{-13}{420}L_e & 0 \\
0 & 0 & 0 & \frac{I_p}{6A} & 0 & 0 \\
0 & 0 & \frac{13}{420}L_e & 0 & \frac{-3}{420}L_e^2 & 0 \\
0 & \frac{-13}{420}L_e & 0 & 0 & 0 & \frac{-3}{420}L_e^2
\end{bmatrix}
\tag{A.7}
$$

$$
m_{22} = \rho A L_e \begin{bmatrix}
\frac{1}{3} & 0 & 0 & 0 & 0 & 0 \\
0 & \frac{156}{420} & 0 & 0 & 0 & \frac{-22}{420}L_e \\
0 & 0 & \frac{156}{420} & 0 & \frac{22}{420}L_e & 0 \\
0 & 0 & 0 & \frac{I_p}{3A} & 0 & 0 \\
0 & 0 & \frac{22}{420}L_e & 0 & \frac{4}{420}L_e^2 & 0 \\
0 & \frac{-22}{420}L_e & 0 & 0 & 0 & \frac{4}{420}L_e^2
\end{bmatrix}
\tag{A.8}
$$

A.2
Mass Matrix of the Tip

The nodal displacement vector at node C is

$$
d^C = \{d_x^C, d_y^C, d_z^C, \theta_x^C, \theta_y^C, \theta_z^C\}^T
\tag{A.9}
$$

The kinetic energy of the tip is expressed as

$$
T_t = \frac{1}{2} m_t \|V_G\|^2 + \frac{1}{2}\left(I_x^G \dot{\theta}_x^{C2} + I_y^G \dot{\theta}_y^{C2} + I_z^G \dot{\theta}_z^{C2}\right)
\tag{A.10}
$$

where m_t is the mass of the tip, I_x^G, I_y^G and I_z^G are the moments of inertia of the tip about the centroid G, around the x, y, and z-axes, $\dot{\theta}_x^C$, $\dot{\theta}_y^C$, and $\dot{\theta}_z^C$ are the angle velocities. In the coordinate system shown in Fig. 7.20, the velocity vector of the tip at the mass center G can be expressed as

$$V_G = \left(\dot{d}_x^C - h_G\dot{\theta}_y^C - e_t\dot{\theta}_z^C\right)\mathbf{i} + \left(\dot{d}_y^C + h_G\dot{\theta}_x^C\right)\mathbf{j} + \left(\dot{d}_z^C + e_t\dot{\theta}_x^C\right)\mathbf{k} \tag{A.11}$$

where $(\mathbf{i}, \mathbf{j}, \mathbf{k})$ is the unit vector of the coordinate system. Substitution of (A.11) into (A.10) leads to

$$T_t = \frac{1}{2}m_t\left[\left(\dot{d}_x^C - h_G\dot{\theta}_y^C - e_t\dot{\theta}_z^C\right)^2 + \left(\dot{d}_y^C + h_G\dot{\theta}_x^C\right)^2 + \left(\dot{d}_z^C + e_t\dot{\theta}_x^C\right)^2\right]$$
$$+ \frac{1}{2}\left(I_x^G\dot{\theta}_x^2 + I_y^G\dot{\theta}_y^2 + I_z^G\dot{\theta}_z^2\right) \tag{A.12}$$

The mass matrix of the tip can be obtained as

$$m_t = \frac{\partial^2 T_t}{\partial\dot{d}^C\partial(\dot{d}^C)^T}$$

$$= \begin{bmatrix} m_t & 0 & 0 & 0 & -m_th_G & -m_te_t \\ 0 & m_t & 0 & m_th_G & 0 & 0 \\ 0 & 0 & m_t & m_te_t & 0 & 0 \\ 0 & m_th_G & m_te_t & I_x^G + m_th_G^2 + m_te_t^2 & 0 & 0 \\ -m_th_G & 0 & 0 & 0 & I_y^G + m_th_G^2 & m_th_Ge_t \\ -m_te_t & 0 & 0 & 0 & m_th_Ge_t & I_z^G + m_te_t^2 \end{bmatrix} \tag{A.13}$$

A.3
Additional Stiffness and Mass Matrices Under Linear Tip–Sample Interaction

The stiffness and damping matrices due to the linear tip–sample interaction are

$$k_{ts} = \tag{A.14}$$

$$\begin{bmatrix} \tilde{c}^2k_t + \tilde{s}^2k_n & 0 & \tilde{c}\tilde{s}(k_t - k_n) & \tilde{c}\tilde{s}e_t(k_t - k_n) & -l(\tilde{c}^2k_t + \tilde{s}^2k_n) & -e_t(\tilde{c}^2k_t + \tilde{s}^2k_n) \\ & k_{lat} & 0 & lk_{lat} & 0 & 0 \\ & & \tilde{c}^2k_n + \tilde{s}^2k_t & e_t(\tilde{c}^2k_n + \tilde{s}^2k_t) & \tilde{c}\tilde{s}l(k_n - k_t) & \tilde{c}\tilde{s}e_t(k_n - k_t) \\ & & & \tilde{c}^2e_t^2k_n + l^2k_{lat} + \tilde{s}^2e_t^2k_t & \tilde{c}\tilde{s}le_t(k_n - k_t) & \tilde{c}\tilde{s}e_t^2(k_n - k_t) \\ & \text{symmetric} & & & l^2(\tilde{c}^2k_t + \tilde{s}^2k_n) & le_t(\tilde{c}^2k_t + \tilde{s}^2k_n) \\ & & & & & e_t^2(\tilde{c}^2k_t + \tilde{s}^2k_n) \end{bmatrix}$$

$$c_{ts} = \tag{A.15}$$

$$\begin{bmatrix} \tilde{c}^2c_t + \tilde{s}^2c_n & 0 & \tilde{c}\tilde{s}(c_t - c_n) & \tilde{c}\tilde{s}e_t(c_t - c_n) & -l(\tilde{c}^2c_t + \tilde{s}^2c_n) & -e_t(\tilde{c}^2c_t + \tilde{s}^2c_n) \\ & c_{lat} & 0 & lc_{lat} & 0 & 0 \\ & & \tilde{c}^2c_n + \tilde{s}^2c_t & e_t(\tilde{c}^2c_n + \tilde{s}^2c_t) & \tilde{c}\tilde{s}l(c_n - c_t) & \tilde{c}\tilde{s}e_t(c_n - c_t) \\ & & & \tilde{c}^2e_t^2c_n + l^2c_{lat} + \tilde{s}^2e_t^2c_t & \tilde{c}\tilde{s}le_t(c_n - c_t) & \tilde{c}\tilde{s}e_t^2(c_n - c_t) \\ & \text{symmetric} & & & l^2(\tilde{c}^2c_t + \tilde{s}^2c_n) & le_t(\tilde{c}^2c_t + \tilde{s}^2c_n) \\ & & & & & e_t^2(\tilde{c}^2c_t + \tilde{s}^2c_n) \end{bmatrix}$$

in which $\tilde{c} = \cos\alpha$, $\tilde{s} = \sin\alpha$. If $e_t = 0$, (A.14) and (A.15) are reduced to

$$
k_{\mathrm{ts}} =
\begin{bmatrix}
\widetilde{c}^2 k_{\mathrm t} + \widetilde{s}^2 k_{\mathrm n} & 0 & \widetilde{cs}(k_{\mathrm t}-k_{\mathrm n}) & 0 & -l(\widetilde{c}^2 k_{\mathrm t}+\widetilde{s}^2 k_{\mathrm n}) & 0 \\
 & k_{\mathrm{lat}} & 0 & lk_{\mathrm{lat}} & 0 & 0 \\
 & & \widetilde{c}^2 k_{\mathrm n}+\widetilde{s}^2 k_{\mathrm t} & 0 & \widetilde{csl}(k_{\mathrm n}-k_{\mathrm t}) & 0 \\
 & & & l^2 k_{\mathrm{lat}} & 0 & 0 \\
 & \text{symmetric} & & & l^2(\widetilde{c}^2 k_{\mathrm t}+\widetilde{s}^2 k_{\mathrm n}) & 0 \\
 & & & & & 0
\end{bmatrix}
\tag{A.16}
$$

$$
c_{\mathrm{ts}} =
\begin{bmatrix}
\widetilde{c}^2 c_{\mathrm t} + \widetilde{s}^2 c_{\mathrm n} & 0 & \widetilde{cs}(c_{\mathrm t}-c_{\mathrm n}) & 0 & -l(\widetilde{c}^2 c_{\mathrm t}+\widetilde{s}^2 c_{\mathrm n}) & 0 \\
 & c_{\mathrm{lat}} & 0 & lc_{\mathrm{lat}} & 0 & 0 \\
 & & \widetilde{c}^2 c_{\mathrm n}+\widetilde{s}^2 c_{\mathrm t} & 0 & \widetilde{csl}(c_{\mathrm n}-c_{\mathrm t}) & 0 \\
 & & & l_{\mathrm t}^2 c_{\mathrm{lat}} & 0 & 0 \\
 & \text{symmetric} & & & l^2(\widetilde{c}^2 c_{\mathrm t}+\widetilde{s}^2 c_{\mathrm n}) & 0 \\
 & & & & & 0
\end{bmatrix}
\tag{A.17}
$$

References

1. Albrecht TR, Grütter P, Horne D, Rugar D (1991) Frequency Modulation Detection Using High-Q Cantilevers for Enhanced Force Microscopy Sensitivity. J Appl Phys 69:668–673
2. Albrecht TR, Quate CF (1987) Atomic Resolution Imaging of a Nonconductor by Atomic Force Microscopy. J Appl Phys 62:2599–2602
3. Amelio S, Goldade AV, Rabe U, Scherer V, Bhushan B, Arnold W (2001) Measurements of Elastic Properties of Ultra-Thin Diamond-Like Carbon Coatings Using Atomic Force Acoustic Microscopy. Thin Solid Films 392:75–84
4. Arinero R, Lévêque G (2003) Vibration of the Cantilever in Force Modulation Microscopy Analysis by a Finite Element Model. Rev Sci Instrum 74:104–111
5. Bhushan B (2005) Introduction to Nanotribology and Nanomechanics, Springer, Berlin Heidelberg New York
6. Bhushan B (2006) Springer Handbook of Nanotechnology, 2nd edn, Springer, Berlin Heidelberg New York
7. Bhushan B, Kasai T (2004) A Surface Topography-Independent Friction Measurement Technique Using Torsional Resonance Mode in an AFM. Nanotechnology 15:923–935
8. Bhushan B, Qi J (2003) Phase Contrast Imaging of Nanocomposites and Molecularly Thick Lubricant Films in Magnetic Media. Nanotechnology 14:886–895
9. Binning G, Gerber Ch (1987) Stoll E, Albrecht TR, and Quate CF, Atomic Resolution with Atomic Force Microscope. Europhys Lett 3:1281–1286
10. Butt HJ, Jaschke M (1995) Calculation of Thermal Noise in Atomic Force Microscopy. Nanotechnology 6:1–7
11. Caron A, Rabe U, Reinstädtler M (2004) Imaging Using Lateral Bending Modes of Atomic Force Microscopy Cantilevers. Appl Phys Lett 85:6398–6400
12. Chen N, Bhushan B (2005) Morphological, Nanomechanical and Cellular Structural Characterization of Human Hair and Condition Distribution Using Torsional Resonance Mode with an Atomic Force Microscopy. J Microscopy 220:96–112
13. Chen J, Workman RK, Sarid D, Höper R (1994) Numerical simulations of a scanning force microscope with a large-amplitude vibrating cantilever. Nanotechnology 5:199–204
14. Cleveland JP, Anczykowski B, Schmid AE, Elings VB (1998) Energy Dissipation in Tapping-Mode Atomic Force Microscopy. Appl Phys Lett 72:2613–2615
15. Derjaguin BV, Muller VM, Toporov YP (1975) Effect of Contact Deformations on the Adhesion of Particles. J Colloid Interface Sci 53:314–326

16. Dupas E, Gremaud G, Kulik A, Loubet JL (2001) High-Frequency Mechanical Spectroscopy with an Atomic Force Microscope. Rev Sci Instrum 72:3891–3897
17. Fujisawa S, Yokoyama K, Sugawara Y, Morita S (1998) Analysis of Experimental Load Dependence of Two-Dimensional Atomic-Scale Friction. Phys Rev B 58:4909–4916
18. García R, Pérez R (2002) Dynamic Atomic Force Microscopy Methods. Surf Sci Rep 47:197–301
19. Giessibl FJ (1995) Atomic Resolution of the Silicon $(111) - 7 \times 7$ Surface by Atomic Force Microscopy. Science 267:68–71
20. Giessibl FJ (1997) Forces and Frequency Shifts in Atomic-Resolution Dynamic-Force Microscopy. Phys Rev B 56:16010–16015
21. Gorman DJ (1975) Free Vibration Analysis of Beams and Shafts, Wiley, New York
22. Hölscher H, Schwarz UD, Wiesendanger R (1996) Simulation of a Scanned Tip on a NaF (001) Surface in Friction Force Microscopy. Europhys Lett 36:19–24
23. Hölscher H, Schwarz UD, Wiesendanger R (1997) Modelling of the Scan Process in Lateral Force Microscopy. Surf Sci 375:395–402
24. Hölscher H, Schwarz UD, Zwörner O, Wiesendanger R (1998) Consequence of the Stick-Slip Movement for the Scanning Force Microscopy Imaging of Graphite. Phys Rev B 57:2477–2481
25. Huang L, Su C (2004) A Torsional Resonance Mode AFM for In-Plane Tip Surface Interfaces. Ultramicroscopy 100:277–285
26. Hurley DC, Shen K, Jenett NM, Turner JA (2003) Atomic Force Acoustic Microscopy Methods to Determine Thin-Film Elastic Properties. J Appl Phys 94:2347–2354
27. Johnson KL (1985) Contact Mechanics, Cambridge University Press, UK
28. Johnson KL, Kendall K, Roberts AD (1971) Surface Energy and the Contact of Elastic Solids. Proc R Soc London Ser A 324:301–313
29. Johnson KL, Woodhouse J (1998) Stick-Slip Motion in the Atomic Force Microscopy. Tribol Lett 5:155–160
30. Kasai T, Bhushan B, Huang L, Su C (2004) Topography and Phase Imaging Using the Torsional Resonance Mode. Nanotechnology 15:731–742
31. Kitamura S, Iwatsuki M (1995) Observation of Silicon Surfaces Using Ultrahigh-Vacuum Noncontact Atomic Force Microscopy. Jpn J Appl Phys 35:L668–L671
32. Lee SI, Howell SW, Raman A, Reifenberger R (2002) Non-Linear Dynamics of Microcantilevers in TappingMode Atomic Force Microscopy: A Comparison between Theory and Experiment. Phys Rev B 66:115409 1–10
33. Maivald P, Butt HJ, Gould SAC, Prater CB, Drake B, Gurley JA, Elings VB, Hansma PK (1991) Using Force Modulation to Image Surface Elasticities with the Atomic Force Microscopy. Nanotechnology 2:103–106
34. Marti O, Colchero J, Mlynek J (1990) Combined Scanning Force and Friction Microscopy of Mica. Nanotechnology 1:141–144
35. Marti O, Drake B, Hansma PK (1987) Atomic Force Microscopy of Liquid-Covered Surfaces: Atomic Resolution Images. Appl Phys Lett 51:484–486
36. Mate CM, McClelland GM, Erlandsson R, Chiang S (1987) Atomic-Scale Friction of a Tungsten Tip on a Graphite Surface. Phys Rev Lett 59:1942–1945
37. Meyer G, Amer N (1990) Simultaneous Measurement of Lateral and Normal Forces with an Optical-Beam-Deflection Atomic Force Microscopy. Appl Phys Lett 57:2089–2091
38. Mizes HA, Park SI, Harrison WA (1987) Multiple-Tip Interpretation of Anomalous Scanning-Tunneling-Microscopy Images of Layered Materials. Phys Rev B 36:R4491–4494
39. MultiMode SPM Instructor Manual, Version 4.31ce (1997) Digital Instruments, Santa Barbara, CA

40. Rabe U, Arnold W (1994) Acoustic Microscopy by Atomic Force Microscopy. Appl Phys Lett 64:1493–1495
41. Rabe U, Janser K, Arnold W (1996) Vibrations of Free and Surface-Coupled Atomic Force Microscope Cantilevers: Theory and Experiment. Rev Sci Instrum 67:3281–3293
42. Rabe U, Turner J, Arnold W (1998) Analysis of the High-Frequency Response of Atomic Force Microscope Cantilevers. Appl Phys A 66:S227–S282
43. Rabe U, Amelio S, Kester E, Scherer V, Hirsekorn S, Arnold W (2000) Quantitative Determination of Contact Stiffness Using Atomic Force Acoustic Microscopy. Ultrasonics 38:430–437
44. Rabe U, Amelio S, Kopycinska M, Hirsekorn S, Kempf M, Göken M, Arnold W (2002) Imaging and Measurement of Local Mechanical Material Properties by Atomic Force Acoustic Microscopy. Surf Interf Anal 33:65–70
45. Reinstädtler M, Kasai T, Rabe U, Bhushan B, Arnold W (2005a) Imaging and Measurement of Elasticity and Friction Using the TRmode. J Phys D Appl Phys 38:R269–R282
46. Reinstädtler M, Rabe U, Scherer V, Hartmann U, Goldade A, Bhushan B, Arnold W (2003) On the Nanoscale Measurement of Friction Using Atomic-Force Microscope Cantilever Torsional Resonances. Appl Phys Lett 82:2604–2606
47. Reinstädtler M, Rabe U, Goldade A, Bhushan B, Arnold W (2005b) Investigating Ultra-Thin Lubricant Layers Using Resonant Friction Force Microscopy. Tribol Int 38:533–541
48. Ruan J, Bhushan B (1994) Atomic-Scale and Microscale Friction Studies of Graphite and Diamond Using Friction Force Microscopy. J Appl Phys 76:5022–5035
49. San Paulo A, García R (2001) Tip-Surface forces, Amplitude, and Energy Dissipation in Amplitude-Modulation (TappingMode) Force Microscopy. Phys Rev B 64:193411 1–4
50. Sasaki N, Kobayashi K, Tsukada M (1996) Atomic-Scale Friction Image of Graphite in Atomic-Force Microscopy. Phys Rev B 54:2138–2149
51. Scherer V, Arnold W, Bhushan B (1999) Lateral Force Microscopy Using Acoustic Friction Microscopy. Surf Interf Anal 27:578–587
52. Scott WW, Bhushan B (2003) Use of Phase Imaging in Atomic Force Microscopy for Measurement of Viscoelastic Contrast in Polymer Nanocomposites and Molecularly-Thick Lubricant Films. Ultramicroscopy 97:151–169
53. Song Y, Bhushan B (2005) Quantitative Extraction of In-Plane Surface Properties Using Torsional Resonance Mode of Atomic Force Microscopy. J Appl Phys 97:083533 1–5
54. Song Y, Bhushan B (2006a) Dynamic Analysis of Torsional Resonance Mode of Atomic Force Microscopy and its Application to In-Plane Surface Property Extraction. Microsyst Technol 12:129–230
55. Song Y, Bhushan B (2006b) Simulation of Dynamic Modes of Atomic Force Microscopy Using a 3D Finite Element Model. Ultramicroscopy 106:847–873
56. Song Y, Bhushan B (2006c) Coupling of Lateral Bending and Torsion in Torsional Resonance and Lateral Excitation Modes of Atomic Force Microscopy. J Appl Phys 99:094911 1–12
57. Song Y, Bhushan B (2006d) Atomic-scale topographic and friction force imaging and cantilever dynamics in friction force microscopy. Phys Rev B, in press
58. Stark RW, Schitter G, Startk M, Guckenberger R, Stemmer A (2004) State-Space Model of Freely Vibrating and Surface-Coupled Cantilever Dynamics in Atomic Force Microscopy. Phys Rev B 69:085412 1–9
59. Tamayo J, García R (1998) Relationship between Phase Shift and Energy Dissipation in Tapping-Mode Scanning Force Microscopy. Appl Phys Lett 73:2926–2928
60. Turner JA (2004) Non-linear Vibrations of a Beam with Cantilever-Hertzian Contact Boundary Conditions. J Sound Vib 275:177–195
61. Turner JA, Wiehn JS (2001) Sensitivity of Flexural and Torsional Vibration Modes of Atomic Force Microscopy Cantilevers to Surface Stiffness Variations, Nanotechnology 12:322–330

62. Turner JA, Hirsekorn S, Rabe U, Arnold W (1997) High-Frequency Response of Atomic-Force Microscope Cantilevers. J Appl Phys 82:966–979
63. Wang L (1998) Analytical Descriptions of the Tapping-Mode Atomic Force Microscopy Response. Appl Phys Lett 73:3781–3783
64. Wang L (1999) The Role of Damping in Phase Imaging in TappingMode Atomic Force Microscopy. Surf Sci 429:178–185
65. Wright OB, Nishiguchi N (1997) Vibration Dynamics of Force Microscopy: Effect of Tip Dimensions. Appl Phys Lett 71:626–628
66. Yamanaka K, Nakano S (1996) Ultrasonic Atomic Force Microscope with Overtone Excitation of Cantilever. Jpn J Appl Phys 35:3787–3792
67. Yamanaka K, Nakano S (1998) Quantitative Elasticity Evaluation by Contact Resonance in an Atomic Force Microscopy. Appl Phys A 66:S313–S317
68. Yamanaka K, Tomita E (1995) Lateral Force Modulation Atomic Force Microscopy for Selective Imaging of Friction Force. Jpn J Appl Phys 34:2879–2882
69. Yamanaka K, Ogiso H, Kolosov O (1994) Ultrasonic Force Microscopy for Nanometer Resolution Substrate Imaging. Appl Phys Letts 64:178–180

8 Combined Scanning Probe Techniques for In-Situ Electrochemical Imaging at a Nanoscale

Justyna Wiedemair · Boris Mizaikoff · Christine Kranz

Abbreviations and Symbols

a	Inner radius of a ring-shaped electrode
AC	Alternating current
AFM	Atomic force microscopy/microscope
AgQRE	Ag quasireference electrode
APD(s)	Avalanche photodiode(s)
ATP	Adenosin 5'-triphosphate
b	Outer radius of a ring-shaped electrode
BEM	Boundary element method
c^*	Concentration of a redox mediator in bulk solution
CD-SP	Constant-distance mode scanning potentiometry
CF	Cystic fibrosis
CNT(s)	Carbon nanotube(s)
CVD	Chemical vapor deposition
CVP	Chemical vapor polymerization
D	Diffusion coefficient of a redox mediator
d	Distance between the UME and the sample surface
DC	Direct current
DNA	Deoxyribonucleic acid
DRIE	Deep-reactive ion etching
EBL	Electron beam lithography
ECAFM	Electrochemical atomic force microscopy/microscope
ECSTM	Electrochemical scanning tunneling microscopy/microscope
EIS	Electrochemical impedance spectroscopy
F	Faraday constant
FDM	Finite difference method
FEM	Finite element method
FIB	Focused ion beam
FMA	Hydroxyl methyl ferrocene
FMA^+	Ferrocenium methylhydroxide
FMN	Flavin mononucleotide
FRET	Fluorescence (foerster) resonance energy transfer
GOD	Glucose oxidase
h	Height of a cone-shaped electrode
HARS	High aspect ratio silicon

HEX	Hexokinase
HOPG	Highly-ordered pyrolytic graphite
HRP	Horseradish peroxidase
HUVEC	Human umbilical vein endothelial cells
IDA	Interdigitated array
λ	Wavelength
I_{diff}	Diffusion-limited steady state current
LIGA	Litographie, Galvanoformung, Abformung (Lithography, Electroforming, Molding)
MWNT(s)	Multiwalled nanotube(s)
n	Number of transferred electrons per converted molecule
NAD	Nicotinamide adenine dinucleotide
NAD^+	Nicotinamide adenine dinucleotide, oxidized form
NADH	Nicotinamide adenine dinucleotide, reduced form
NO	Nitric oxide
NSOM	Near-field scanning optical microscopy/microscope
OM	Optical microscopy/microscope
PECVD	Plasma-enhanced chemical vapor deposition
PEG	Poly(ethylene glycol)
PEM	Photoelectrochemical microscopy/microscope
PET	Polyethylene terephthalate
PPS	Pitting precursor
RIE	Reactive ion etching
r_0	Electroactive radius of a UME
SAM(s)	Self-assembled monolayer(s)
SCE	Saturated calomel electrode
SCM	Scanning confocal microscopy/microscope
SCLM	Scanning chemiluminescence microscopy/microscope
SECM	Scanning electrochemical microscopy/microscope
SEM	Scanning electron microscopy/microscope
SICM	Scanning ion conductance microscopy/microscope
SKP	Scanning Kelvin probe
SKPFM	Scanning Kelvin probe force microscopy/microscope
SMFS	Single-molecule fluorescence spectroscopy
SPM	Scanning probe microscopy/microscope
SSCM	Scanning surface confocal microscopy/microscope
SSV	Steady state voltammetry
STM	Scanning tunneling microscopy/microscope
SWNT(s)	Single-walled nanotube(s)
TIRF	Total internal reflection fluorescence
TIRFM	Total internal reflection fluorescence microscopy/microscope
TLSTM	Thin layer scanning tunneling microscopy/microscope
UHV	Ultrahigh vacuum
UME	Ultramicroelectrode
VLP(s)	Virus-like particle(s)

8.1
Overview

Scanning probe microscopy (SPM) techniques are gaining importance in spatially and temporally resolved investigations of electrochemical processes, which are of substantial interest in fundamental research, industrial applications, and, increasingly, the biological/biomedical field. Redox processes and the kinetics of the involved active species addressed with electrochemical techniques are of fundamental importance in areas ranging from biochemical signaling processes to material sciences, including, e.g., fuel cell technology, catalysis, electroanalytical sensors, and environmental processes. The challenge of gaining insight and detailed understanding on in-situ surface and interface processes demands for surface sensitive measurement techniques providing information at an atomic or molecular level, which usually are difficult to address by ex-situ ultrahigh vacuum (UHV)-based surface techniques. This gap is bridged by scanning probe microscopies providing in-situ information on structural changes at a nanoscopic scale in real time.

Shortly after the introduction of scanning tunneling microscopy (STM) in 1982 by Binnig and Rohrer [1], the first in-situ STM investigations in electrolyte solutions were performed [2–4]. The transition of STM from a vacuum tool to an in-situ technique in liquid environments was a key step for SPM techniques transforming into standard surface science tools for studying electrochemical processes. Electrochemical applications of STM are a logical consequence of the (semi)conductive surface required for maintaining a tunneling current between tip and sample, which is generated within the angstrom-sized gap between a sharp tip and the sample if a potential is applied. Independent control of the electrochemical substrate potential and the tip [5, 6] with respect to a reference electrode enables in-situ investigations on electrochemical reactions at potentiostatic conditions, which is frequently referred to as electrochemical STM (ECSTM). Consequently, ECSTM has developed into a routine tool for probing the structure and dynamics of electrode surfaces. For more information, the reader is referred to detailed reviews pertaining this SPM technique [3, 7–11]. During the last decades, STM and its derivatives atomic force microscopy (AFM) [12], near-field scanning optical microscopy (NSOM) [13], scanning ion conductance microscopy (SICM) [14], and scanning electrochemical microscopy (SECM) [4, 15, 16] have successfully been applied for probing electrochemical processes at the solid–liquid and liquid–liquid interface. AFM and ECAFM are considered routine tools for studying the interfacial region between charged surfaces, such as electrodes, or biological membranes and electrolyte solution [17–22]. In contrast to ECSTM, which requires insulation of the STM probe such that only a small area at the tip is exposed, ECAFM is technically less challenging and routinely applied to a wide variety of samples immersed in solution. Furthermore, the latter technique enables force spectroscopy probing the electrostatic double layer, which provides high-resolution information on surface charges [23–25].

While ECSTM and ECAFM are based on imaging structural and electronic changes at a (biased) macroscopic sample surface, SECM introduced the opportunity of combining microelectrochemistry and electroanalytical techniques with imaging, by scanning a microelectrode probe across the sample surface. Hence, since its introduction by Bard and Engstrom in 1986 [4, 15, 16], SECM has gained increasing importance among the scanning probe techniques for the investigation of in-situ electrochemical processes at the solid–liquid and liquid–liquid interface. Measurements providing information on local (electro)chemical reactivity are based on the disturbance of steady state Faraday currents at a scanned microelectrode in dependence on the surface properties. For a more extensive review the reader is referred to reference [26].

While nowadays SPM techniques are considered routine tools, a common drawback is the limited availability of surface information with high chemical specificity. Furthermore, the benefit of high lateral resolution is frequently accompanied by the fact that the obtained information is limited to a small fraction of the entire sample surface. Hence, the correlation between laterally-resolved information obtained by SPM, and macroscopic properties derived from bulk electrochemical techniques might not be representative for the overall surface reactivity. Shortly after the introduction of SPM, first attempts to increase the information space have been published demonstrating the combination of individual scanning probe techniques, such as AFM-STM [27, 28] and STM-NSOM [29–32] as a viable tool for surface chemistry. Noteworthy are also the combinations of AFM with optical imaging techniques such as NSOM [33] and scanning confocal laser microscopy [34, 35], which are particularly attractive for in-situ studies at biological specimen. Among the most prominent representatives of combined SPM techniques is NSOM-AFM, which is available as a commercial instrument [36].

This chapter focuses on combined scanning probe techniques with particular relevance to electrochemical imaging at the nanoscale. Although spatially induced changes of Faradic currents are emphasized, potentiometric and ion conductance approaches will also be discussed. An overview on state-of-the-art applications of combined SPM techniques will focus on life sciences, with particular attention to recent developments in combining AFM and SECM for high-resolution electrochemical imaging at a nanoscale.

8.2
Combined Techniques

Despite the superior lateral resolution provided by many SPM techniques, limited chemical specificity remains among the challenges in promoting SPM to an even broader user community. Recently, chemical specificity was added to the topographical information by force recognition imaging, which involves modification of AFM tips with smart linker chemistries facilitating selective binding to recognition molecules. While this powerful technique relies on molecular tailoring of the chemical functionality [37], alternative approaches aim at increasing the information space for SPM techniques by combining electroanalytical imaging techniques with complementary scanning probe microscopies. Electrochemical identification of (electro)active species is of particular significance in life sciences, as many fundamental molecular biological processes are based on redox chemistry. Figure 8.1 provides a general scheme of the predominant combined SPM-SECM probe concepts.

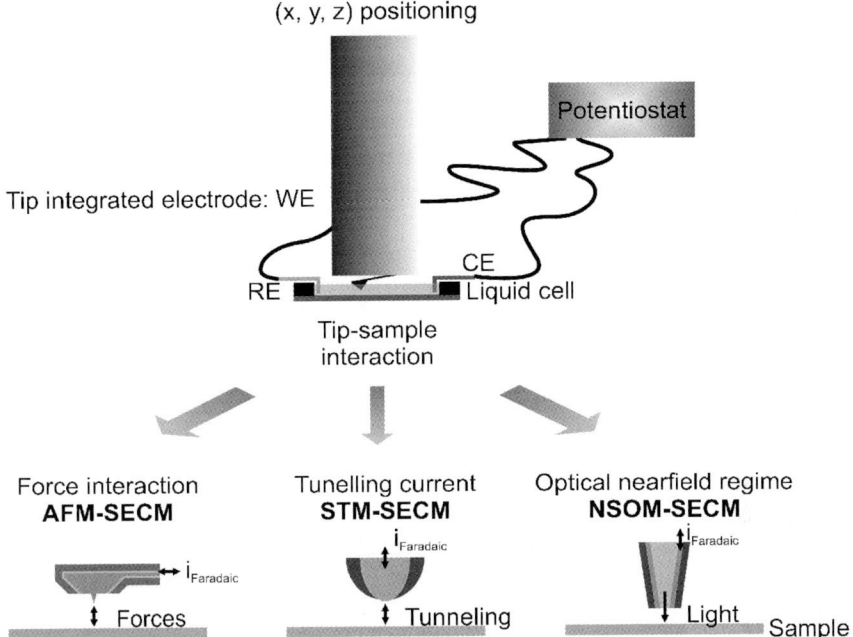

Fig. 8.1. Schematic setup of combined SECM-based SPM techniques. Besides the electrochemical signal, different types of physical principles govern the tip–sample interaction and are responsible for signal generation. The tube-shaped (x,y,z) piezopositioner in the *upper part* of the figure refers to a typical top-down AFM setup with flexible space for an electrochemical cell (WE – working electrode, RE – reference electrode, CE – counter electrode)

Combined techniques based on AFM platforms are the most prevalent approaches given their broad applicability, and will be discussed in detail in this chapter. AFM is among the most versatile and widely applied SPM techniques, almost independent of the nature of the investigated sample. Bard and co-workers published some of the first combined electrochemical SPM experiments describing a dissolution process using a biased Pt-coated cantilever inducing the dissolution of ferrocyanide crystals. Topological changes during the dissolution process were simultaneously monitored by the conductive AFM tip [38].

Far fewer applications of combined electrochemical techniques using STM platforms have been reported [39–42]. The detection of localized ion fluxes combined with SPM measurements was introduced by Hansma and co-workers [14, 43], and was further expanded by Korchev's group towards a routine tool for investigations at live cell surfaces [44–47]. Potentiometric probes, which are traditionally less applied in SECM experiments due to the difficulty in current-less positioning, are receiving increased attention based on combined approaches due to alternative positioning mechanisms, as laterally-resolved pH measurements or the determination of biologically relevant ions such as Ca^{2+} are of substantial importance for many biological systems.

8.2.1
Integration of Electrochemical Functionality

In general, the motivation for integrating electrochemical functionality into SPM techniques derives from two driving aspects: (1) expanding the information content from an SPM experiment, and (2) improving the performance of existing SECM technology. Although the first instrumental combinations were pioneered by scientists with SECM background [48, 49], more recently, leading experts in biological applications of AFM adopted the modification of conventional AFM probes by integrating a defined electroactive area at the apex of the tip [50].

Historically, the main thrust for integrating SECM functionality into SPM tips was the limited spatial resolution of conventional SECM experiments compared to AFM and STM. In a basic SECM experiment, the probe (usually an ultramicroelectrode (UME) with an electroactive diameter in the range of 1 to 25 µm) is positioned in close proximity to the sample surface as a function of the recorded Faraday current, and laterally scanned in constant height across the sample surface [51]. SECM measurements can be performed in feedback mode [52] with an artificial redox mediator added to the solution, or in generation-collection mode [53] with the probe sensing (electro)chemically active species generated at or released from the sample surface. Similar to related SPM techniques, the achievable lateral resolution is dependent on the dimensions of the probe. Hence, there is evident argumentation for using nanometer-sized electrodes in SECM experiments providing improved lateral resolution. Furthermore, scanning the electrode at constant height across the sample surface results in a convolution of the electrochemical response and the topographical information. Based on the required distance between probe and sample surface (roughly the diameter of the nanometer-sized probe), precise positioning at the nanoscale independent of the Faraday current is a pre-requisite enabling the probe following the topology of the sample surface (constant distance SECM mode). Early approaches using Faraday current independent positioning of the tip were adapting shear-force-dependent positioning of electrodes derived from NSOM using either optical detection [54], tuning fork detection [55], or piezo-based actuator-detector systems [56]. However, routine imaging using nanoelectrodes in shear-force mode SECM has yet to be achieved.

Signal generation in laterally-resolved electrochemical imaging is based on changes in the flux of the converted molecular species and, hence, its local concentration. For locally mapping changes of flux or concentration pertinent to the electroactive species, the electrochemical probe is positioned at a defined distance close to the sample surface. Dependent on the design of the combined electrochemical SPM technique, the sample may need to be scanned twice: if the tip–sample interaction is based on direct contact between probe and surface, and the electroactive area extends to the apex of the probe, in a first scan the topography is recorded. In a second scan after retracting the combined SPM probe the electrochemical information is mapped positioning the tip in close proximity and following the height profile retrieved in the first scan. Hence, a constant distance between the probe and the sample surface is maintained. However, especially for biological samples or during monitoring of fast processes at the solid–liquid interface, it is desirable to simultaneously retrieve the topographical and electrochemical information correlated in space and time during a single scan. Hence, alternative combined probe designs with an electroactive area recessed from the apex of the tip have been implemented [49], or probe positioning avoiding direct contact between electroactive area and sample surface has been devised. In either case, of a quantitative description for the electrochemical response precise knowledge on the distance between the electroactive probe and the sample surface is indispensable.

8.2.2
Combined Techniques Based on Force Interaction

8.2.2.1
SECM

AFM is among the most widely used SPM techniques almost independent of the nature of the investigated sample providing an advantageous platform for the combination with SECM (AFM-SECM). Additionally, well-established microfabrication of AFM cantilevers [57] enables batch processing of silicon and silicon nitride cantilevers at comparatively low costs and with high reproducibility. High spatial resolution and versatility renders AFM in combination with SECM of particular interest for electrochemical imaging at the nanoscale. The main benefit of merging AFM with SECM is the direct correlation of structural information with chemical surface activity at excellent lateral resolution [58]. In principle, the combination AFM-SECM requires the development of combined scanning probes integrating AFM and SECM functionality into a single SPM tip. For combined electrochemical and topographical measurements, it is advisable to locate the AFM instrument in a Faraday cage. Usually, a three-electrode setup comprising a reference electrode, an auxiliary electrode (platinum wire), and the combined AFM-SECM tip as working electrode is applied in an AFM liquid cell. However, the sensitivity of the (bi)potentiostats implemented in most commercial AFM instruments equipped for electrochemical experiments is insufficient for recording currents at the sub-nA range. Hence, to date most published results on combined AFM-SECM experiments were obtained with external (bi)potentiostats feeding the signal output into the data acquisition board of the AFM for visualizing correlated electrochemical and topographical data. Recently, the first commercial AFM-SECM instrument with an integrated SECM software module was introduced to the market [59]. AFM-SECM measurements have been demonstrated in contact mode or dynamic mode AFM operation, and feedback mode or generator-collector SECM mode. However, dynamic mode AFM imaging of soft samples in liquid requires cantilever force constants on the order of one hundreth of 1 N/m, which cannot be achieved with some of the proposed combined scanning probe designs. Within the last few years, combined AFM-SECM has been developed into an acknowledged novel analytical surface science tool providing sufficient resolution for both techniques [58].

In principle, shear-force mode SECM can be considered a variation of non-contact mode AFM. Hydrodynamic forces dampen the vibration amplitude of a tapered micro- or nanoelectrode, if the probe is in close proximity to the sample surface. A feedback loop relays this information into a tip positioning software routine of the SECM maintaining a constant distance between the tip and the sample surface. The first experimental realization using optical shear force detection has been published by Ludwig et al. [54]. Nowadays, tuning fork-type resonators first applied in NSOM [60], and piezo-based actuator-detector systems [56] are preferred due to less demanding instrumental efforts [55,56,61–65]. Shear-force positioning is well-received in SECM for routine current-independent positioning of micrometer-sized electrodes. While conventional nanoelectrodes suitable for SECM experiments are described in the literature, routine imaging in shear-force mode SECM remains to be demonstrated. Considering that the entire diameter of a well-insulated nano-electrode usually exceeds the curvature of conventional AFM tips and accounting for non-contact operation, the achievable lateral resolution for the sample topography cannot compete with AFM. Furthermore, even though the tip can be accurately positioned with shear-force mode control, the precise absolute distance between tip and sample is unknown, which adds complexity and uncertainty to a quantitative description of the electrochemical response.

Nonetheless, shear-force mode SECM operation has eliminated many of the initial limitations of conventional SECM based on current-dependent positioning of the electrochemical probe, enabling positioning of and imaging with miniaturized biosensors and potentiometric probes (e.g., ion selective electrodes) [66–68], along with simultaneous topographical and electrochemical imaging.

8.2.2.2
SICM

The fundamental principle in SICM is based on scanning an electrolyte-filled micropipette housing an electrode with apertures ranging from 50 nm to 1.5 μm across the surface of a sample immersed into an electrolytic solution. Distance control is based on a feedback loop controlling the ion current developing across the pipette aperture, if a constant voltage between the electrode within the pipette and an external reference electrode is applied. Especially, for biological applications, such as imaging living cells, ion channels, and subcellular structures, where many processes are controlled or initiated by fluxes of ions, this non-invasive in-situ technique has significant potential [14,47,69–74]. A comprehensive review on SICM was recently published within this book series [75]. Proksch et al. [43] demonstrated a combined AFM-SICM system based on force-regulated positioning of the pipette. In principle, a bent micropipette is applied as a dual-function probe, which allows positioning similar to a conventional AFM cantilever evaluating the deflection signal while simultaneously serving as SICM probe. However, the advantage of non-invasive imaging is usually lost in this combination. Recently, alternatives to these distance control modes using AC recording of the current flow [76,77], or shear-force-based probe positioning [78] have been introduced.

8.2.3
Combined Techniques Based on Tunneling Current

Based on the physical near-field interaction principle of controlling the tunneling current, ECSTM-SECM is usually performed in a sequential mode. After recording the topography in solution, the tip is usually retracted to a designated distance correlated with the appropriate working distance of the electroactive area of the combined probe, enabling SECM imaging. The first combination of ECSTM-SECM was demonstrated by Williams et al. [79] investigating pitting corrosion at stainless steel surfaces. However, this approach was based on comparatively large electrodes limiting the achievable lateral resolution to the micrometer range. Wittstock and co-workers used the same principle with a modified tip design providing an electroactive area of 4 nm [40]. Similar to combined AFM-SECM measurements, commercial AFM-STM systems combined with a bipotentiostat controlling the four-electrode setup, and a pre-amplifier for ECSTM and SECM operation are adapted for combined ECSTM-SECM experiments. Due to the superior lateral resolution of STM, this combination has potential for simultaneous measurements at a molecular level. However, applicability to live biological specimen or rapidly changing samples is still limited due to the sequential nature of the image acquisition by first STM and then SECM operation.

8.2.4
Combined Techniques Based on Optical Near-Field Interaction

Near-field scanning optical microscopy (NSOM) provides a versatile tool for optical measurements in the near-field regime to the sample surface at resolutions below the diffraction limit. Smyrl and co-workers [80] published one of the first approaches towards combined NSOM-SECM. Again, the combination is realized by modifying the optical SPM probe with a conducting electrode layer insulated against the solution. In NSOM, radiation from a laser source is guided through an aperture smaller than dictated by Abbe's limit (e.g., conical optical fiber tip) to the sample surface, while the tip is scanned across the sample surface, and collected by an optical microscope objective or by a focusing mirror onto a photomultiplier or similar optical detection device. To date, most combinations are based on home-built SECM systems, which

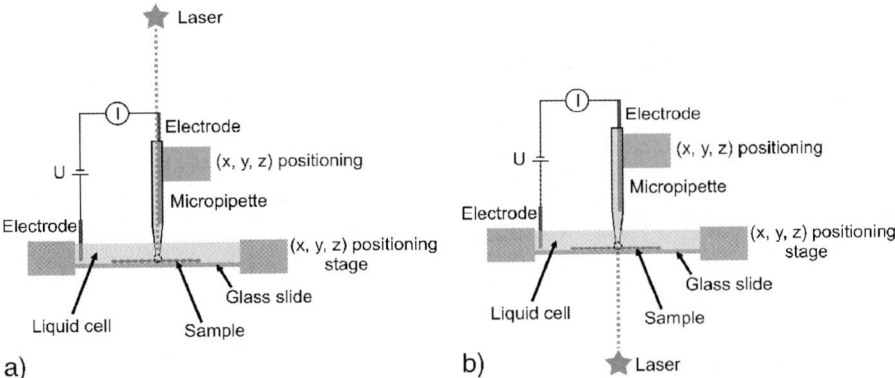

Fig. 8.2. Combined optical-SICM techniques. Schematic of combined NSOM-SICM (**a**) and fluorescence/optical-SICM (**b**). The sample is mounted on a (x,y,z) positioning stage and scanned during imaging, whereas the micropipette and optical path remain static during scanning. In case of NSOM-SECM, the micropipette is replaced by a combined optical/electrochemical tapered fiber

are extended by a light source, mirrors, and an optical detection device facilitating optical measurements, or a commercially available NSOM system is combined with an electrochemical cell and a (bi)potentiostat, which controls the electrochemical reaction producing an output signal that is fed into the data acquisition board of the NSOM instrument.

The combination of SICM and NSOM has substantial potential for the investigation of interfacial processes (NSOM-SICM). In this approach, the NSOM probe is modified to house the electrode immersed into electrolyte solution, which enables simultaneous recording of optical images during mapping of selected ion concentrations. By coupling laser radiation into the aluminum-coated NSOM-SICM pipette and controlling the distance by maintaining constant ion current, simultaneous optical and topographical imaging can be achieved [45]. Figure 8.2 shows a typical setup used for optical microscopy-SECM or optical microscopy-SICM, respectively.

8.2.5
Theory

The possibility of relating experimental SECM data to a quantitative theoretical description is a major asset of SECM in contrast to other SPM techniques. Diffusion of a redox mediator towards the electroactive area is affected by the sample morphology, the interfacial reactions at the sample surface, the distance d between the UME and the sample surface, and by the geometry of the UME and its insulating shielding, when the probe is in close proximity to the sample surface. These parameters can be used as input data for mathematical models and numerical tools for a quantitative treatment of the mass transport characteristics in SECM, which have theoretically been described for a variety of experimental conditions and UME geometries [81,82]. Simulations utilizing modifications of the finite difference method (FDM), the finite element method (FEM), and – for more complex three-dimensional simulations – the boundary element method (BEM) [83–97] have been developed. Based on BEM [92, 98], Sklyar et al. [93] provided the first comprehensive theoretical description of integrated AFM-SECM micro- and submicroframe electrodes recessed from the apex of the tip, and quantitatively modeled their behavior during combined AFM-SECM measurements in feedback mode [99]. Kottke and Fedorov developed numerical simulations for integrated frame electrodes describing advective and transient effects based on BEM [100]. Based on previous dissolution experiments using conductive AFM probes [38, 101, 102], Unwin and Macpherson developed a model for mass transport at non-insulated metal-coated AFM probes using FEM [103].

8.2.6
Combined Probe Fabrication

Independent of the physical near-field interaction, following pre-requisites have to be fulfilled for realizing useful multifunctional high-resolution probes: (1) the apex of the tip has to be in the nanometer regime ensuring high-resolution topographical imaging (for some SPM techniques, e.g., STM, ideally atomically sharp), (2) a well-defined geometry of the electroactive area enabling quantitative treatment of the electrochemical response, and (3) perfect insulation of the entire probe avoiding leakage currents leading to erroneous electrochemical information. The first probes used for combined electrochemical scanning probe measurements were ECSTM tips. In principle, the conductive probe has to be insulated avoiding interfering Faraday currents at the tip, which is achieved by exposing an electroactive area only at the very end of the sharp tip.

Combined scanning probes compromising an amperometric electrode are characterized similar to conventional SECM UMEs by scanning electron microscopy (SEM), steady state voltammetry (SSV), or by recording approach curves. The diffusion-limited plateau current retracted from cyclic voltammograms depends on the shape and size of the electroactive area, and can be expressed for different geometries by the following equations [104, 105]:

$$I_{diff} = 2\pi n F D c^* r_o \qquad \text{Hemisphere} \tag{8.1}$$

$$I_{diff} = 4\pi n F D c^* r_o \qquad \text{Disk} \tag{8.2}$$

$$I_{diff} = n F D c^* l_o \qquad \text{Ring} \tag{8.3}$$

with

$$l_o = \frac{\pi^2 (a+b)}{\ln[16(a+b)/(b+a)]} \qquad a/b < 1.25 \tag{8.4}$$

$$I_{diff} = 4\pi n F D c^* r_o (1 + q H^p) \qquad \text{Finite cone} \tag{8.5}$$

with

$$q = 0.3661, \ p = 1.14466 \text{ and } H = h/r \ ,$$

where n is number of electrons transferred, F is the Faraday constant, D is the diffusion coefficient of the redox mediator, r_o is the radius of the UME and c^* is the concentration of the redox mediator in bulk solution.

Approach curves performed by approaching the tip in the z-direction towards an insulating or conducting surface while recording the Faraday current are frequently subject to numerical fitting procedures providing useful information about the tip geometry and insulation quality.

8.2.6.1
Manual Probe Preparation

In principle, combined ECSTM-SECM probes are fabricated similar to conventional ECSTM tips. In recent years, conductive STM probes were usually fabricated by controlled electrochemical etching of metal wires or alloy materials such as iridium, tungsten, etc. Etching is performed with etching solutions of different compositions producing high aspect ratio tips with curvatures < 10 nm. The metallic wire is immersed in an electrolyte solution and a potential is applied between the wire and a counter electrode initiating the electrochemical etching process [106–110]. As the etching process proceeds, the lower part of the wire separates, and a sharp tip is formed at the breaking junction. Insulation materials for ECSTM probes include SiO$_2$, glass [4], and polymers [40, 41, 108, 111–114]. Most recently, electrophoretic paints, which can be electrochemically deposited and cross-linked by temperature treatment, have emerged as preferred insulation layer [40, 41, 111–114]. A detailed review on aspects of STM tip fabrication is provided by Melmed [115].

Figure 8.3 shows a schematic of ECSTM and ECSTM-SECM probes along with SEM images of probes used for combined measurements. Treutler et al. fabricated combined ECSTM-SECM with a slightly modified geometry as shown in Fig. 8.3a right panel. For ECSTM a geometry as shown in Fig. 8.3a left and middle panel with a portion of several micrometers of the sharp tip non-insulated is sufficient, since a potential window can be selected eliminating interfering Faraday currents, however, ECSTM-SECM experiments in feedback mode require a much smaller defined electroactive area. In feedback mode SECM experiments, large background currents, which would superimpose the signal in positive feedback mode, have to be avoided. Furthermore, at conventional ECSTM tips diffusion blocking by passivated samples would not occur. Hence, Wittstock and co-workers sacrificed the option of atomic resolution topographical imaging by using less pointed tips, where the insulation layer was less recessed compared to conventional ECSTM tips. Consequently, the resulting ECSTM-SECM probe provides more similarity to a non-ideal nanoscopic disk electrode than a pointed ECSTM probe facilitating SECM experiments.

Manually fabricated AFM-SECM probes are derived from etched metal wires (usually gold or platinum), which are bent after the etching process and shaped like AFM cantilevers. Again, appropriate insulation is the crucial aspect for fabricating useful electrochemical probes. Following the insulation procedures with electrophoretic paint introduced for ECSTM tips [111], similar approaches are most commonly applied to cantilever-shaped nanoelectrodes. The first commercially available bent microelectrodes sealed into glass were introduced by Nanonics Imaging Ltd (Israel) in the 1990s. Macpherson et al. [116] adapted the etching process derived from fabricating nanoelectrodes, and insulated the cantilever-shaped probes with anodic electrophoretic paint exposing a conically-shaped nanoelectrode due to shrinking of the coating during the curing process at 200 °C (Fig. 8.3b, left). This procedure results in effective tip radii ranging from 50 nm to 2.5 μm within a batch of 50 fabricated tips [48]. Abbou et al. [117] published a similar approach for spherical and conical gold nanoelectrodes by melting the metal at the end of an etched wire within an electrical arc between the gold electrode and a tungsten wire (Fig. 8.3b, right). In contrast to the approach by Macpherson et al. [48, 118], they developed a protocol for insulation with cathodic electrophoretic paint, which coats the entire probe and is less prone to shrinking during the curing process. Exposure of the electroactive area was achieved by applying a voltage pulse with an amplitude of a few kilovolts to the coated microwire. The resulting electric field localized at the tip exposed spherical and conical electrodes with radii ranging from 150 to 550 nm. Both approaches enable high-resolution electrochemical imaging based on the exposed nanometer scale electroactive area, however, since the electrode is at the apex of the combined probe, limitations during simultaneous topographical and electrochemical imaging may occur dependent on the nature of the sample [119], as previously discussed in this chapter. A problem still not solved for these approaches is the mediocre reproducibility of the fabrication process, and the lack of control for defined electrode geometries [58].

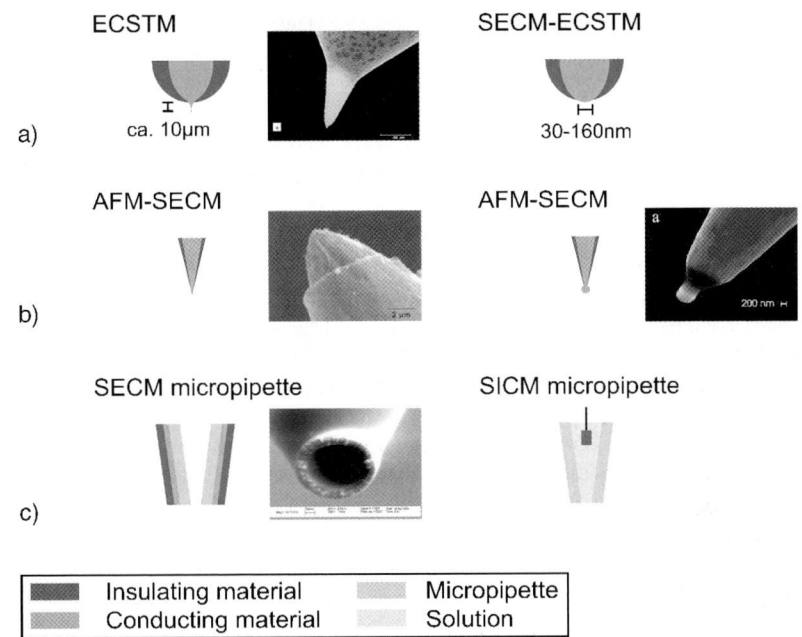

Fig. 8.3. Different types of manually fabricated SPM probes schemes and related SEM images. (**a**) ECSTM (*scale bar* in SEM image: 40 µm) and combined SECM-ECSTM probe. (**b**) Combined AFM-SECM probes with different electrode tip shapes: conical (*scale bar* in SEM image: 2 µm) and spherical (*scale bar* in SEM image: 200 nm). (**c**) SECM micropipette probe with ring-shaped electrode (*scale bar* in SEM image: 300 nm), and SICM micropipette with internal electrode. SEM images are reprinted from [236] copyright 1998 with permission from [48, 117, 126] copyright 2002, 2005 with permission from the American Chemical Society

Combined NSOM-SECM probes can be fabricated by coating fiber-shaped probes with a defined aperture with a metal layer followed by insulation against the electrolyte solution. Thereby, an exposed micro-ring electrode around the aperture of the optical fiber tip is provided. Typical aperture dimensions produced by fiber pulling are in the range of approximately 50 nm with a cone angle of approximately 20°. Shi et al. [120] demonstrated this concept by coating an optical fiber with gold providing a SECM microelectrode. However, this first approach suffered from considerable leakage currents due to imperfections in the insulation layer. More recently, a combined probe suitable for simultaneous optical and electrochemical measurements was published by Bard and co-workers [121], and Maruyama et al. [122]. Both concepts for combined optical and nanometer-sized electrode probes are based on coating a tapered optical fiber with a metal layer followed by insulation with electrophoretic paint. In contrast to the approach of Bard et al. [51] using a laser pipette puller for obtaining a tapered fiber with diameters of the combined probe < 1 µm, more recently published approaches by Xiong et al. [123] use selective chemical etching for fiber tapering to nanometer diameters obtaining conical microelectrodes templated by etched optical fibers. Similar to the fabrication of single optoelectrochemical probes, optoelectrochemical arrays have been described in the literature [124, 125]. While the fabrication process is outlined in detail, to date no results have been published on simultaneous measurements using the combined near-field techniques.

Similar to optical fibers, nanopipettes for SICM (see scheme in Fig. 8.3c, right), and recently for SECM investigations at the liquid–liquid interfaces, are fabricated based on laser-assisted pulling utilizing commercially available pipette pullers equipped with a CO_2 laser for controlled melting of the glass (Fig. 8.3c, middle and left) [126].

Carbon nanotubes (CNTs) have great potential for high-resolution AFM and STM imaging in addition to their promising properties as electrochemical scanning probe tips. Their unique features combining a high aspect ratio, large Young's modulus [127], and subnanometer radius of curvature with chemical stability and mechanical robustness, have identified carbon nanotubes as an interesting alternative to conventional scanning probes [127–138]. Recently, the application of nanotubes as conductive probes and nanoelectrodes has been described [139–143]. In order to use nanotubes as SPM probes, the CNT has to be attached to a conventional scanning probe tip, which is usually realized by adhesive pick-up with an AFM or STM probe (Fig. 8.4a), or by direct growth of CNTs at the tip [144]. Very recently, their unique features as scanning nanoprobes was combined with their potential simultaneously serving as nanoelectrodes [141, 145, 146]. The resulting nanowire has a high aspect ratio providing a beneficial geometry for imaging small and/or deep pore structures. In the following, the most prevalent current approaches

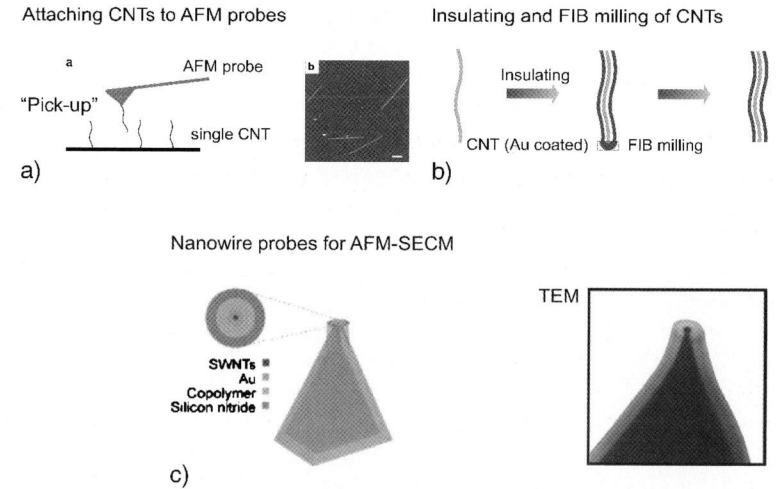

Fig. 8.4. Nanowire probes for AFM-SECM imaging. (**a**) Scheme for attaching vertically aligned carbon nanotubes (CNTs) to AFM probes via the pick-up technique (*a*), AFM image of single wall carbon nanotubes on silicon substrates (*b*) (*scale bar*: 100 nm). (**b**) Insulating of Au-coated CNTs previously attached to AFM probes, and subsequent FIB milling for exposure of the very tip. (**c**) Resulting nanowire-based AFM-SECM probe. The image width of the TEM image is 3.7 μm. AFM images/AFM-SECM probe scheme/TEM images are reprinted from [129, 145], copyright 2001, 2005 with permission from the American Chemical Society

toward electrochemical CNT-SPM tips are summarized. Burt et al. [145] used an assembly of a single-walled nanotube (SWNT) attached to an AFM tip and modified with a conducting metal layer consecutively insulated with poly(oxyphenylene) and silicon nitride, respectively. A defined disk electrode at the end of the insulated carbon nanotube was obtained by focused ion beam (FIB) milling (Fig. 8.4b,c). Alternatively, a multiwalled nanotube (MWNT) was attached to the apex of a SPM tip without further metal deposition, and the entire assembly was insulated with parylene C [146]. A nanoelectrode was exposed by locally removing parylene C via laser vaporization at the apex of the probe. Finally, single-walled nanotubes were attached to an AFM tip and insulated with fluorocarbon polymer [147]. The active electrode was exposed by electrical pulsing.

Macpherson and co-workers demonstrated the first application of combined imaging using a CNT-AFM-SECM probe [145] imaging a Pt substrate with a diameter of $< 2\,\mu m$. Since this approach was based on sputter coating a single-walled carbon nanotube with a conductive metal layer, metal nanowires with diameters of 50 to 100 nm were obtained, which results in a total diameter of the combined probe of 200 to 300 nm after insulation. Hence, high-resolution imaging taking advantage of CNT probes, in particular their high aspect ratio and subnanometer radius of curvature, demands further microstructuring of such tips. With the electroactive area again exposed at the apex of the CNT-SPM probe, some samples may still require consecutive scans for topographical and electrochemical imaging in "lift mode" [119]. However, merging the benefits of carbon nanotube probes for SPM along with sophisticated microstructuring may lead to combined electrochemical scanning probe tips with unique imaging properties in the near future.

8.2.6.2
Batch Microfabrication

Metallized AFM tips have been used for triggering dissolution or deposition processes by application of a potential [38, 101, 102], and imaging the changes in-situ. These probes can be easily produced by sputtering or evaporating a thin metal layer onto commercially available AFM cantilevers [148]. However, these probes are not suitable for imaging at the nanometer scale since lateral contrast on electrochemical information is only attainable if just a small fraction of the cantilever is immersed into the electrolyte solution. In order to obtain a functional probe for combined AFM-SECM measurements, an insulation layer has to be applied to the cantilever and cantilever mount, which covers everything except a defined electroactive area. For reproducibility of the fabrication process at acceptable cost, microfabrication processes are ideally suited providing batches of probes at a wafer level. Several research groups suggested full wafer processes providing combined AFM-SECM probes with conical electrodes at the apex of the combined tips [50, 149–151]. The group of De Rooij developed a standard microfabrication process based on plasma-enhanced chemical vapor deposition (PECVD), reactive ion etching (RIE), and a molding process [57], which is used to fabricate standard silicone nitride probes. In principle, a pyramidal etch-pit was formed in Si(100) followed by Si_xN_y deposition using PECVD. On top of this layer, the conductive layer forming the conically-shaped electrode is deposited. An additional layer of SiO_2 ensures electrical insulation. At the tip apex, the oxide layer was removed by etching resulting in a cantilever and tip that were entirely insulated, except for the tip apex (Fig. 8.5a, upper). In order to ensure high reflectivity on the backside of the cantilever for optical cantilever deflection readout, a chrome-gold layer was deposited, which additionally compensated the mechanical stress of the layered cantilever structure. Similarly, a batch process was developed, which consists of three main fabrication steps [149]. In a first step, high aspect ratio silicon (HARS) tips are shaped combining isotropic etching with an anisotropic deep-reactive ion etching (DRIE) in silicon. In the following, silicon tips are embedded into a silicon nitride layer, the electrode material is patterned, and then passivated with an insulation layer. Similar to the approach of De Rooij and co-workers, a conical electrode on top of the tip is formed by back-etching the insulation layer at the tip side. In order to obtain structures in the nanometer regime for HARS tips, an etch mask technology was developed utilizing FIB milling.

Recently, the same research group demonstrated a batch process for cantilever arrays with conical electrodes at the apex of the individual probes (Fig. 8.5a, lower) [150]. Although a wafer-level fabrication process was demonstrated, to date the routine application of these probes was not realized. A simplified approach was published by Hirata et al. [152] based on commercial silicone nitride cantilevers, which were sputtered with gold and insulated with photoresist. The photoresist was locally removed with a maskless optical micropattern generator. Combined probes with electrodes at the apex of the tip ranging from 50 nm to 6.2 μm have

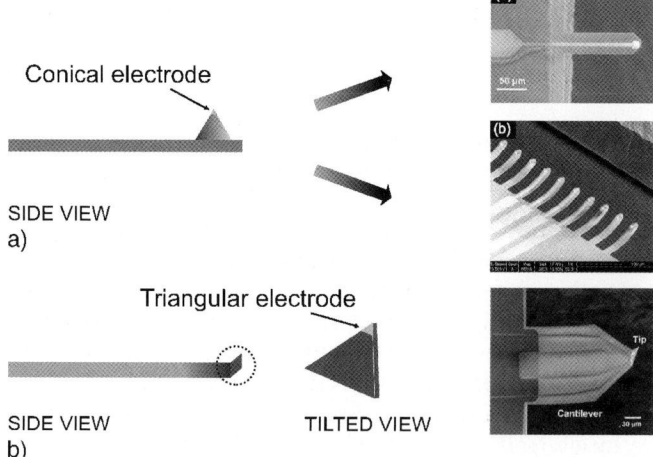

Fig. 8.5. Schemes of batch fabricated AFM-SECM probes along with SEM images. (**a**) The apex of the AFM probe is conducting and serves as conical electrode. The overall shape of the AFM tip can be varied. In (*a*), the tips are pillar-shaped (*scale bar* in SEM image: 50 μm), whereas the tips in (*b*) are pyramidal (*scale bar* in SEM image: 100 μm). (**b**) Triangular electrode is integrated into the probe. The length of the scale bar in the SEM image corresponds to 30 μm. SEM images are reprinted from [150] copyright 2004 with permission from Elsevier Science [237], copyright 2003 with permission from American Institute of Physics [238] copyright 2005 with permission from the American Chemical Society

been prepared by this approach. Again, this process is amenable to fabrication at a wafer scale.

Macpherson and co-workers reported on a batch microfabrication process for combined probes based on electron beam lithography (EBL) [153]. Cantilever and tip are processed from silicon nitride deposited by PECVD onto a 3 inch Si wafer. EBL is used to deposit the individual gold lines, contact pads, and the gold-coated triangular tip (Fig. 8.5b). An additional layer of silicon nitride was applied for insulating the gold lines from the tip electrode to the contact pad. In a last step, the contact pad and the gold-coated tip were exposed by using a resist mask and RIE.

All processes are similar in that the final combined probe provides an electroactive area at the apex of the tip, which may require, depending on the nature of the sample, that in a first step the topography has to be recorded without biasing the integrated electrode, while the electrochemical image is recorded in a consecutive step after retracting the probe from the sample surface. This is a necessary requirement if a potential is applied to the sample, in order to avoid short-circuiting of the probe. Shin et al. [154] developed a batch process, which enables fabrication of a ring electrode recessed from the AFM tip similar to the probes fabricated by FIB-assisted processing. This multistep microfabrication process allows integrating micro-ring electrodes recessed from the apex of silicon carbide tips, which are located at the center of the ring electrode. While the dimensions of the electrodes are still at the micrometer scale (outer diameter approximately 1 to 3 μm), these bifunctional probes are suitable for simultaneously recording topography and Faraday currents independent of the nature of the substrate. Furthermore, this batch fabrication process can be performed at a wafer scale resulting in affordable bifunctional probes.

8.2.6.3
Focused Ion Beam (FIB)-Assisted Fabrication

Focused ion beam milling is among the most promising techniques for three-dimensional nanofabrication based on the distinct advantage of a maskless process with unsurpassed flexibility [155–157]. Although FIB technology is mainly applied in semiconductor industries, nowadays, it has gained broad interest in nanotechnology and nanofabrication, e.g., for structuring of silicon, for fabricating conventional and combined SPM probes either exposing an electroactive area [49, 158, 159] or obtaining well-defined apertures such as for NSOM probes [160–163], or for producing high aspect ratio AFM tips [164–166]. Our research group and collaborators introduced a successful concept for bifunctional AFM-SECM probes based on integrating nano- and submicroelectrodes into AFM tips utilizing microfabrication technologies and FIB milling [49, 158, 159, 167, 168]. In contrast to the majority of approaches featuring the electroactive area at the apex of the probe, our design enables simultaneous recording of topographical and electrochemical properties at any sample surface, independent of the nature of the sample. With the electroactive area recessed from the apex of the tip, the distance between the integrated electrode and the surface is intrinsically constant and precisely defined, as the AFM tip itself serves as spacer for the integrated electrode. The fabrication process utilizes thin film chemical vapor deposition (CVD) and sputter techniques, along with focused ion beam (FIB) micromachining ensuring reproducible fabrication of integrated nanoelectrodes. This design enables integrating an electrode at a precisely defined and deliberately selected distance from the apex of a scanning probe tip, realizing virtually any desired well-defined electrode geometry. Commercially available silicon nitride cantilevers are sputter-coated with a metal electrode layer (e.g., gold, iridium, antimony, platinum, etc.) ranging in thickness from 50 to 150 nm. In a consecutive step, the conductive layer is insulated either by PECVD silicon nitride or sandwiched silicon nitride/silicon oxide deposition, or by chemical vapor polymerization (CVP) of parylene C. Due to its high uniformity and biocompatibility along with only minor effects on the original force constant of the cantilever, parylene C is the preferred choice for many applications. Exposing the electroactive area recessed from the apex of the tip is realized by three-dimensional focused ion beam milling. While the first two fabrication steps can be performed at a wafer scale, to date the FIB milling step has to be performed on a tip-by-tip basis increasing the costs for bifunctional probe fabrication.

As shown in Fig. 8.6, a series of FIB milling steps are performed to reproducibly expose an electrode recessed from the apex of the AFM tip at deliberately selected dimensions. Recent advancements in dual-beam FIB-SEM technology enable bitmap-assisted nanofabrication for significantly accelerated processing, as introduced by FEI Corp. [169]. A bitmap of the desired milling structure can be created by any graphics software, and uploaded into the FIB milling software. Hence, the number of individual milling steps of the original fabrication process is significantly reduced, as simultaneous synchronized milling on both sides of the tip is enabled. For bitmap-assisted fabrication, the size of the electrode is ultimately determined by the position of the first bitmap-based FIB milling step relative to the apex of the tip. The AFM tip is instantaneously reshaped at the center of the frame electrode with the length already correlated to the optimum working distance of the exposed recessed electrode. Several different electrode geometries have successfully been integrated into AFM probes, including frame, ring [158], and disk-shaped electrode structures [170]. Frame electrodes with electrode dimension in the range of 400 to 800 nm are nowadays integrated on a routine basis. However, the resolution of FIB systems promises the integration of nanoelectrodes in the near future. Up to now, disk-shaped integrated nanoelectrodes with diameters as small as 150 nm have been fabricated [170]. A distinct advantage of this technology is the maintained integrity of the basic features of the AFM cantilever and the fact that (electro)chemical and topographical information is inherently recorded at the same time, which is crucial for applications in life sciences, the biomedical field, for corrosion processes, and for other rapidly changing sample systems. The quality of the topographical images is determined by the precision of focused ion milling, which is suitable for the fabrication of ultrasharp AFM tips [165].

An alternative approach for obtaining electrodes recessed from the apex of the tip was recently published by Davoodi et al. [171]. In principle, the approach is similar to the cantilever-shaped nanoelectrodes described by Macpherson et al. [48], however, the electrodes are not etched and the exposed electroactive area is on the micrometer scale. The bent part of the electrode is insulated with epoxy and its end is cut by FIB milling exposing a disk electrode with core Pt electrode dimensions of

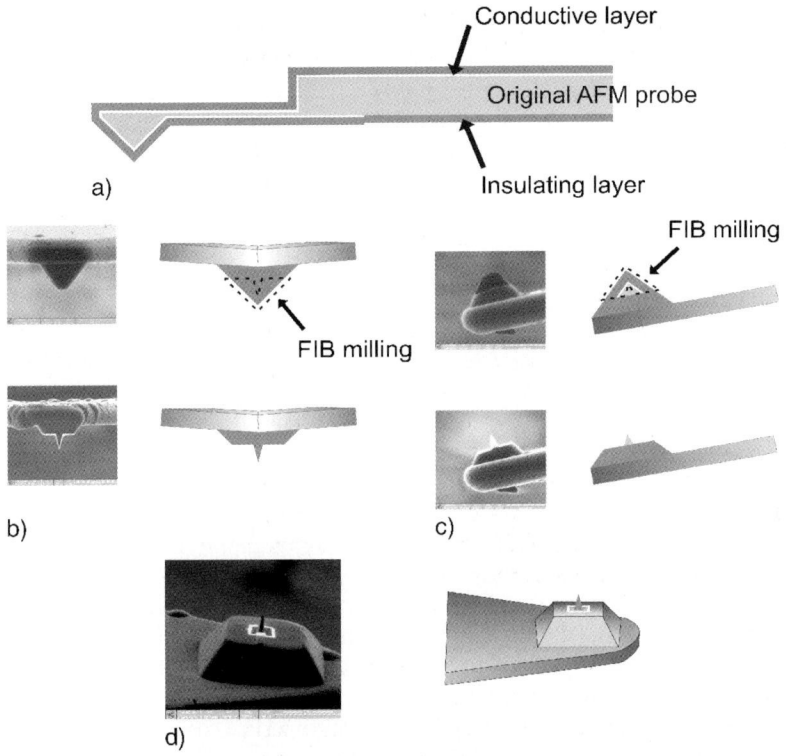

Fig. 8.6. FIB-assisted fabrication of bifunctional AFM-SECM probes. (**a**) Schematic cross section of sputter-coated and subsequently insulated AFM probe. (**b**) SEM and schematic front view of the probe prior to FIB milling (*upper part*) indicating the location for the FIB bitmap mask and after FIB milling (*lower part*). (**c**) SEM and schematic side view of the probe after the first milling step and after a 90° rotation (*upper part*). The *arrow* indicates the part that is removed by FIB milling. (**d**) An integrated AFM-SECM probe is shown

approximately 1 to 5 μm. FIB technology is then applied for shaping an insulating epoxy tip next to the electroactive Pt disk electrode. However, the epoxy AFM tip may be prone to abrasion limiting the application for hard sample surfaces and imaging over extended periods.

8.3
Applications

Approximately a decade after the Nobel Prize in Physics was awarded for STM [1, 172] and the introduction of AFM [12], first combined electrochemical SPM techniques appeared in literature [38, 45, 48, 49, 173]. Since combined SPM techniques are relatively new developments, corresponding efforts are primarily focused on technical improvements and optimization of these techniques along with proving the basic operational concept for imaging of simple model systems.

The following section will briefly summarize state-of-the-art combined electro-chemical SPM imaging of model systems, and consecutively focus on the applications of combined SPM for electrochemical imaging of more complex samples and scientific problems.

8.3.1
Model Systems

Favored model systems studied by combined electrochemical SPM techniques include imaging and/or modifying surfaces with alternating conducting and insulating features, such as grids, sputtered conducting features or electrodes embedded in insulating shielding, membrane structures or crystal surfaces, and molecule layers.

First combined AFM-SECM measurements demonstrated the flux of an electroactive species through a synthetic track-etched membrane. Macpherson and co-workers published several papers using AFM-SECM for imaging diffusion-controlled transport of electroactive target molecules through hydrated (i.e., the solution was confined to only the pores of the membrane) track-etched membranes along with the surface topography in air [173, 174] using a non-insulated conductive AFM probe. Similar experiments were performed at membranes that were fully immersed in liquid [48, 175] by applying a cantilever-shaped bifunctional probe. In their first publication [48], the membrane was studied in the absence and presence of an external concentration gradient of electroactive mediator $IrCl_6^{3-}$, revealing corresponding current decrease (hindered diffusion) and increase (diffusion transport) above the pores of the membrane with pore diameters of 0.6 μm and 1.2 μm, respectively. The images were obtained with cantilever-shaped microelectrodes with radii ranging from 1.1 μm to 275 nm for the electroactive probe. In the latter case [175], a potential was applied throughout the membrane to study iontophoretic transport of the redox mediator, and a quantitative description for the experimentally measured electrochemical signal was provided based on diffusion through a cylindrical pore. Figure 8.7a presents a schematic of the experimental setup, along with obtained topography (Fig. 8.7b) imaged in contact mode, and fixed height current images (Fig. 8.7c) after retracting the probe by ∼ 1 μm. It can be seen that not all of the pores, which have diameters varying from 0.5 to 1.5 μm, seem to be active during membrane transport.

First combined AFM-SECM measurements using microfabricated, integrated AFM-SECM probes with the electroactive area recessed from the apex of the tip were demonstrated by simultaneously mapping topography and electrochemistry of conducting features such as Au gratings [49, 168] or Au rings [176]. Both contributions show experiments conducted in SECM feedback mode, whereas AFM was performed in contact mode [49], as well as contact and dynamic mode operation [176]. The authors demonstrated that although a commercial silicon nitride cantilever with a nominal spring constant of 0.06 N/m was modified with a layered structure generating a bifunctional AFM-SECM tip, the altered force constant still enables dynamic mode operation. Modification with 100 nm Au layer and insulation with parylene C leads to an approximate threefold increase in force constant (personal communication, Wiedemair, Mizaikoff, Kranz, 2006).

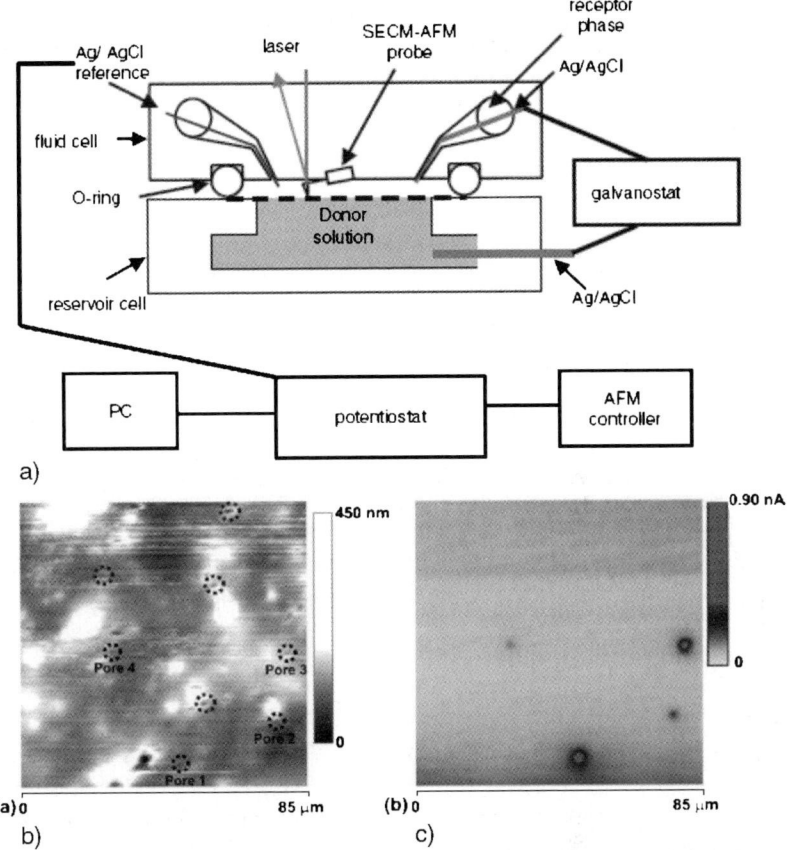

Fig. 8.7. (a) Scheme of the experimental setup for AFM-SECM imaging of iontophoretic transport across a porous polyethylene terephthalate (PET) membrane. Representative AFM-SECM topography **(b)**, and fixed-height current **(c)** images of membrane structure and transport activity are shown. The donor solution contained $0.1 \, \mathrm{mol \, L^{-1}}$ $Fe(CN)_6^{4-}$, $0.1 \, \mathrm{mol \, L^{-1}}$ KCl and $1 \, \mu \mathrm{L/mL}$ surfactant (Triton X-100) to ensure wetting of the membrane, whereas the receptor solution only contained $0.1 \, \mathrm{mol \, L^{-1}}$ KCl. A iontophoretic current of $14.7 \, \mu \mathrm{A}$ was applied across the membrane and the probe was biased at a potential sufficient for the detection of $Fe(CN)_6^{4-}$ (typically $+0.7$ V vs. AgQRE). The topography was recorded in CM AFM, whereas for the current image the probe was kept retracted from the surface at a fixed height of $1 \, \mu \mathrm{m}$. The *dotted labeled circles* (in **(b)**) represent transport active pores, whereas *unlabelled circles* indicate inactive pores. Reprint from [175], copyright 2005 with permission from Elsevier Science

Three-dimensional LIGA microstructures were imaged with Pt nanoelectrodes by SECM with shear-force-based distance control obtaining high-resolution topographical and electrochemical images of the porous three-dimensional structure [64]. However, the authors claim that although shear-force-based distance control was used, positioning and current measurements were increasingly difficult at small electrode diameters. Therefore, a minimal electrode diameter of 450 nm was used for imaging of the LIGA structures, and the improved imaging quality was compared to images recorded with a microelectrode ($d = 10 \, \mu m$).

The surface of bare [119, 145] and modified electrodes [177], as well as the topography and activity of electrode arrays [153] was studied.

Combined scanning electrochemical/optical microscopy (SECM-OM) was also used for simultaneous electrochemical and optical imaging of interdigitated array (IDA) electrodes [121]. However, it should be noted that in this approach the achieved resolution was substantially reduced due to difficulties during probe positioning. The same research group published a subsequent report with improved positioning of the combined optoelectrochemical probe by shear-force and current feedback investigating IDA electrodes, polycarbonate membrane filters, and living diatoms [178]. Maruyama et al. [122] developed nanometer-sized optical fibers for SECM-OM by selective electrochemical etching and simultaneously imaged implantable IDA electrodes with an electrode tip radius in the range of 100 nm.

It should be mentioned that combined AFM-SECM was also used for in-situ studies of processes involved in electrochemically induced crystal dissolution [38, 48, 101, 102]. Dissolution of potassium bromide [38] and potassium ferrocyanide trihydrate [48, 102] crystal surfaces was electrochemically induced, and topographical changes were imaged in-situ by AFM revealing information on local dissolution mechanisms. Dissolution of the cleavage surfaces of calcite single crystals was observed by local electrogeneration of protons at the tip-integrated electrode [101].

Tailored monomolecular layers play an important role in many areas ranging from organic catalysis to architecturing molecular materials for sensing. Hence, the comprehensive investigation of these monolayers with respect to their physical and chemical properties is of substantial importance. Recently, AFM-SECM based on cantilever-shaped nanoelectrodes with spherical tips ≥ 250 nm was applied for the investigation of structure and dynamics of flexible poly(ethylene glycol) (PEG) chains, which were end-grafted to a gold substrate and labeled with a redox moiety [179]. The chains were found to perform a tip-to-substrate cycling motion as a result of oxidation and re-reduction at the electrode and at the substrate surface. Modeling the data enables quantitative analysis of the polymer thickness and diffusion coefficients of the chains. A combination of ECSTM and SECM was used for the tip-induced modification of self-assembled monolayers (SAMs) [40]. Current-distance curves showed that the structure of the SAM was locally disturbed by applying a tunneling current through the combined probe. However, the structure relaxed back to its ordered passivating state throughout the experiment. The SAM was locally removed by probe-induced mechanical damage of the modified surface, and this modification was subsequently imaged. In an extension of this approach, combined ECSTM-SECM probes were used for the localized deposition of Pt dots at SAM covered Au samples [41].

8.3.2
Imaging Enzyme Activity

SECM has been extensively used for imaging and characterization of localized enzyme activity. For comprehensive reviews the reader is referred to [180, 181]. Contributions using combined AFM-SECM [152, 167, 182], SECM-SCLM [183], as well as SECM-SCM [184] techniques for enzyme activity studies can be found in the literature.

The group of Mizaikoff and Kranz [167] reported measurements with integrated AFM-SECM probes for simultaneous topographical and activity imaging of the oxidoreductase enzyme glucose oxidase (GOD) in GC mode of SECM and dynamic mode of AFM. GOD was entrapped in an electrophoretic polymer, which was locally deposited at a microstructured gold-disk array embedded in silicon nitride. The hydrophilic enzyme-containing polymer can be deposited by electrochemically-induced precipitation of the polymer suspension [185] generating a micropattern of isolated enzyme containing polymer spots. Since the generated enzyme spots have a high water content and are very soft, successful imaging was performed in dynamic mode AFM, which is advantageous for imaging of soft biological samples or polymers due to the reduced force impact and contact time [186, 187]. GOD catalyzes the conversion of glucose to glucono lactone (see scheme in Fig. 8.8a). H_2O_2 is generated as a byproduct and can be directly detected via oxidation at the tip-integrated electrode, which was biased at a potential of +0.65 V (vs. AgQRE). Hence, only when the enzyme is active and the substrate (glucose) is present in solution, an increase of the current recorded at the tip can be obtained due to localized production of H_2O_2 when the tip is scanned across the GOD containing polymer spots (see Fig. 8.8a). In absence of glucose, the current recorded at the electrode during AFM tapping mode imaging of the enzyme-containing polymer spots is negligible and reveals no electrochemical features. The periodicity of the pattern in the current image corresponds well with the topography provided by the integrated AFM tip.

FIB microfabricated AFM-SECM probes were also used to study the enzymatic activity of immobilized horseradish peroxidase HRP [182]. A self-assembled thiol monolayer with functionalized headgroups, which was previously chemisorbed to an Au/Si_xN_y micropatterned substrate, was used for covalent attachment of a glutaraldehyde/enzyme mixture leading to the formation of a cross-linked protein gel. HRP catalyzes the reduction of H_2O_2 to H_2O (see scheme in Fig. 8.8b). Hydroxyl methyl ferrocene (FMA, containing Fe^{2+}) was added to the solution as an electron donor and oxidized during the enzymatic conversion to ferrocenium methylhydroxide (FMA$^+$, containing Fe^{3+}). The AFM probe was biased at 0.05 V (vs. AgQRE) in order to detect FMA$^+$ that is continuously generated during the enzymatic reaction and, hence, can be directly correlated with the enzyme activity. Simultaneously recorded topography and current images were obtained in presence/absence of the substrate H_2O_2. In the presence of H_2O_2, an electrochemical pattern corresponding to the topography of the Au/Si_xN_y pattern was observed (see Fig. 8.8b), whereas if no substrate was present the current did not change significantly above the enzyme-containing spots.

Hirata et al. [152] used a similar approach for the measurement of enzyme activity at biosensor surfaces. However, the combined probe bore the electroactive area at the apex of the probe. GOD was immobilized in a cross-linked polyion complex layer (poly-L-lysine and poly(4-styrenesulfonat) composite) at the surface of graphite (HOPG) electrodes. In this experimental setup the AFM-SECM probe was operated in AFM non-contact mode enabling non-invasive measurements of the sample surface. The AFM-SECM tip was manually modified with a magnetic bead (10–30 µm), which provided the possibility for magnetic excitation at the respective

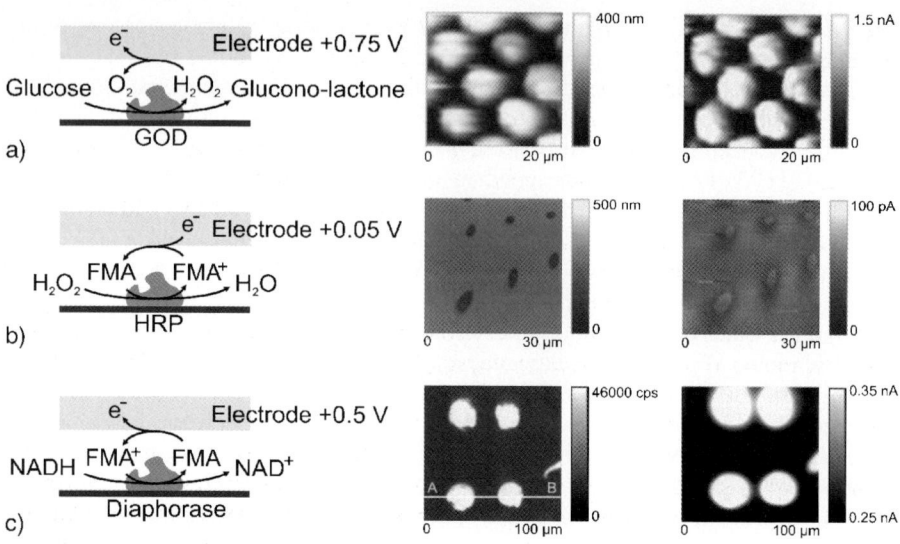

Fig. 8.8. Combined SPM techniques for imaging of enzyme activity/enzyme reactions and simultaneously-obtained images. (**a**) GOD catalyzes the conversion of glucose to glucono lac-tone. H_2O_2 is generated as a byproduct and can be detected at a biased electrode ($+0.65$ V vs. AgQRE). The topography of the periodic enzyme containing polymer structure and current increase due to GOD activity were simultaneously imaged in a glucose solution (50 mmol L^{-1} glucose in 0.1 mol L^{-1} phosphate buffer, pH 7.4) with dynamic mode AFM. (**b**) HRP cat-alyzes the conversion of H_2O_2 to H_2O; the byproduct FMA$^+$ can be reduced at the electrode ($+0.05$ V vs. AgQRE). Simultaneously-recorded height and current images of a periodic HRP micropattern in a solution of 0.5 mmol L^{-1} H_2O_2, 2 mmol L^{-1} FMA and 0.1 mol L^{-1} KCl in phosphate buffer (0.1 mol L^{-1}, pH 7) are shown. (**c**) The conversion of NADH to NAD$^+$ is catalyzed by diaphorase. The enzyme reaction can be detected at an electrode via elec-trochemical conversion of FMA to FMA$^+$ (at $+0.5$ V vs. AgQRE). Additionally the active center of diaphorase emits strong fluorescence, which can be detected by the SCFM system. Simultaneously-obtained optical and electrochemical images of diaphorase spots on glass sub-strates are presented. The measurements were performed in a solution of 5 mmol L^{-1} NADH, 0.5 mmol L^{-1} FMA and 0.1 mol L^{-1} Na$_2$HPO$_4$. Reprinted from [167], copyright 2003 with permission from Wiley [62, 182], copyright 2001, 2004 with permission from Elsevier Science

resonance frequency. The distance control in non-contact operation is based on monitoring the change of the oscillation frequency, and maintaining constant distance using phase-locked loop electronics. The probe was biased at $+0.9$ V (vs. SCE) to oxidize enzymatically produced H_2O_2. If the substrate glucose was present in solution, an increase in oxidation current was observed with local current variations indicating the distribution of active enzyme spots, whereas no significant current changes could be determined for the control experiment without glucose solution. In an additional control, the substrate electrode was also biased at $+0.9$ V (vs. SCE) leading to an immediate conversion of H_2O_2 at the substrate, which resulted in a reduced current at the AFM-SECM probe.

Besides combination with force microscopy, SECM in combination with opti-cal techniques was used to image enzyme activity. Matsue and co-workers [188] used a combination of scanning electrochemical-scanning chemiluminescence microscopy (SECM-SCLM) for simultaneous mapping of chemiluminescence and topography of HRP immobilized on a glass slide. An UME was used for imaging the sample surface and the current along with the photoncounting intensity, which was correlated with the probe position to give a two-dimensional image. The electrode was biased at -1 V (vs. AgQRE) for the reduction of O_2 to H_2O_2, which in turn was used by the immobilized enzyme for the oxidation of luminol to yield an excited luminescent state. Therefore, the two-dimensional plot of the chemiluminescence signal provides information on enzyme activity, whereas the current image was used for displaying the sample topography.

Another approach for imaging enzyme activity was performed using a combined scanning electrochemical-scanning confocal microscope (SECM-SCM) for the formation and characterization of diaphorase patterns at aminosilanized glass substrates [189]. In this setup the microelectrode is positioned above the focus of the laser. During the subsequent measurement only the sample is scanned, while the optical path and electrode remain static. Patterning of the structures was performed by different approaches including manual deposition of enzyme/glutaraldehyde solution spots, partially laser-induced, and by local electrochemical deactivation of uniformly covered, spin-coated enzyme surfaces. Diaphorases are flavin-bound enzymes catalyzing the oxidation of the reduced form of di- and triphosphopyridine nucleotides (NADH, NADPH) in presence of an electron acceptor. A ferrocene derivate (FMA) was used as electron acceptor. The electrode was biased at 0.5 V (vs. AgQRE) to oxidize FMA to FMA^+. In close proximity of the microelectrode to the enzyme-spotted surface, FMA^+ diffuses to the surface and diaphorase catalyzes the oxidation of NADH to NAD^+ using FMA^+ as acceptor molecule, which is regenerated to FMA during the enzymatic conversion (see Fig. 8.8c). Simultaneously the sample was irradiated with a laser ($\lambda = 473$ nm) inducing strong fluorescence of the active center of the enzyme (flavin mononucleotide, FMN), which was detected by the SCM. Hence, simultaneous SECM-SCM measurements were performed demonstrating laterally-resolved complementary information on enzyme activity by SECM and location of the active enzyme center by SCM, as shown in Fig. 8.8c. Although the lateral resolution achieved with this approach was still in the micrometer range, it clearly demonstrates the potential for increasing the information space by combining laterally-resolved electrochemical measurements with scanning confocal microscopy.

8.3.3
AFM Tip-Integrated Biosensors

Miniaturized amperometric biosensors are gaining increasing interest, as they offer many advantages by combining the inherent selectivity of biological recognition elements with well-established transduction principles. Many relevant processes in biochemistry and molecular biology are based on oxidation or reduction of molecules. Hence, techniques for obtaining laterally-resolved information on coupled oxidation-reduction processes with correlation in space and time are of particular interest. Few approaches have been reported using miniaturized enzyme electrodes as probes in conventional SECM due to the common problem of positioning the miniaturized biosensor above the sample surface [190]. Although dual-enzyme approaches or shear-force detection can be used [66, 191–193], several challenges remain in particular if soft biological samples are imaged. An elegant approach to circumvent these problems could be demonstrated by integrating a miniaturized biosensor into a combined AFM-SECM probe [170, 194]. With the electroactive area recessed from the apex of the combined probe, the reshaped AFM tip acts as a spacer between sample surface and the tip-integrated miniaturized biosensor, and prevents deterioration of both. This approach is not limited to integrating amperometric microbiosensors, as shown in Fig. 8.9 for alternative immobilization schemes and, hence, can be extended to the integration of a wide variety of electrochemical sensors.

The first example integrating an amperometric glucose sensor into combined AFM-SECM probes was demonstrated by Kueng et al. [170]. With the integrated microbiosensor recessed from the apex of the AFM tip, simultaneous mapping of topography and biochemical activity is enabled. GOD was immobilized at the

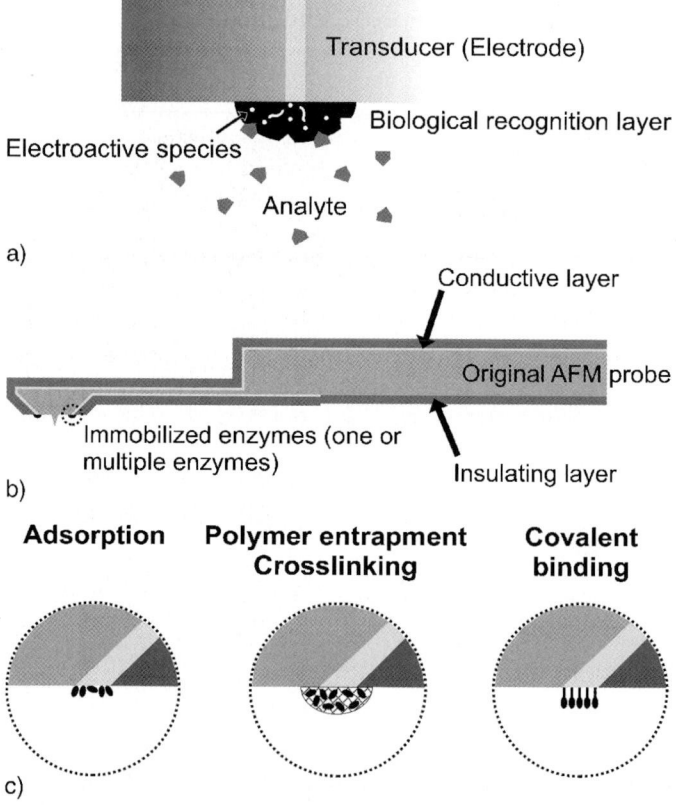

Fig. 8.9. (a) General scheme of amperometric biosensors. The highly specific interaction of an analyte to a biological recognition layer produces an electroactive species, which can be subsequently detected at a biased electrode. **(b)** Scaling of enzyme based amperometric biosensors to bifunctional AFM-SECM probes. **(c)** Immobilization strategies for enzyme-containing biorecognition layers at electrode surfaces

AFM-SECM probe by entrapping it into an electrophoretic polymer matrix [185]. Glucose diffusion through a polycarbonate membrane mimicking cellular membrane transport was used as a model substrate for demonstrating the versatility as shown in Fig. 8.10. Glucose diffuses from the lower compartment to the upper compartment through pores (diameter of 200 nm) and can be detected by biasing the integrated electrode at +0.65 V (vs. AgQRE) in order to oxidize H_2O_2, which was produced as a byproduct of the enzymatic reaction.

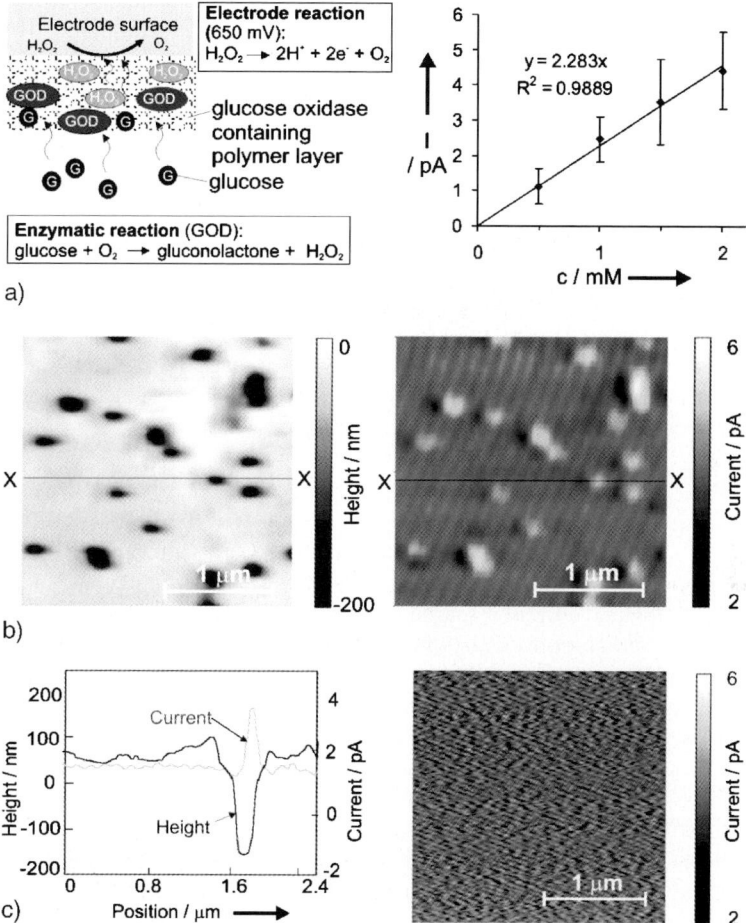

Fig. 8.10. (**a**) *Left panel*: Reactions for glucose detection with a biosensor based on glucose oxidase (GOD). *Right panel*: Glucose calibration of an AFM-tip-integrated biosensor (edge length: 800 nm) fabricated by entrapment of glucose oxidase within a polymer film. Simultaneously recorded. (**b**) *Left panel*, height and *right panel*, current images of glucose diffusion through a porous polycarbonate membrane (pore size: 200 nm); images recorded in AFM dynamic mode. (**c**) *Left panel*: Exemplary corresponding line scans of height and current. *Right panel*: Current image recorded without glucose in the donor compartment. Reprinted from [195], copyright 2005 with permission from Wiley-VCH Verlag

Figure 8.10b shows representative simultaneously obtained topographical and electrochemical images. In the presence of glucose, an increased current due to diffusion of glucose was observed above the pores of the membrane. After calibrating the integrated microbiosensor, which resulted in highly linear response function (R^2 of 0.9889) in the concentration range of 0.5 to 2.5 mM glucose (see Fig. 8.10a, right), the obtained data for the glucose transport can be quantitatively described. Based on the knowledge of the electrode geometry and the absolute distance between the electrochemical probe and the sample surface, the concentration of glucose was determined by applying a simple hemispherical single pore model [195], and by assuming the flux of glucose away from the pore opening is dominated by radial diffusion in absence of appreciable convective flows. The glucose transport rate can be obtained from solving the continuity equation using Fick's law at appropriate boundary conditions, and the measured glucose concentration was estimated to be 0.66 ± 0.13 mM, which corresponds well to the theoretically estimated concentration of 0.62 mM. Again, control experiments verified that without glucose negligible currents were measured (Fig. 8.10c, right) during control experiments. The successful integration of a microbiosensor into AFM-SECM probes could be also demonstrated by immobilizing HRP via self-assembled thiol monolayers.

A more complex biosensor architecture was shown using multiple types of immobilized enzymes for implementing an adenosine 5′-triphosphate (ATP) biosensor based on the co-immobilization of glucose-oxidase (GOD) and hexokinase (HEX) [196]. The sensing scheme involves the competitive reaction for glucose (substrate) between GOD and HEX with ATP as a co-substrate:

$$\text{glucose} + O_2 \rightarrow \text{gluconic acid} + H_2O_2 \quad \text{Catalyzed by glucose oxidase} \quad (8.6)$$

$$\text{glucose} + \text{ATP} \rightarrow \text{glucose-6-P} + \text{ADP} \quad \text{Catalyzed by hexokinase .} \quad (8.7)$$

Miniaturization of this sensing scheme to dual-microelectrodes was successfully established [193], and current research in the research group of Mizaikoff and Kranz focuses on the integration of an ATP biosensor into AFM-SECM probes for the detection of ATP metabolism at live epithelial cells. ATP is a key molecule involved in the molecular mechanism of cystic fibrosis (CF) [197], in muscle contraction, and in the generation of pain signals and/or ion channels. Hence, there is a significant interest for localized detection of ATP at cellular levels with high spatial and temporal resolution. AFM-SECM tip-integrated biosensors seem to be uniquely suitable solutions for this task.

An alternative approach for positioning microsensors in SECM experiments was demonstrated by constant-distance SECM based on shear-force mode positioning. Etienne et al. [67] described the application of a constant-distance mode scanning potentiometry (CD-SP) approach with Ca^{2+} ion selective microsensors for measurements at calcite crystals and at calcium carbonate shells of *Mya arenaria*. Although the obtained lateral resolution is in the micrometer range due to the size of the applied probe, these probes present an interesting approach for combined force/electrochemical studies. Laterally-resolved measurements of nitric oxide (NO) are of immense importance in many physiological processes, such as neurotransmission, immune response, and vasodilatation. Hence, tackling this challenging task by laterally-resolved determination of NO would have a major impact on cell physiological investigations. First approaches developing microsensors for NO detection have recently been published by Schuhmann and co-workers [198–206].

Pailleret et al. [204] prepared NO microsensors by in-situ modification, which were subsequently scanned across a layer of adherently growing human umbilical vein endothelial cells (HUVEC). Isik et al. [207] used dual-electrode-based NO sensors for positioning at the sample surface. In this approach one electrode remains unmodified enabling independent positioning of the sensor. However, to date combined measurements presenting topographical and simultaneous electrochemical data of NO release have not been demonstrated. Recently, an alternative approach was published combining electrochemical and optical measurements for the simultaneous detection of intracellular production and diffusion of NO into the extracellular space [208].

8.3.4
Combined SPM for Imaging of Living Cells

SPM techniques and in particular AFM have a promising potential for imaging living biological systems with high-resolution at physiological conditions. However, one critical aspect of this approach is the constant interaction of the scanning probe with the cell surface leading to possible deformation and perturbation of the soft cell membrane. Additionally, living cells may have mechanosensitive ion channels, which can respond to the mechanical stimulation by the applied force. Therefore non-invasive approaches such as SICM [14] or NSOM [13, 209] are interesting alternatives. In 2000, Korchev et al. [45] reported the development of a hybrid SICM-NSOM technique for studying living rabbit cardiac myocyte cells. Combined SICM-NSOM provides an alternative distance control mechanism enabling separation of the ion flux signal from the topographical information. In the presented work, laser radiation ($\lambda = 532$ nm) was coupled into a coated SICM micropipette with an internal diameter of ~ 500 nm, and was scanned in the near-field regime across the sample surface. Simultaneous optical and topographical images were acquired showing a characteristic striated pattern that corresponds to the sarcomeric structure of the cells. The achieved probe-sample distance was significantly larger than in shear-force detection NSOM. However, the authors claim that the distance between capillary and sample can be significantly decreased, by decreasing the aperture of the micropipette. Additionally Korchev and co-workers [47] described an improvement of the distance control mechanism via modulation methods based on periodically changing the distance between tip and surface via oscillation, while keeping the overall distance constant. The distance between micropipette and sample is modulated, and this AC contribution of the current signal is used via an integrated feedback loop making the measurement insensitive to changes in ionic strength or DC drifts. These improvements along with the combination with laser confocal microscopy enabled simultaneous measurements of the motion of contracting cardiac myocytes, and of localized intracellular calcium concentrations. Although combining SICM and NSOM seems to be an interesting approach for imaging cell chemistry, technical difficulties such as significant losses of light intensity emerging at the very front of the tip along with a decrease in topographical resolution due to the metallic coating need to be resolved. Bruckbauer et al. [44] described an alternative approach for generating an intense near-field light source for combined SICM-NSOM measurements. The nanoscopic light source is generated by a chemical reaction between the calcium indicator fluo-3^{5-} and Ca^{2+} [210] (see Fig. 8.11a):

$$\text{fluo-}3^{5-} + 2Ca^{2+} \rightarrow \text{fluo-}3(Ca)_2^-\ . \tag{8.8}$$

The generated complex is highly fluorescent. An applied electric field releases fluo-3^{5-}, which is confined to the interior of an uncoated capillary, and the electrical field gradient forces Ca^{2+}, which is located in the solution, into the capillary (see scheme in Fig. 8.11a). Hence, a mixing zone is formed at the capillary aperture enabling localized complex formation and eliminating problems with photobleaching due to the continuous supply of reagents. Diffusion characteristics along with the photophysical properties of this novel chemical light source were characterized [44]. Figure 8.11b shows a microscopic image of the light source in the pipette with $160\,\mu W$ laser illumination. This novel concept was applied for imaging of biological systems [211]. Fixed sea-urchin sperm flagella, which have a known size and shape [212] were imaged to determine the optical and topographic resolution. Then, the technique was extended to imaging of live A6 cells at physiological conditions. The simultaneously-obtained SICM and NSOM images of sperm flagellae were in good agreement, and the diameter of the flagellum showed the expected size ($\sim 100\,nm$). SICM images obtained during imaging of live A6 cells were consistent with observations from previous experiments [69].

The same group published a novel combined technique, scanning surface confocal microscopy (SSCM) based on SICM and SCM [213]. The interaction of fluorescently-labeled polyoma virus-like particles (VLPs) with living cells was studied, and single particles could be detected. VLPs are nanospheres consisting of 360 molecules of a single viral coat protein (VP1), which interact with and carry plasmid DNA into cells [214,215]. Using SSCM the position of fluorescently-labeled features could be correlated with the topography of the surface. Background fluorescence is minimized in contrast to conventional confocal microscopy, since only the sample surface is scanned.

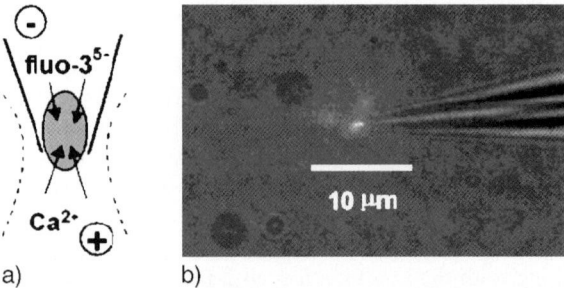

a)   b)

Fig. 8.11. Novel SICM-SNOM light source based on the formation of a fluorescent complex between fluo-3 and calcium. (**a**) Schematic representation of the light source at the apex of the micropipette. Due to the applied electromagnetic field, fluo-3^{5-} is forced out of the pipette while Ca^{2+} is forced into the pipette. (**b**) Microscope image of the light source at the micropipette with $160\,\mu W$ laser illumination. Reprinted from [44], copyright 2002 with permission from the American Chemical Society

Combined SECM-fluorescence imaging was demonstrated for live cell imaging in order to distinguish between non-transformed human breast cells and metastatic breast cells due to their redox activity [216]. Based on previous experiments [217] demonstrating substantially different behavior of the redox mediator recycling at different types of mammalian cells, combined measurements were performed. Fields of metastatic and non-transformed breast cells were plated together, with the former cells pre-loaded with fluorescent particles. It was possible to differentiate between both cell types even if they were grown in dense cell layers. Lee et al. [178] used combined SECM-OM with current feedback for controlling the tip–sample distance to obtain simultaneous SECM and optical images of living unicellular organisms (diatoms). In case of soft and sticky biological samples, using a constant current for the feedback mechanism is advantageous to shear-force sensing due to the reduced interaction force as the tip is not oscillated. However, the current feedback mode is restricted to samples comprising only insulating or only conducting materials. Unfortunately, the obtained resolution was not sufficiently high to resolve structures at the diatom surface. Although the lateral resolution is still in the micrometer range for the electrochemical response and no laterally-resolved optical information was obtained, the demonstrated investigation has significant relevance for future developments in combined electrochemical/optical techniques at the nanoscale.

8.3.5
Measurement of Local pH Changes

Laterally-resolved measurement of pH profiles or gradients is important as pH changes are often involved in enzymatic reactions, metabolic processes of microorganisms and corrosion processes.

Boldt et al. [218] described an approach by combining scanning electrochemical microscopy with single-molecule fluorescence spectroscopy (SECM-SMFS). During SMFS the optical detection volume has to be confined to several femtoliters and analyte concentrations in the nanomolar range, which can be achieved by two approaches, total internal reflection fluorescence microscopy (TIRFM) and confocal excitation and reflection, respectively.

Although, SMFS is not within the focus of this chapter, it is noteworthy since the combination appears versatile and highly sensitive for localized pH measurements. UMEs were used for electrochemical generation of pH fluctuations, which were detected by avalanche photodiodes (APDs) for single-photon counting through changes in the fluorescence of pH-sensitive dyes. UMEs and nanoelectrodes with diameters ranging from 5 μm to 35 nm were fabricated and used for generating a pH gradient, which was optically mapped with freely diffusing SNARF-1 dextran fluorescence dye. After positioning of the UME ($r = 2\,\mu$m) or nanoelectrode ($r = 35$ nm) above the TIRFM field, a pH gradient was generated by applying 1.4 V or -0.8 V (vs. AgQRE) in buffered solution containing 10 mM NO_2^- in borate buffer or in MES buffer, respectively. The insets in Fig. 8.12a, left panel for the UME experiment and Fig. 8.12a, right panel for the nanoelectrode experiment, respectively, show the CCD images of the two-dimensional $F_I/_{II}$ in the TIRFM. The line scans in Fig. 8.12a left and right panel demonstrate the generated pH gradient in dependence of the distance between tip and coverslip surface. Furthermore, Boldt et al. [218] investigated single

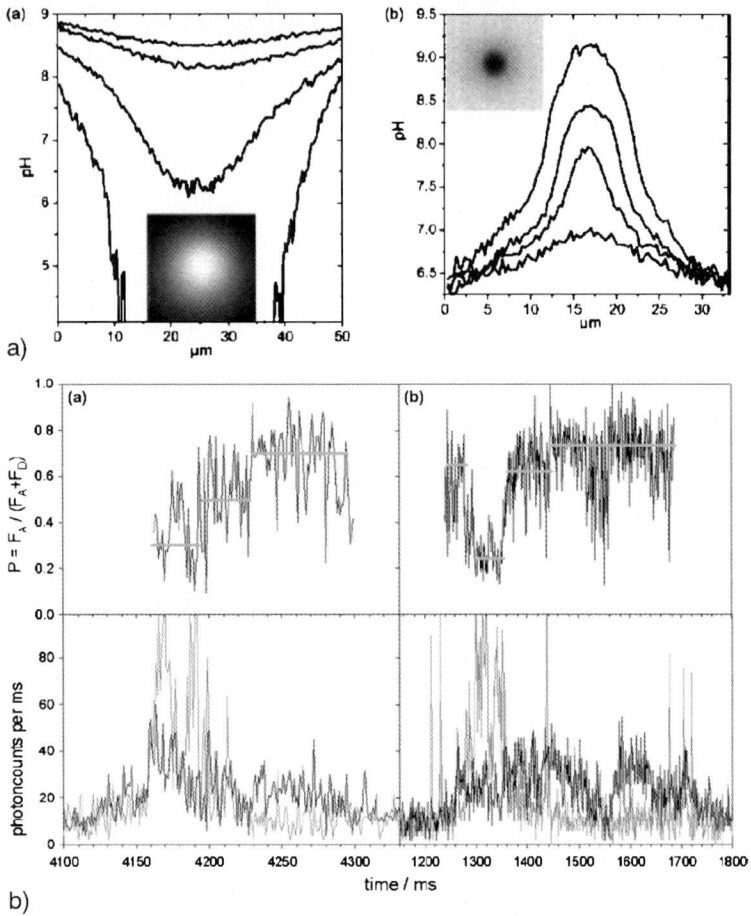

a)

b)

Fig. 8.12. (a) Surface pH profiles generated with a 2 μm radius (*upper*) and with a 35 nm radius (*lower*) SECM tip in dependence of the surface distance; pH profile monitored by TIRFM. Each pH profile consists of an average of at least 10 detection ratios FI/II. The *insets* illustrate the FI/II ratio in TIRF fields of 50 × 50 μm with an electrode distance to the surface of 6 μm and 33 × 33 μm with a distance of 1 μm, respectively. The *left panel* shows pH profiles recorded during the generation of protons at 1.4 V vs. AgQRE starting from pH 9.0 at various distances (from *top* to *bottom*): 12, 9, 6, and 3 μm. The *right panel* shows pH profiles recorded during water electrolysis and generation of OH$^-$ at -0.8 V vs. AgQRE starting from pH 6.3 at various distances (from *top* to *bottom*): 1, 2, 3, and 4 μm. **(b)** Photon bursts and corresponding proximity factors obtained from single-molecule FRET measurements on single EF0F1-ATP synthases in liposomes. ATP synthesis during SECM-induced ΔpH generation at a SECM tip with 5 μm radius. The *lower panel* shows photon count rates of the donor channel (FD, *gray lines*) and the acceptor channel (FA, *black lines*). The *upper panel* shows the calculated proximity factor P (*black lines*) and different level means (*grey lines*). Reprinted from [218], copyright 2004 with the permission from the American Chemical Society

membrane-bound F0F1-ATP synthases in freely diffusing lipid vesicles during catalysis powered by a transmembrane proton gradient after SECM-induced pH changes (Fig. 8.12b). Depending on whether ATP synthase synthesizes or hydrolyses ATP the protons are pumped in reverse directions through the transmembrane portion. Additionally the head subunit of ATP synthase rotates in opposite directions depending on whether ATP is synthesized or hydrolyzed. By labeling two subunits on the ATP synthase the rotation of the subunits during synthesis/hydrolysis resulted in distinct fluorescence resonance energy transfer (FRET) patterns that could be indirectly correlated with the direction of the proton flow.

Miniaturized potentiometric pH sensors have been reported in the literature [219–221]. Combined STM-pH measurement were published by Siegenthaler and co-workers [39]. Thin layer STM (TLSTM) combining an annular iridium oxide microelectrode with an STM probe was used for the investigation of the H^+ exchange behavior of electropolymerized polyaniline films. Iridium oxide microelectrodes [221] are used as generators and scavengers for H^+/OH^-. In the presented TLSTM experiments, the Ir/IrOx microelectrode was applied for injecting and removing protons into a polyaniline layer along with the detection of pH changes occurring during potential controlled film oxidation or reduction. Changes in topography due to partial film reduction/oxidation were simultaneously recorded with STM. Additionally, potential controlled film reduction/oxidation revealed areas where proton exchange was observed. The presented concept can be expanded to a variety of pH-dependent phase formation/adsorption processes, such as fundamental mechanisms correlated with corrosion especially when the size of the pH-electrode can be significantly reduced.

8.3.6
Corrosion Studies

Corrosion processes can be defined as localized (electro)chemical reactions between a metal and its surrounding medium leading to material deterioration. A pending goal in corrosion science is the development of fundamental molecular level understanding of processes occurring at surfaces and interfaces. Hence, in-situ SPM techniques providing complementary information with lateral resolution at a molecular level are of particular need. Besides in-situ spectroscopic techniques such as surface-enhanced IR and Raman spectroscopy, mainly electrochemical impedance spectroscopy (EIS), scanning Kelvin probe (SKP), and scanning probe microscopic techniques including AFM and SECM are used to investigate corrosion processes [222–227]. SECM is of particular interest given the unique possibility to probe localized electrochemical reactions. SECM and PEM (photoelectrochemical microscopy) was combined and used for the identification of pitting precursor (PPS) sites at titanium samples [228]. PEM is based on scanning an optical fiber across the surface of a semiconductor and recording the induced photoelectrochemical current. Pitting precursor sites could be imaged simultaneously by SECM/PEM using a metal-coated optical fiber. However, the achieved resolution was poor for both techniques considering the aperture of the fiber and the dimension of the probe, which consists of a metal-coated optical fiber that is subsequently insulated. Although, SECM-PEM probes were fabricated [229] with optical core diameters of 1–2 μm and electrode diameters of approximately

3 μm, their electrochemical characterization performed by cyclic voltammetry revealed poor electrical insulation or high surface roughness, which renders them of limited utility for combined experiments. In a subsequent study, miniaturized probes were successfully fabricated [230, 231], however the concept was not yet applied to SECM-PEM measurements.

Pan and co-workers [171] presented an integrated EC-AFM-SECM technique for in-situ investigations of aluminum alloys in NaCl solutions. After determining corrosion potential, current density in passive regions, and the breakdown potential by conventional cyclic polarization of macroscopic Al samples, the EC-AFM-SECM studies were performed. SECM experiments were performed in generation-collection mode involving the following reaction scheme:

$$Al \rightarrow Al^{3+} + 3e^- \tag{8.9}$$

$$I_3^- + 2e^- \rightarrow 3I^- . \tag{8.10}$$

Although the bifunctional probe provides the electroactive area recessed from the apex of the polymer tip, combined measurements were conducted in a sequential mode. After recording the topography the tip was retracted from the sample surface and the electrochemical information was recorded. The sample was anodically polarized in presence of the I^-/I_3^- redox couple, and localized Al corrosion (anodic dissolution) started to occur at certain sites. In the case of Al dissolution I_3^- is simultaneously reduced leading to an increase in I^- concentration, which can be locally detected (via oxidation) at the biased AFM tip-integrated electrode. Hence, surface areas where Al dissolution occurs result in high oxidation current detected at the combined probe. In the case of the surface biased at a small anodic potential (50 mV), large dissolution sites could be revealed in the current image, although there was no significant change in the surface topography (see Fig. 8.13a). If the sample is polarized at values near the breakdown potential (300 mV), the topography reveals many small defects in the surface, which is corroborated by a higher background current in the electrochemical image (see Fig. 8.13b). An explanation may be that many active sites are closely spaced and, hence, cannot be resolved with the current bifunctional probes. If the sample is biased in the passive region (100 mV), deposition of corrosion products along with increased currents can be observed (see Fig. 8.13c). The obtained results indicate that intermetallic particles are involved in local dissolution of Al. The same research group investigated the corrosion behavior of duplex stainless steel using STM and SECM in a sequential mode [232]. Typically in-situ STM images were recorded while biasing the surface at different potentials and observing the resulting topographical changes. Additionally in a few experiments after STM imaging the probe was retracted from the sample and immediately used for SECM investigation at the previously depicted surface.

It should be mentioned that scanning Kelvin probe force microscopy (SKPFM) is frequently applied in corrosion studies. Its capability of mapping surface potentials along with simultaneously imaging the surface topography renders SKFM a valuable tool in corrosion science [233, 234]. However, potential measurements are typically performed in air and hence, this technique is beyond the scope of this chapter.

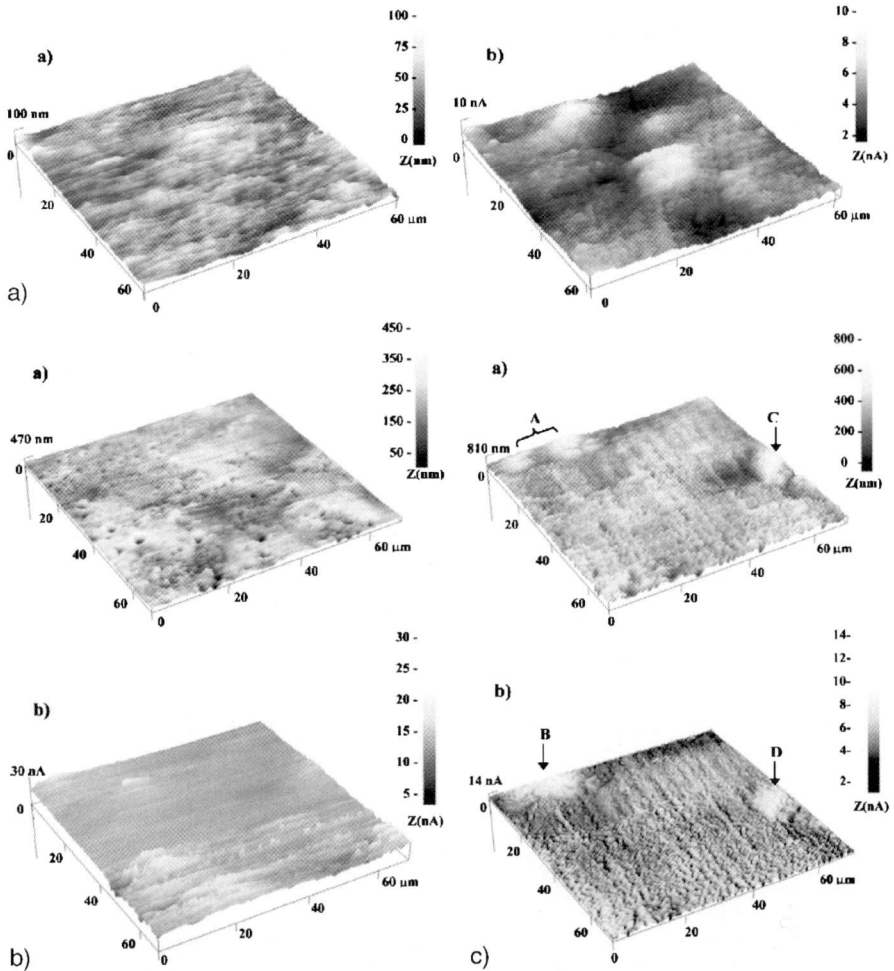

Fig. 8.13. EC-AFM-SECM images of aluminum alloys (AA1050) in 10 mmol L^{-1} NaCl and 5 mmol L^{-1} KCl at 50 mV (**a**), 300 mV (**b**), and 100 mV (**c**) anodic polarization. The probe was held at +750 mV (vs. AgQRE). Topography Fig. 8.12a (*left panel*) and Fig. 8.12b,c (*upper panel*) and electroactivity Fig. 8.12a (*right panel*) and Fig. 8.12b,c (*lower panel*) maps are presented. Reprinted from [171], copyright 2005 with the permission from the Electrochemical Society

8.4
Outlook: Further Aspects of Multifunctional Scanning Probes

Investigations on electrochemical surface processes relevant to bioelectrochemistry, biocorrosion, fuel cell research, electrochemical nanotechnology, sensor technology, and in material sciences have been performed with scanning probe microscopies providing insight on fundamental processes at a molecular level. Very recently, combined scanning probe techniques were introduced, which allow the multiparametric investigation of dynamic processes correlated in time and space and, hence, expand the information content achievable with individual SPM techniques. While in recent years substantial progress was achieved in fabricating the next generation of combined scanning probes such as carbon nanotubes, combined SPM probes, batch fabricated SPM probes with integrated nanoelectrodes at the tip, or FIB-milled bifunctional probes, to date most applications are still limited to model systems such as interdigitated electrode arrays or mapping localized diffusion through artificial pores. Novel probe designs such as trifunctional scanning probes integrating AFM-SECM-NSOM functionality or multiple electrodes into a single tip can certainly be envisaged with ongoing progress in micro- and nanofabrication technology. Furthermore, smart modification of the integrated electrodes with biorecognition membranes [170], or the possibility of integrating potentiometric sensors has yet to be achieved. Without doubt, microfabrication will continue to have a major impact on reliable and reproducible fabrication of scanning probe tips. However, their application for real scientific or industrial problems utilizing sophisticated probe designs has yet to be fulfilled.

Whereas in recent years main focus in this field was on combination techniques facilitating the positioning of nanometer-sized electroactive probes and, hence, deconvolute topographical and electrochemical information, researchers in this field realized the immense potential of combining electrochemical detection with optical detection at a time- and space-resolved scale. First examples using this powerful combination were demonstrated, although frequently only one of the combined techniques provides the promised lateral resolution. Nonetheless, the opportunity of combining powerful optical techniques such as near-field spectroscopy, or confocal laser scanning microscopy with micro- and nanoelectrochemistry may have the potential to revolutionize scanning probe microscopy in the near future.

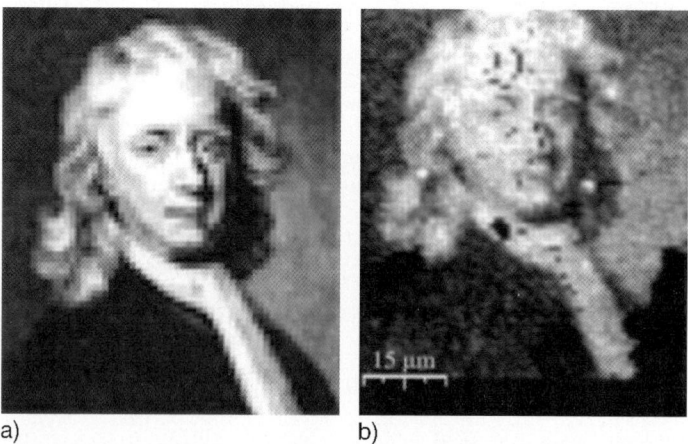

a)   b)

Fig. 8.14. Fluorescence images of DNA deposited onto a streptavidin-coated glass surface. (**a**) An image of Sir Isaac Newton downsized with Adobe Photoshop to 75×62 pixels. (**b**) An image of Sir Isaac Newton reproduced with a barreled pipette. The image was written with alexa 647-labeled DNA with 1 mm pixels (printed area is 75 mm^2). Reprinted from [239], copyright 2005 with the permission from Wiley

Besides imaging of surface processes or kinetic studies, micro- and nanoelectrochemistry driven by miniaturized electrochemical SPM probes can also be applied for surface modification, which was recently demonstrated by Korchev and coworkers [235] (see Fig. 8.14). Furthermore, the combination with well-established cell physiological techniques such as patch-clamp, as demonstrated by combined SICM/patch-clamp studies, provide the potential of addressing complex, live biological systems [73].

While there is increasing interest in localized electrochemical measurements outside the so-called SECM community, it may still take another decade until micro- and nanoelectrochemistry based on combined scanning probe microscopies will be a routine surface analytical tool such as AFM or STM.

Acknowledgements. The authors acknowledge support for their research on integrated AFM-SECM from the National Science Foundation (project # 0216368, Biocomplexity/IDEA and # 0343029, CRIF), the National Institute of Health (# EB000508), the Microelectronic Research Center (MiRC) at Georgia Tech, the FIB^2 Center at Georgia Tech, and FEI Corporation (Hillsboro, Oregon). Furthermore, A. Lugstein and E. Bertagnolli at the Institute of Solid State Electronics, Vienna University of Technology, are greatly acknowledged for their collaboration in the initial stages of AFM-SECM probe fabrication. Finally, Jean-Francois Masson is thanked for critical discussions and helpful input during preparation of this manuscript.

References

1. Binnig G, Rohrer H, Gerber C, Weibel E (1982) Appl Phys Lett 40:178
2. Sonnenfeld R, Hansma PK (1986) Science 232:211
3. Sonnenfeld R, Schneir J, Hansma PK (1990) Mod Asp Electrochem 21:1
4. Liu HY, Fan FF, Lin CW, Bard AJ (1986) J Am Chem Soc 108:3838
5. Lustenberger P, Rohrer H, Christoph R, Siegenthaler H (1988) J Electroanal Chem 243:225
6. Wiechers J, Twomey T, Kolb DM, Behm RJ (1988) J Electroanal Chem 248:451
7. Tao NJ, Li CZ, He HX (2000) J Electroanal Chem 492:81
8. Christensen PA (1992) Chem Soc Rev 21:197
9. Moffat TP (2003) Encyclopedia Electrochem 3:393
10. Siegenthaler H (1992) In: Wiesendanger R, Guntherodt HJ (eds) Scanning Tunneling Microscopy II. Springer Series in Surface Science, Vol. 28. Springer, Berlin Heidelberg New York
11. Kolb DM, Simeone FC (2005) Electrochim Acta 50:2989
12. Binnig G, Quate CF, Gerber C (1986) Phys Rev Lett 56:930
13. Betzig E, Lewis A, Harootunian A, Isaacson M, Kratschmer E (1986) Biophys J 49:269
14. Hansma PK, Drake B, Marti O, Gould SA, Prater CB (1989) Science 243:641
15. Engstrom RC, Weber M, Wunder DJ, Burgess R, Winquist S (1986) Anal Chem 58:844
16. Engstrom RC, Meaney T, Tople R, Wightman RM (1987) Anal Chem 59:2005
17. Uosaki K, Koinuma M (1998) In: Lorenz WJ, Plieth W (eds) Electrochemical Nanotechnology. Wiley, Weinheim, p 253
18. Kienberger F, Moser R, Schindler H, Blaas D, Hinterdorfer P (2001) Single Mol 2:99
19. Lindsay SM, Lyubchenko YL, Tao NJ, Li YQ, Oden PI, DeRose JA, Pan J (1993) J Vac Sci Technol A 11:808
20. Kindt JH, Sitko JC, Pietrasanta LI, Oroudjev E, Becker N, Viani MB, Hansma HG (2002) Methods Cell Biol 68:213

21. Perfetti P, Cricenti A, Generosi R (2000) Surf Rev Lett 7:411
22. Takano H, Kenseth JR, Wong S-S, O'Brien JC, Porter MD (1999) Chem Rev 99:2845
23. Tao N (2003) Encyclopedia Electrochem 2:221
24. Hillier AC, Kim S, Bard AJ (1996) J Phys Chem B 100:18808
25. Raiteri R, Butt H-J (1995) J Phys Chem 99:15728
26. Bard AJ, Mirkin MV (eds) (2001) Scanning Electrochemical Microscopy, Marcel Dekker, New York
27. Ermakov AV, Garfunkel EL (1994) Rev Sci Instrum 65:2853
28. Bryant PJ, Miller RG, Yang R (1988) Appl Phys Lett 52:2233
29. Nakajima K, Micheletto R, Mitsui K, Isoshima T, Hara M, Wada T, Sasabe H, Knoll W (1999) Appl Surf Sci 144–145:520
30. Nakajima K, Jacobsen V, Noh J, Isoshima T, Hara M (2001) RIKEN Rev 38:52
31. Nakajima K, Micheletto R, Mitsui K, Isoshima T, Hara M, Wada T, Sasabe H, Knoll W (1999) Jpn J Appl Phys, Part 1 38:3949
32. Hecht B, Heinzelmann H, Pohl DW (1995) Ultramicroscopy 57:228
33. Kim J, Song K-B, Park K-H (2002) Jpn J Appl Phys, Part 1 41:1903
34. Hillner PE, Walters DA, Lal R, Hansma HG, Hansma PK (1995) J Microsc 1:127
35. Kodama T, Ohtani H, Arakawa H, Ikai A (2004) Jpn J Appl Phys, Part 1 43:4580
36. Jauss A, Koenen J, Weishaupt K, Hollricher O (2002) Single Mol 3:232
37. Hinterdorfer P, Gruber HJ, Kienberger F, Kada G, Riener C, Borken C, Schindler H (2002) Coll Surf B23:115
38. Macpherson JV, Unwin PR, Hillier AC, Bard AJ (1996) J Am Chem Soc 118:6445
39. Ammann E, Beuret C, Indermuhle PF, Kotz R, De Rooij NF, Siegenthaler H (2001) Electrochim Acta 47:327
40. Treutler TH, Wittstock G (2003) Electrochim Acta 48:2923
41. Sklyar O, Treutler TH, Vlachopoulos N, Wittstock G (2005) Surf Sci 597:181
42. Williams DE, Hutton RS (1996) Bull Electrochem 12:121
43. Proksch R, Lal R, Hansma PK, Morse D, Stucky G (1996) Biophys J 71:2155
44. Bruckbauer A, Ying L, Rothery AM, Korchev YE, Klenerman D (2002) Anal Chem 74:2612
45. Korchev YE, Raval M, Lab MJ, Gorelik J, Edwards CRW, Rayment T, Klenerman D (2000) Biophys J 78:2675
46. Gorelik J, Gu Y, Spohr HA, Shevchuk AI, Lab MJ, Harding SE, Edwards CRW, Whitaker M, Moss GWJ, Benton DCH, Sanchez D, Darszon A, Vodyanoy I, Klenerman D, Korchev YE (2002) Biophys J 83:3296
47. Shevchuk AI, Gorelik J, Harding SE, Lab MJ, Klenerman D, Korchev YE (2001) Biophys J 81:1759
48. Macpherson JV, Unwin PR (2000) Anal Chem 72:276
49. Kranz C, Friedbacher G, Mizaikoff B, Lugstein A, Smoliner J, Bertagnolli E (2001) Anal Chem 73:2491
50. Frederix PLTM, Gullo MR, Akiyama T, Tonin A, De Rooij NF, Staufer U, Engel A (2005) Nanotechnol 16:997
51. Bard AJ, Fan FF, Pierce DT, Unwin PR, Wipf DO, Zhou F (1991) Science 253:68
52. Kwak J, Bard AJ (1989) Anal Chem 61:1221
53. Kwak C, Anson J, Lee FC (1991) Anal Chem 63:1501
54. Ludwig M, Kranz C, Schuhmann W, Gaub HE (1995) Rev Sci Instrum 66:2857
55. James PI, Garfias-Mesias LF, Moyer PJ, Smyrl WH (1998) J Electrochem Soc 145:L64
56. Katemann BB, Schulte A, Schuhmann W (2003) Chem Eur J 9:2025
57. Albrecht TR, Akamine S, Carver TE, Quate CF (1990) J Vac Sci Technol A 8:3386
58. Gardner CE, Macpherson JV (2002) Anal Chem 74:576A
59. Windsor Scientific (2006) http://www.windsorscientific.co.uk

60. Karrai K, Grober RD (1995) Appl Phys Lett 66:1842
61. Buchler M, Kelley SC, Smyrl WH (2000) Electrochem Solid State Lett 3:35
62. Oyamatsu D, Hirano Y, Kanaya N, Mase Y, Nishizawa M, Matsue T (2003) Bioelectrochemistry 60:115
63. Hirano Y, Mase Y, Oyamatsu D, Yasukawa T, Shiku H, Matsue T (2004) Chem Sensors 20:754
64. Katemann BB, Schulte A, Schuhmann W (2004) Electroanalysis 16:60
65. Etienne M, Schulte A, Schuhmann W (2004) Electrochem Commun 6:288
66. Hengstenberg A, Kranz C, Schuhmann W (2000) Chem Eur J 6:1547
67. Etienne M, Schulte A, Mann S, Jordan G, Dietzel ID, Schuhmann W (2004) Anal Chem 76:3682
68. Turyan I, Etienne M, Mandler D, Schuhmann W (2005) Electroanalysis 17:538
69. Korchev YE, Gorelik J, Lab MJ, Sviderskaya EV, Johnston CL, Coombes CR, Vodyanoy I, Edwards CRW (2000) Biophys J 78:451
70. Gorelik J, Shevchuk AI, Frolenkov GI, Diakonov IA, Lab MJ, Kros CJ, Richardson GP, Vodyanoy I, Edwards CRW, Klenerman D, Korchev YE (2003) Proc Natl Acad Sci USA100:5819
71. Korchev YE, Negulyaev YA, Edwards CRW, Vodyanoy I, Lab MJ (2000) Nature Cell Biol 2:616
72. Korchev YE, Bashford CL, Milovanovic M, Vodyanoy I, Lab MJ (1997) Biophys J 73:653
73. Gorelik J, Zhang Y, Shevchuk AI, Frolenkov GI, Sanchez D, Lab MJ, Vodyanoy I, Edwards CRW, Klenerman D, Korchev YE (2004) Mol Cell Endocr 217:101
74. Zhang Y, Gorelik J, Sanchez D, Shevchuk A, Lab M, Vodyanoy I, Klenerman D, Edwards C, Korchev Y (2005) Kidney Int 68:1071
75. Schäffer TE, Anczykowski B, Fuchs H (2006) Scanning Ion Conductance Microscopy in Bushan B, Fuchs H, Hosaka S (eds) Applied Scanning Probe Methods. Springer, Berlin Heidelberg New York, in press
76. Pastre D, Iwamoto H, Liu J, Szabo G, Shao Z (2001) Ultramicroscopy 90:13
77. Mann SA, Hoffmann G, Hengstenberg A, Schuhmann W, Dietzel ID (2002) J Neurosci Meth 116:113
78. Nitz H, Kamp J, Fuchs H (1998) Probe Microsc 1:187
79. Williams DE, Mohiuddin TF, Zhu YY (1998) J Electrochem Soc 145:2664
80. Casillas N, James P, Smyrl WH (1995) J Electrochem Soc 142:L16
81. Mirkin MV, Horrocks BR (2000) Anal Chim Acta 406:119
82. Bard AJ, Fan FRF, Mirkin MV (1994) Electroanal Chem 18:243
83. Selzer Y, Mandler D (2000) Anal Chem 72:2383
84. Selzer Y, Mandler D (1999) Electrochem Commun 1:569
85. Unwin PR, Bard AJ (1991) J Phys Chem 95:7814
86. Kwak J, Bard AJ (1989) Anal Chem 61:1221
87. Ferrigno R, Girault HH (2000) J Electroanal Chem 492:1
88. Harriman K, Gavaghan DJ, Houston P, Suli E (2000) Electrochem Commun 2:157
89. Bartlett PN, Taylor SL (1998) J Electroanal Chem 453:49
90. Nann T, Heinze J (2003) Electrochim Acta 48:3975
91. Fulian Q, Fisher AC, Denuault G (1999) J Phys Chem B 103:4393
92. Fulian Q, Fisher AC, Denuault G (1999) J Phys Chem B 103:4387
93. Sklyar O, Wittstock G (2002) J Phys Chem B 106:7499
94. Sklyar O, Ufheil J, Heinze J, Wittstock G (2003) Electrochim Acta 49:117
95. Zhou F, Unwin PR, Bard AJ (1992) J Phys Chem 96:4917
96. Zoski CG, Aguilar JC, Bard AJ (2003) Anal Chem 75:2959
97. Amphlett JL, Denuault G (1998) J Phys Chem B 102:9946

98. Sarler B, Mavko B, Kuhn G (1993) In: Wrobel LC, Brebbia CA (eds) Computational Methods for Free and Moving Boundary Problems, vol. 1, 2nd edn. Elsevier Science, Oxford, p 373–398

99. Sklyar O, Kueng A, Kranz C, Mizaikoff B, Lugstein A, Bertagnolli E, Wittstock G (2005) Anal Chem 77:764

100. Kottke PA, Fedorov AG (2005) J Electroanal Chem 583:221

101. Jones CE, Unwin PR, Macpherson JV (2003) Chem Phys Chem 4:139

102. Jones CE, Macpherson JV, Unwin PR (2000) J Phys Chem B 104:2351

103. Holder MN, Gardner CE, Macpherson JV, Unwin PR (2005) J Electroanal Chem 585:8

104. Zoski CG, Mirkin MV (2002) Anal Chem 74:1986

105. Zoski CG (2002) Electroanalysis 14:1041

106. Mao BW, Ye JH, Zhuo XD, Mu JQ, Fen ZD, Tian ZW (1992) Ultramicroscopy 42–44:464

107. Zhang B, Wang E (1994) Electrochim Acta 39:103

108. Chen ZF, Wang E (1994) Electroanalysis 6:672

109. Quaade UJ, Oddershede L (2002) Europhys Lett 57:611

110. Kerfriden S, Nahle AH, Campbell SA, Walsh FC, Smith JR (1998) Electrochim Acta 43:1939

111. Bach CE, Nichols RJ, Beckmann W, Meyer H, Schulte A, Besenhard JO, Jannakoudakis PD (1993) J Electrochem Soc 140:1281

112. Zhu L, Claude-Montigny B, Gattrell M (2005) Appl Surf Sci 252:1833

113. Gueell AG, Diez-Perez I, Gorostiza P, Sanz F (2004) Anal Chem 76:5218

114. Bach CE, Nichols RJ, Meyer H, Besenhard JO (1994) Surf Coat Technol 67:139

115. Melmed AJ (1991) J Vac Sci Technol B 9:601

116. Lee YH, Tsao GT, Wankat PC (1978) Ind Eng Chem Fund 17:59

117. Abbou J, Demaille C, Druet M, Moiroux J (2002) Anal Chem 74:6355

118. Slevin CJ, Gray NJ, Macpherson JV, Webb MA, Unwin PR (1999) Electrochem Commun 1:282

119. Macpherson JV, Unwin PR (2001) Anal Chem 73:550

120. Shi G, Garfias-Mesias LF, Smyrl WH (1998) J Electrochem Soc 145:2011

121. Lee Y, Bard AJ (2002) Anal Chem 74:3626

122. Maruyama K, Ohkawa H, Ogawa S, Ueda A, Niwa O, Suzuki K (2006) Anal Chem 78:1904

123. Xiong H, Guo J, Kurihara K, Amemiya S (2004) Electrochem Commun 6:615

124. Wang CL, Creasy KE, Shaw BR (1991) J Electroanal Chem 300:365

125. Szunerits S, Walt DR (2003) Chem Phys Chem 4:186

126. Walsh DA, Fernandez JL, Mauzeroll J, Bard AJ (2005) Anal Chem 77:5182

127. Wong SS, Joselevich E, Woolley AT, Cheung CL, Lieber CM (1998) Nature 394:52

128. Hafner JH, Cheung CL, Woolley AT, Lieber CM (2001) Progr Biophys Mol Biol 77:73

129. Hafner JH, Cheung CL, Oosterkamp TH, Lieber CM (2001) J Phys Chem B 105:743

130. Nguyen CV, Ye Q, Meyyappan M (2005) Meas Sci Technol 16:2138

131. Solares SD, Matsuda Y, Goddard WA, III (2005) J Phys Chem B 109:16658

132. Solares SD, Esplandiu MJ, Goddard WA, III, Collier CP (2005) J Phys Chem B 109:11493

133. Deng Z, Yenilmez E, Leu J, Hoffman JE, Straver EWJ, Dai H, Moler KA (2004) Appl Phys Lett 85:6263

134. Zhang L, Ata E, Minne SC, Hough P (2004) IEEE Int Conf MEMS Systems, 17th Technical Digest, Maastricht, Netherlands, 25–29 January 2004, p 438

135. Nguyen CV, So C, Stevens RM, Li Y, Delziet L, Sarrazin P, Meyyappan M (2004) J Phys Chem B 108:2816

136. Mizutani W, Choi N, Uchihashi T, Tokumoto H (2001) Jpn J Appl Phys Part 1 40:4328

137. Shingaya Y, Nakayama T, Aono M (2002) Physica B: Cond Matter 323:153

138. Xiao B, Albin S (2005) Int J Nanosci 4:437

139. Li J, Ng HT, Cassell A, Fan W, Chen H, Ye Q, Koehne J, Han J, Meyyappan M (2003) Nano Lett 3:597
140. Koehne J, Li J, Cassell AM, Chen H, Ye Q, Ng HT, Han J, Meyyappan M (2004) J Mater Chem 14:676
141. Wilson NR, MacPherson JV (2003) Nano Lett 3:1365
142. Wilson NR, Cobden DH, Macpherson JV (2002) J Phys Chem B 106:13102
143. Heller I, Kong J, Heering HA, Williams KA, Lemay SG, Dekker C (2005) Nano Lett 5:137
144. Hafner JH, Cheung CL, Lieber CM (1999) J Am Chem Soc 121:9750
145. Burt DP, Wilson NR, Weaver JMR, Dobson PS, Macpherson JV (2005) Nano Lett 5:639
146. Patil A, Sippel J, Martin GW, Rinzler AG (2004) Nano Lett 4:303
147. Esplandiu MJ, Bittner v G, Giapis KP, Collier PC (2004) Nano Lett 4:1873
148. Oshea SJ, Atta RM, Welland ME (1995) Rev Sci Instr 66:2508
149. Fasching RJ, Tao Y, Prinz FB (2004) Proc SPIE Int Soc Opt Eng 5342:53
150. Fasching RJ, Tao Y, Prinz FB (2005) Sens Actuators B 108:964
151. Akiyama T, Gullo MR, De Rooij NF, Tonin A, Hidber H-R, Frederix PLTM, Engel A, Staufer U (2004) Jpn J Appl Phys Part 1 43:3865
152. Hirata Y, Yabuki S, Mizutani F (2004) Bioelectrochemistry 63:217
153. Dobson PS, Weaver JMR, Holder MN, Unwin PR, Macpherson JV (2005) Anal Chem 77:424
154. Shin H, Kranz C, Mizaikoff B, Hesketh P (2006) Anal Chem, submitted
155. Tseng AA (2004) J Micromech Microeng 14: R15
156. Reyntjens S, Puers R (2001) J Micromech Microeng 11:287
157. Petit D, Faulkner CC, Johnstone S, Wood D, Cowburn RP (2005) Rev Sci Instrum 76:026105/1
158. Lugstein A, Bertagnolli E, Kranz C, Mizaikoff B (2002) Surf Int Anal 33:146
159. Lugstein A, Bertagnolli E, Kranz C, Kueng A, Mizaikoff B (2002) Appl Phys Lett 81:349
160. Lehrer C, Frey L, Petersen S, Sulzbach T, Ohlsson O, Dziomba T, Danzebrink HU, Ryssel H (2001) Microelectron Eng 57–58:721
161. Muranishi M, Sato K, Hosaka S, Kikukawa A, Shintani T, Ito K (1997) Jpn J Appl Phys Part 2 36: L 942
162. Krogmeier JR, Dunn RC (2001) Appl Phys Lett 79:4494
163. Pilevar S, Edinger K, Atia W, Smolyaninov I, Davis C (1998) Appl Phys Lett 72:3133
164. Ando Y, Nagashima T, Kakuta K (2000) Tribol Lett 9:15
165. Ximen H, Russell PE (1992) Ultramicroscopy 42–44:1526
166. Dziomba T, Danzebrink HU, Lehrer C, Frey L, Sulzbach T, Ohlsson O (2001) J Microsc 202:22
167. Kueng A, Kranz C, Lugstein A, Bertagnolli E, Mizaikoff B (2003) Angew Chem Int Ed 42:3238
168. Kranz C, Mizaikoff B, Lugstein A, Bertagnolli E (2002) ACS Symp Ser 811:320
169. FEI Corporation (2006) http://www.feicompany.com
170. Kueng A, Kranz C, Lugstein A, Bertagnolli E, Mizaikoff B (2005) Angew Chem Int Ed 44:3419
171. Davoodi A, Pan J, Leygraf C, Norgren S (2005) Electrochem Solid State Lett 8:B21
172. Binnig G, Rohrer H, Gerber C, Weibel E (1982) Phys Rev Lett 49:57
173. Jones CE, Macpherson JV, Barber ZH, Somekh RE, Unwin PR (1999) Electrochem Commun 1:55
174. Macpherson JV, Jones CE, Barker AL, Unwin PR (2002) Anal Chem 74:1841
175. Gardner CE, Unwin PR, Macpherson JV (2005) Electrochem Commun 7:612
176. Kueng A, Kranz C, Mizaikoff B, Lugstein A, Bertagnolli E (2003) Appl Phys Lett 82:1592
177. Macpherson JV, Gueneau de Mussy J-P, Delplancke J-L (2002) J Electrochem Soc 149:B306

178. Lee Y, Ding Z, Bard AJ (2002) Anal Chem 74:3634
179. Abbou J, Anne A, Demaille C (2004) J Am Chem Soc 126:10095
180. Horrocks BR (2001) In: Bard AJ (ed) Scanning Electrochemical Microscopy. Marcel Dekker, New York, p 445
181. Wittstock G (2001) Fres J Anal Chem 370:303
182. Kranz C, Kueng A, Lugstein A, Bertagnolli E, Mizaikoff B (2004) Ultramicroscopy 100:127
183. Diao G, Wang P (2001) Bull Electrochem 17:505
184. Hirano Y, Mitsumori Y, Oyamatsu D, Nishizawa M, Matsue T (2003) Biosens Bioelectron 18:587
185. Kurzawa C, Hengstenberg A, Schuhmann W (2002) Anal Chem 74:355
186. Putman CAJ, van der Werf KO, De Grooth BG, Van Hulst NF, Greve J (1994) Appl Phys Lett 64:2454
187. Delain E, Michel D, Le Grimellec C (2000) Morphologie 84:25
188. Zhou H, Kasai S, Matsue T (2001) Anal Biochem 290:83
189. Oyamatsu D, Kanaya N, Shiku H, Nishizawa M, Matsue T (2003) Sens Actuators B 91:199
190. Horrocks BR, Schmidtke D, Heller A, Bard AJ (1993) Anal Chem 65:3605
191. Wei C, Bard AJ, Nagy G, Toth K (1995) Anal Chem 67:1346
192. Yasukawa T, Kaya T, Matsue T (1999) Anal Chem 71:4637
193. Kueng A, Kranz C, Mizaikoff B (2005) Biosens Bioelectron 21:346
194. Kueng A, Kranz C, Lugstein A, Bertagnolli E, Mizaikoff B (2005) Methods in Molecular Biology. Humana, Totowa NJ, p 403
195. Bath BD, Phipps JB, Scott ER, Uitto OD, White HS (2001) Stud Surf Sci Catal 132:1015
196. Kueng A, Kranz C, Mizaikoff B (2004) Biosens Bioelectron 19:1301
197. Riordan JR, Rommens JM, Kerem B, Alon N, Rozmahel R, Grzelczak Z, Zielenski J, Lok S, Plavsic N, Chou JL (1989) Science 245:1066
198. Wartelle C, Schuhmann W, Bloechl A, Bedioui F (2005) Electrochim Acta 50:4988
199. Ryabova V, Schulte A, Erichsen T, Schuhmann W (2005) Analyst 130:1245
200. Isik S, Berdondini L, Oni J, Blochl A, Koudelka-Hep M, Schuhmann W (2005) Biosens Bioelectron 20:1566
201. Diab N, Oni J, Schuhmann W (2005) Bioelectrochemistry 66:105
202. Castillo J, Isik S, Blochl A, Pereira-Rodrigues N, Bedioui F, Csoregi E, Schuhmann W, Oni J (2005) Biosens Bioelectron 20:1559
203. Oni J, Pailleret A, Isik S, Diab N, Radtke I, Bloechl A, Jackson M, Bedioui F, Schuhmann W (2004) Anal Bioanal Chem 378:1594
204. Pailleret A, Oni J, Reiter S, Isik S, Etienne M, Bedioui F, Schuhmann W (2003) Electrochem Commun 5:847
205. Oni J, Diab N, Radtke I, Schuhmann W (2003) Electrochim Acta 48:3349
206. Groppe M, Thanos S, Schuhmann W, Heiduschka P (2003) Anal Bioanal Chem 376:797
207. Isik S, Etienne M, Oni J, Bloechl A, Reiter S, Schuhmann W (2004) Anal Chem 76:6389
208. Pereira-Rodrigues N, Zurgil N, Chang S-C, Henderson JR, Bedioui F, McNeil CJ, Deutsch M (2005) Anal Chem 77:2733
209. Pohl DW, Denk W, Lanz M (1984) Appl Phys Lett 44:651
210. Minta A, Kao JPY, Tsien RY (1989) J Biol Chem 264:8171
211. Rothery AM, Gorelik J, Bruckbauer A, Yu W, Korchev YE, Klenerman D (2003) J Microsc 209:94
212. Darszon A, Labarca P, Nishigaki T, Espinosa F (1999) Physiol Rev 79:481
213. Gorelik J, Shevchuk A, Ramalho M, Elliott M, Lei C, Higgins CF, Lab MJ, Klenerman D, Krauzewicz N, Korchev Y (2002) Proc Natl Acad Sci USA 99:16018
214. Forstova J, Krauzewicz N, Sandig V, Elliott J, Palkova Z, Strauss M, Griffin BE (1995) Human Gene Therapy 6:297

215. Krauzewicz N, Cox C, Soeda E, Clark B, Rayner S, Griffin BE (2000) Gene Therapy 7:1094
216. Feng W, Rotenberg SA, Mirkin MV (2003) Anal Chem 75:4148
217. Liu B, Rotenberg SA, Mirkin MV (2000) Proc Natl Acad Sci USA 97:9855
218. Boldt F-M, Heinze J, Diez M, Petersen J, Boersch M (2004) Anal Chem 76:3473
219. Matsumura Y, Kajino K, Fujimoto M (1980) Membrane Biochem 3:99
220. Giaume C, Korn H (1983) Science 220:84
221. Glab S, Hulanicki A, Edwall G, Ingman F (1989) Crit Rev Anal Chem 21:29
222. Rohwerder M, de Boeck A, Ogle K, Rehnisch O, Reier T, Stellnberger KH, Steinbeck G, Wormuth R, Stratmann M (2001) Galvatech '2001, 5th International Conference on Zinc and Zinc Alloy Coated Steel Sheet, Brussels, Belgium, 26–28 June 2001, p 585
223. Hornung E, Rohwerder M, Stratmann M (2001) Proc Electrochem Soc 2001 22:618
224. Rohwerder M, Stratmann M, Leblanc P, Frankel GS (2006) Anal Meth Corr Sci Eng: 605
225. Rohwerder M, Stratmann M (2000) Proc Electrochem Soc 1999 28:302
226. Amirudin A, Thierry D (1995) Progr Org Coat 26:1
227. Grundmeier G, Stratmann M (2005) Ann Rev Mater Res 35:571
228. James P, Casillas N, Smyrl WH (1996) J Electrochem Soc 143:3853
229. Shi G, Garfias-Mesias LF, Smyrl WH (1998) J Electrochem Soc 145:2011
230. Garfias-Mesias LF, Smyrl WH (1999) Electrochim Acta 44:3651
231. Garfias-Mesias LF, Smyrl WH (1999) J Electrochem Soc 146:2495
232. Pan J, Femenia M, Leygraf C (2000) Proc Electrochem Soc 1999 28:131
233. Guillaumin V, Schmutz P, Frankel GS (2001) J Electrochem Soc 148: B163
234. de Wit JHW (2004) Electrochim Acta 49:2841
235. Rodolfa KT, Bruckbauer A, Zhou D, Schevchuk AI, Korchev YE, Klenerman D (2006) Nano Lett 6:252
236. Kazinczi R, Szocs E, Kalman E, Nagy P (1998) Appl Phys A:S535
237. Akiyama T, Gullo MR, De Rooij NF, Staufer U, Tonin A, Engel A, Frederix PLTM (2003) AIP Conf Proc 696:166
238. Dobson Phillip S, Weaver John MR, Holder Mark N, Unwin Patrick R, Macpherson Julie V (2005) Anal Chem 77:424
239. Rodolfa KT, Bruckbauer A, Zhou D, Korchev YE, Klenerman D (2005) Angew Chem Int Ed 44:6854

9 New AFM Developments to Study Elasticity and Adhesion at the Nanoscale

Robert Szoszkiewicz · Elisa Riedo

Abbreviations and Symbols

α	Coefficient
δ	AFM tip–sample deformation depth
$\Delta \overline{F}$	Changes of the ultrasonic force
Λ	Dislocation density at the surface
a	Radius of a contact area between an AFM tip and a sample
A	Ultrasonic amplitude
AFAM	Acoustic force atomic microscopy
AFM	Atomic force microscopy or atomic force microscope
AH	Adhesion hysteresis
AH_{el}	Elastic part of adhesion hysteresis
AH^{rough}	Adhesion hysteresis in a contact between rough surfaces
BCP	Burnham–Colton–Pollock contact mechanics
CapH	Capillary forces hysteresis
d	AFM tip–sample distance
D	Fractal dimension
d_d	Dislocation length
d_o	Given indentation
DMT	Derjaguin–Muller–Toporov contact mechanics
d_s	Transducer's amplitude in SLAM
d_{tot}	Total normal or lateral displacement of an AFM tip (the sum of the contact deformation plus the cantilever bending)
$E_{1,2}$	Young's modulus of the AFM tip or the sample
E^*	Reduced Young's modulus between the AFM tip and the sample
F	Force
F_{all}	Sum of all interaction forces between an AFM tip and a sample
F_{adh}	AFM tip–sample adhesion force
F_o	Given force between an AFM tip and a sample
$F_{pull-off}$	Pull-off force (maximum tensile force at which an AFM tip and a sample are still in contact)
F_R	Frictional force
F_R^{rough}	Frictional force in a contact between rough surfaces
F_{SP}	AFM set point (load force set at tip–sample contact)
f-d curve	Force-distance curve
$G_{1,2}$	Shear modulus of the tip, or the sample

G^*	reduced shear modulus between the tip and the sample
H	Hurst exponent
HF	High frequency
HFM	Heterodyne force microscopy
HOPG	Highly-oriented pyrolitic graphite
JKR	Johnson–Kendall–Roberts contact mechanics
$k_{contact}$	Normal or lateral stiffness of an AFM tip in contact with a surface
$k_{contact}^{l}$	Lateral stiffness of an AFM tip in contact with a surface
$k_{contact}^{n}$	Normal stiffness of an AFM tip in contact with a surface
k_{lever}	Normal or lateral stiffness of a free AFM tip
k_{level}^{n}	Normal stiffness of a free AFM tip
l_c	Contact length ($l_c = 2a$)
LF	Low frequency
l_i	AFM image size
NT	Nanotube
R	Reduced AFM tip–sample curvature radius
SLAM	Scanning local acceleration microscopy
SMM	Scanning microdeformation microscopy
S_q	RMS roughness at the acquired AFM images size l_i
UFM	Ultrasonic force microscopy
v	Poisson ratio
w_{t-s}	AFM tip–sample adhesion energy
z_c	Cantilever's response amplitude in SLAM
z_{HF}	Cantilever response amplitude at the excitation frequency in AFAM

9.1
Introduction

The development of new materials with the size of a few nanometers has opened a new field of scientific and technological research. Nanomaterials such as carbon nanotubes, oxide nanobelts and semiconductor nanowires are promising building blocks in future integrated nanoelectronic and photonic circuits, nanosensors, interconnects and electromechanical nanodevices. The goal is to develop faster and better communication systems and transports, as well as smarter and smaller nanodevices for biomedical applications. To reach these objectives it is crucial to have knowledge and ability to control the mechanical behavior of these nanoobjects. In general, the mechanical properties of the materials at the nanoscale are not well-understood. The experimental challenge is to have an instrumentation that allows us to image and manipulate a nanoobject, characterize its atomic structure and measure forces of the order of nanonewtons. From the theoretical side, developing a theory of elasticity and friction at the nanoscale is an intriguing task that lies at the crossover between the atomic level and the continuum.

In this chapter we will describe recently developed methods to measure the elastic and adhesive properties of the materials at the nanoscale. Local elastic properties are usually studied by both static and dynamic AFM methods. The static methods include force-distance curves together with AFM-based nanoindentation tests as well as various local realizations of the classical triple point contact tests. The dynamic techniques utilize AFM cantilevers vibrating either at low frequencies, e.g., of a few kHz (modulated nanoindentation), or at ultrasonic frequencies. We will focus our description on two AFM-based dynamic methods: modulated nanoindentation and ultrasonic microscopy.

The chapter is divided in three parts: in Sect. 9.2 we describe continuum contact mechanics theories and their limitations. In the Sect. 9.3 we present the modulated nanoindentation method and we show how this technique is used to obtain high-resolution force versus indentation curves and the elastic modulus of nanomaterials. In the Sect. 9.4 we describe the use of ultrasonic AFM techniques to study local elasticity. We also highlight an application of ultrasonic AFM to investigate the local adhesive properties and their relation with friction at the tip–sample contact.

9.2
Contact Mechanics Theories and Their Limitations

The theories used to describe the contact between idealized macroscopic bodies are called continuum contact mechanics theories. These theories do not consider the local mechanical properties of the materials, but they treat each object in contact as a continuum. For this reason many questions can arise on the validity of these theories when we are dealing with nanocontacts. Here, we will present a brief compilation of classical continuum contact mechanics theories and we will discuss briefly the limitations of these theories in the last part of this section.

Hertz theory dates back to 1882 [1]. This approach assumes that surface forces and adhesion can be neglected and that the elastic deformation due to the contact can be modeled taking into consideration only the geometry of the two contacting bodies and their elastic macroscopic properties. More advanced and general theories for bodies with axisymmetric shape versus a soft elastic surface have been developed by Sneddon [2]. When surface forces such as friction and adhesion are negligible Hertz and Sneddon models can be used, otherwise, other theories need to be considered.

The Derjaguin–Muller–Toporov (DMT) theory expands Hertz by including van der Waals forces between the bodies in the contact region [3]. DMT is applicable to AFM experiments where the reduced AFM tip is small, and the adhesion between the surfaces is small. JKR theory, named after Johnson, Kendall and Roberts, takes into account only short-range forces, such as adhesion, inside the contact region and neglects long-range forces outside of it [4]. For AFM experiments, this theory can be employed for highly adhesive systems when the tip has a low stiffness and large radius. A very complete contact mechanics theory was developed by Maugis and Pollock, and it deals with a wide range of behaviors [5]. By varying a single parameter this theory can go continuously from DMT to JKR, and can be applied to compliant, large and adhesive bodies as well as small rigid materials with low adhesion.

None of the afore-mentioned theories accounts for the AFM tip–sample forces outside the contact area. These forces are either semiempirically estimated (i.e., in the Burnham–Colton–Pollock theory [6, 7]), or calculated by first principles molecular dynamics simulations [8–10]. There is also an enormous bibliography on analytical and numerical studies about contact between varying geometries.

Contact mechanics theories assume no plastic deformation and no viscoelastic phenomena. These are accounted for through finite element studies for bodies of different geometries in contact. With typical AFM tip–sample contact radii of the order of a few nanometers and with loads of the order of tens of nanonewtons, the pressure into the tip–sample contact area is of the order of GPa. High pressure at the contact exceeds the bulk value of macroscopic tensile/compressive strengths (i.e., of about 250 MPa for mica), and the material can deform plastically. The compressive/tensile strengths depend, however, on the density of dislocations into the concerned volume. The average dislocation length d_d is about 100 nm, roughly 1000 times the interatomic distance. The dislocation density at the surface Λ reads: $\Lambda = d_d^{-2}$. Thus, the dislocations are very rare in nanovolumes, and the values of the local tensile/compressive strengths approach their theoretical limits of about 10 percent of the corresponding Young's moduli E (e.g., $E = 70$ GPa for mica). Therefore, in typical AFM scanning conditions, the tip–sample interactions are mostly within the elastic limit. However, too high AFM tip–sample approaching speed and too high loads induce local plastic deformation at the surface [6].

Another important element not typically considered is surface roughness. The existence of roughness complicates tremendously the calculation of the real area of contact, because the contact occurs at several points where the asperities of the surfaces meet. How to incorporate roughness to contact mechanics has been studied for a long time. A recent review on the historical and modern understanding of the true contact area was written by Gao et al. [11]. There is extensive research on stochastic methods to treat contact among rough surfaces [12, 13]. There are also analytical models [14]. In particular, Greenwood and Williamson have assumed the surface as covered by hemispherical asperities with the same radii [15]. Whitehouse and Archard have extended that model by allowing random radii of curvature [16]. Nayak has introduced the random process theory with input parameters such as the distribution of asperities heights, the density of asperities, the mean surface gradient, and the mean curvature of asperities [17]. We will not treat this topic here, but we suggest some interesting reading, found in [18–26].

There is a lingering question as to whether continuum contact mechanics should be used at the nanoscale. At what point does the discreteness of matter start to play such an important role that continuum mechanics can no longer be used? There are some molecular dynamics simulations studying at what point continuum mechanics breaks down. Recently, Luan and Robbins [27] conducted molecular dynamics simulations to test whether the Hertz theory is still valid at the nanoscale. They found that the atomic discreteness at the bulk does not seem to affect the tip–sample contact area as much as roughness does. They concluded that continuum mechanics may underestimate the area of SPM contacts by up to 100% at small loads, though the error decreases with increasing load and radius. Miesbauer et al. [28] conducted molecular dynamics simulations of the contact between two NaCl nanocrystals. After comparing with the Hertz theory, they concluded that this theory can be employed to describe systems with sizes down to 5 nm. Furthermore, several friction force microscopy experiments support the validity of continuum contact mechanics at the nanoscale [29–31].

9.3
Modulated Nanoindentation

9.3.1
Force-Indentation Curves

The ideal and simplest way to measure the elasticity of nanoobjects is to indent an atomic force microscope tip in the nanoobject, while measuring force versus indentation curves. Measurements of force versus indentation curves are particularly challenging on nanoobjects because to insure the elastic regime, we usually have to measure forces of a few nanonewtons versus displacements of a few angstroms with subangstrom resolution.

An AFM force-distance (f-d) curve [32] is a measure of the normal force acting between an AFM tip and a sample surface as a function of the distance between the same tip and the undeformed sample surface. Thus, when the tip starts to be in contact with the sample, a f-d curve is a force versus indentation curve. During an AFM experiment the force is detected through the bending of the AFM cantilever, while the distance is given by the AFM scanner vertical position as the tip approaches the sample (Fig. 9.1). It is thus clear that f-d curves need to be calibrated [32] accounting on the cantilever's bending and the cantilever's spring constant. Recently, some convenient methods have been developed to calibrate the cantilever's spring constants based on thermal noise (see Sader et al. [33], Higgins et al. [34], Proksch et al. [35], and references therein, as well as the review written by Burnham [36] and Gibson [37]). The errors in f-d calibration come from many sources: detection of the cantilever's position and detection of the forces, coupling between torsion and vertical cantilever's bending, hysteresis and resonance of a piezoscanner, etc. Furthermore, Heim et al. [38] as well as Hutter [39] have recently reported on some systematic errors in f-d curves due to usual cantilever tilt with respect to the surface.

Some instrumental developments have pursued the improvement of the force-distance sensitivity through: (1) use of segmented scanners with lower hysteresis,

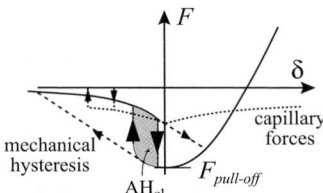

Fig. 9.1. An example of a force-distance curve obtained with an AFM. The f-d curves represent forces between an AFM tip and a sample as a function of the tip–sample indentation (positive value of δ) or separation (negative value of δ). The *arrows* mark that the curve is intrinsically hysteretic, e.g., different for approaching and retracting to/from the sample. The *shaded area* corresponds to the elastic part of adhesion hysteresis (AH$_{el}$), but there are other hysteretic effects present as well. Mechanical hysteresis results from a sudden cantilever snap-in and snap-out to/from the contact, when a gradient of existing forces exceeds a spring constant of a cantilever. Capillary forces hysteresis is associated with the presence of water on the sample. The pull-off force $F_{pull-off}$ is marked as well

(2) separation and minimization of X, Y, Z scanners to obtain better signal to noise ratio, and (3) use of low-coherence lasers to minimize any interference patterns on the f-d curves. Manyes et al. [40] with indentations smaller than one nanometer have obtained local hardness from the onset of a plastic deformation between diamond AFM tips and a Si surface. By improving the electronics to the limit and building scanners with very large resonance frequencies, Rost et al. [41] show that force-distance imaging can be obtained with video rates. Still however, the comparative force-distance measurements with the same AFM configuration are the most meaningful and trusted.

Recently several attempts have been pursued to attain more reliable methods to measure force versus indentation curves. Special attention has been dedicated in finding a technique to investigate the radial elasticity of carbon nanotubes (NTs). Some authors have proposed to vibrate an AFM cantilever in non-contact mode with small vibration amplitudes (a few angstroms) while it is scanned across the sample [42]. Because of van der Waals forces between the tip and the tube, the normal force can be considered in first approximation proportional to the amplitude of the cantilever vibrations. However, in this case it is extremely difficult to obtain quantitative results. In fact, this method requires a good description of tip–sample interactions and a good model of the cantilever vibrations taking into accounting the geometry of the problem and the different forces acting on the cantilever [43].

Another AFM-based method, modulated nanoindentation, has been demonstrated to be a powerful means to measure quantitatively the shear and radial elasticity of nanoobjects [44].

Modulated nanoindentation (Fig. 9.2) consists in indenting an AFM tip in a sample while laterally (parallel to the substrate) or normally (perpendicular to the substrate) oscillating the sample. The idea is to measure the slope of the f-d curves in each point (F_o, d_o). Here, we will describe a modulated nanoindentation experiment on a NT lying on a flat, hard substrate. The oscillations have to be very small, about one angstrom, so that the tip sticks on the nanoobject, i.e., on the NT. During the stick regime, the normal or lateral force necessary to move the NT vertically or laterally with respect to the cantilever support coincides with the force needed to elastically stretch two springs in series [45, 46]: the cantilever, with (normal or lateral) stiffness k_{lever}, and the tip–sample contact with (normal or lateral) stiffness k_{contact}. Mathematically we can write

$$\partial F / \partial d_{\mathrm{tot}} = (1/k_{\mathrm{lever}} + 1/k_{\mathrm{contact}})^{-1} \,, \tag{9.1}$$

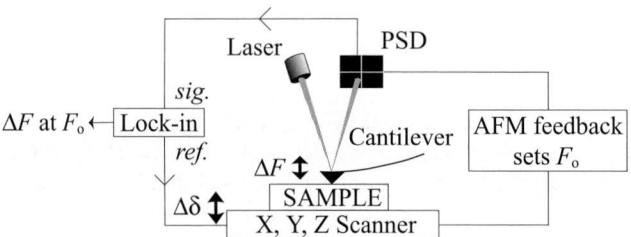

Fig. 9.2. Experimental setup for the modulated nanoindentation method

where F is the normal or lateral force and d_{tot} is the total normal or lateral displacement. Thus, the value of d_{tot} is the sum of the contact deformation plus the cantilever bending. Since k_{lever} is known, a measure of $\partial F/\partial d_{tot}$ at different normal loads, F_o, gives allows for the obtainment of $k_{contact}$ as a function of F_o.

The value of $\partial F/\partial d_{tot}$ is experimentally measured by means of a lock-in amplifier. The lock-in amplifier modulates vertically or laterally the sample position Δd_{tot} by sending an appropriate signal to the piezotube, and it simultaneously measures the signal ΔF coming from the AFM (see Fig. 9.2). For the experiments on the NT the authors have chosen $\Delta d_{tot} = 1.3 \text{Å}$, and they have worked at a frequency of 3.111 kHz. From normal modulation measurements they have obtained the normal contact stiffness $k_{contact}^n$ as a function of F_o, which contains all the information related to the radial elasticity of the NT. Besides, in lateral modulation experiments, we measure the lateral contact stiffness $k_{contact}^l$ versus F_o, which provides information on the shear elastic properties of the nanoobject.

In the following, first we describe how to derive force versus indentation curves from the measurements of $k_{contact}^n$. Second, we show how continuum contact mechanics is used to extract from $k_{contact}^n$ and $k_{contact}^n$ the radial Young's modulus and the axial shear modulus of the NT. In the analogy to Hook's law of elastic deformation, for normal modulation experiments we have $F = k_{contact}^n \delta$, where δ is the indentation of the tip in the NT. By integrating this equation we obtain

$$\int_{F(0)}^{F(d)} \left(\frac{1}{k_{contact}^n} \right) dF = \int_0^d d\delta \,. \tag{9.2}$$

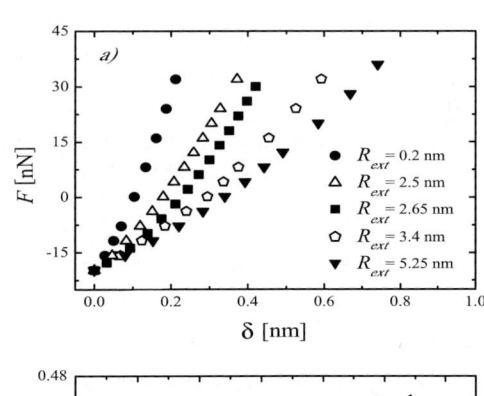

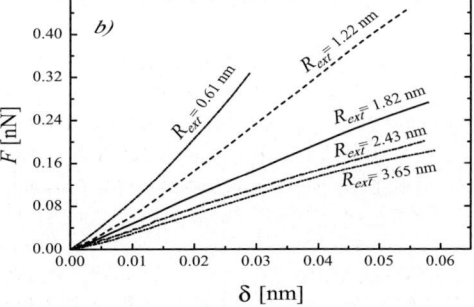

Fig. 9.3. Normal force as a function of indentation for NT of different radii, obtained by integration of experimental $1/k_{contact}$ versus F curves using the trapeze method (**a**). Theoretical normal force as a function of indentation for NT of different radii (different number of layer N), obtained by simulating the indentation of a rigid sphere in a NT (**b**)

By substituting (9.1) into (9.2) we obtain from the experimentally measured $\partial F/\partial d_{\text{tot}}$ the F versus δ curves

$$\int_{F(0)}^{F(d)} \left(\frac{\Delta d_{\text{tot}}}{\Delta F} - \frac{1}{k_{\text{lever}}^{\text{n}}} \right) dF = \int_0^d d\delta , \tag{9.3}$$

where $k_{\text{level}}^{\text{n}}$ is the normal stiffness of a free AFM tip. In Fig. 9.3 we show the indentation curves obtained from experiments and by means of molecular dynamics simulations. Both experiments and theory show that the radial stiffness of carbon NT increases by decreasing the tube's external radius, e.g., by decreasing the internal radius. It is interesting to note that experiments and calculations agree in finding the same 3/2-power law dependence of F on indentation, $F \propto \delta^{3/2}$, which is expected in the framework of the Hertz contact mechanics [5].

9.3.2
Elastic Moduli

The elastic modulus of a material is usually extracted from force versus indentation curves by using the following relationship:

$$E^* = \frac{\sqrt{\pi}}{2} \frac{\partial F}{\partial \delta} \frac{1}{\sqrt{A}} \tag{9.4}$$

where E^* is the effective modulus of elasticity defined by $E^* = [(1 - v_1^2)/E_1 + (1 - v_2^2)/E_2]^{-1}$ with $v_{1,2}$ and $E_{1,2}$ the Poisson's ratio and Young's moduli of the indenter and sample and A is the contact area, respectively. In (9.4), $\frac{\partial F}{\partial \delta}$ is $k_{\text{contact}}^{\text{n}}$ and is measured during the experiments, whereas to determine A, we need to model the contact between the indenter, i.e., the AFM tip and the sample. The simplest way to do it is to use continuum contact mechanics models.

Molecular dynamics simulations show that continuum contact mechanics have some limits in the case of a nanometer size sphere in contact with a NT. Errors up to 30% have been calculated. However, the 3/2-power law dependence found in experimental and theoretical indentation curves is a strong indication of the reliability of the Hertz theory for our system. Following the continuum contact mechanics theories, for small normal and lateral forces, the contact area and the contact radius a [45, 46] for a sphere indenting a flat surface are given by

$$A = \pi a^2 \tag{9.5a}$$

$$a = R^{1/3} (F_{\text{SP}} + F_{\text{adh}})^{1/3}/(E^*)^{1/3} , \tag{9.5b}$$

where R is the sphere (tip) radius, F_{sp} is the load force set at the tip–sample contact, F_{adh} is the tip–sample adhesion force. Furthermore, we can show that the normal and lateral contact stiffness are directly proportional to the contact radius a [5]

$$k_{\text{contact}}^{\text{n}} = 2E^* a \tag{9.6b}$$

$$k_{\text{contact}}^{\text{l}} = 2G^* a , \tag{9.6b}$$

where G^* is the effective shear modulus defined by $G^* = [(2 - v_1)/G_1 + (2 - v_2)/G_2]^{-1}$ with $G_{1,2}$ being the shear moduli of the sphere and the flat surface. We underline that (9.5a) and (9.6b) are equivalent to (9.4), but now we also have access to the area of contact through (9.5b). Furthermore (9.5) and (9.6) can also be applied to the shear elasticity measurement. If the geometry in consideration is different from that one of sphere-plane, e.g., sphere-cylinder geometry as in the case of an AFM tip indenting a nanotube, relationships slightly different from those presented in (9.5) and (9.6) need to be considered. A detailed treatment of contact stiffness in different geometries can be found in [5].

In general, under the assumptions of continuum elasticity theories, the dependence of the contact stiffness versus the external force between two elastic solids in contact is an analytical function of the external force, the tip and the sample elastic moduli, and some geometrical parameters, e.g., the tip radius. By fitting the experimental $k_{contact}$ versus F curves obtained by modulated nanoindentation with the appropriate analytical function, it is possible to obtain the normal and transversal modulus of the nanoobject.

Figure 9.4 shows the experimental values of $k_{contact}^n$ as a function of normal loads for a carbon NT. These curves have been fitted with the appropriate equations [44] for the sphere-cylinder geometry. We underline that in this fitting procedure the only free fitting parameter is the radial Young's modulus of the NT, which thus can be extracted without ambiguity. In Fig. 9.5 we show the radial modulus found by means

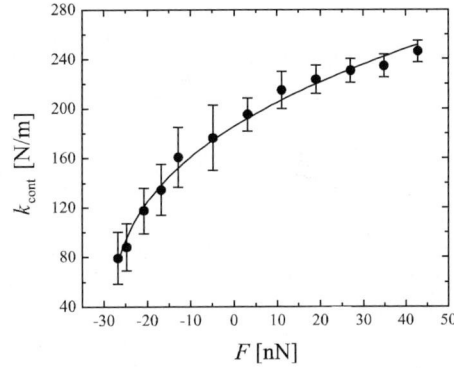

Fig. 9.4. Experimental normal contact stiffness versus normal indentation force for a 3 nm tube radius

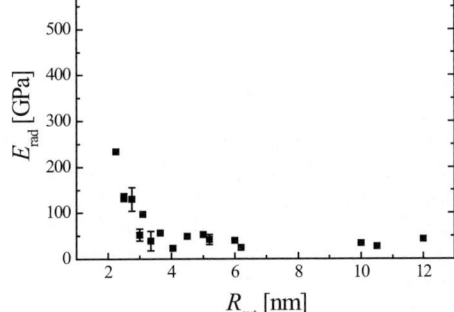

Fig. 9.5. Experimental values of the radial Young's modulus of carbon NTs as a function of external radius as obtained from normal modulation experiments. *Error bars* correspond to mean errors on tubes of the same diameter

of this method as a function of the NT radius for NTs with external to internal radius ratio equal to 2.

9.4
Ultrasonic Methods at Local Scales

Classical acoustic microscopy [47] is a very well-developed non-destructive method for evaluating the material's elasticity on a large scale because its resolution is limited, almost as in the case of an optical microscopy, up to about one micron. However, it has been reported that near-field techniques can greatly overcome physical barriers for microscope resolution [48–50]. Therefore, at local scales, the near-field combination of ultrasound with AFM has been widely applied. The experimental techniques include scanning microdeformation microscopy (SMM), acoustic force atomic microscopy (AFAM), scanning local acceleration microscopy (SLAM), heterodyne force microscopy (HFM), and ultrasonic force microscopy (UFM).

Also, an interesting recent application of the UFM technique has been proposed by Szoszkiewicz et al. [51]. With a modified UFM setup, they have been able to measure the elastic part of the adhesion hysteresis at micro- and nanoscales, and relate these measurements with corresponding friction in the case of fine-polished metallic samples as well as HOPG and mica [52–54]. This approach sets an ultimate way to measure local frictional forces, as will be described in the Sect. 9.4.3.

9.4.1
Brief Description of Ultrasonic Methods

The SMM method was developed by Cretin et al. [55, 56]. Here, a sapphire tip is mounted at the end of a silicon cantilever that is embedded on the piezoelectric transducer. Looped with a specific amplifier, this assembly constitutes an electromechanical oscillator oscillating on the first flexural mode of the cantilever and generating microdeformations at the sample surface. The strain field generated by the cantilever induces an acoustic wave in the sample. In the original setup [55] the acoustic wave was detected with a piezoelectric transducer placed underneath the sample. In the improved setup (Fig. 9.6), which works in reflection, the tip contacts the sample, and any variations in the local elastic properties of the sample induce

Fig. 9.6. The principles of the SMM technique. A sapphire tip is mounted at the end of a silicon cantilever embedded in the piezotransducer. Looped with an amplifier it is an oscillator at the first flexural mode of the cantilever. The strain field induces microdeformations of the sample and frequency change of the oscillator, which are detected by a frequency counter. Heterodyne interferometer measures the true amplitude of the cantilever's vibrations

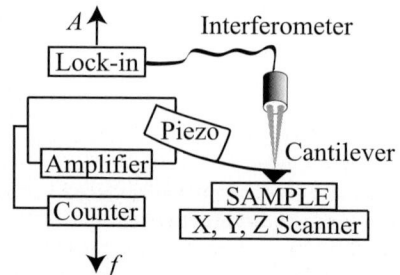

variations of the resonance frequency of the oscillator. A counter detects these frequency variations, and a heterodyne interferometer detects the cantilever amplitude. The SMM technique produces an image of the subsurface defects or of the local elastic properties of the sample. Local elastic properties are inferred from a fit of the resonance spectrum (amplitude versus frequency) of the clamped (through contact stiffness) cantilever beam to the measured resonance spectrum of the cantilever. Due to high curvature radii of sapphire tips, the original resolution was limited to a few microns. Nowadays the tips can be crafted with finer precision, i.e., using the Focused Electron Beam (FEB) or the Focused Ion Beam (FIB) techniques [57,58].

AFAM (Fig. 9.7) was originally developed by Rabe and Arnold [59]. The cantilever, in contact with the sample, is actuated by small amplitude vibrations of a few angstroms at one of its high-frequency (MHz) contact resonances (flexural or torsional). Unless a sample is extremely compliant (i.e., cells), high-frequency cantilever modes provide an enhanced sensitivity to contact stiffness [60,61]. In AFAM, the cantilever response amplitude at the excitation frequency z_{HF} provides the local tip–sample elasticity, and the quasistatic cantilever's position changes provide topography of a sample. Insufficient AFM frequency bandwidth propelled the usage of a knife-edge detector [59] to measure the value of z_{HF} for higher ultrasonic modes. Later, it was realized that together with a reference sample of known elasticity both first and second modes of the cantilever's vibrations are sufficient to determine the tip–sample contact stiffness and an unknown elasticity of the sample. Nowadays, with a proper choice of AFM cantilevers, AFAM utilizes heterodyne down frequency converters or just commercial AFM systems without a need for knife-edge detectors [61–63]. Of the ultrasonic techniques, AFAM is probably the most intensively studied. The effects of microscope tip radius changes due to abrasion [62], humidity [64], and geometry of the tip [65] have been discussed. It has been also reported that attachment of a concentrated mass at the cantilever's extremity improves the AFAM's sensitivity [66].

The SLAM technique (Fig. 9.8) was developed by Burnham et al. [67]. Here, a transducer placed underneath the sample actuates it in the kHz range with amplitude d_s. The cantilever's response z_{HF} at the excitation frequency is detected by a standard beam-bounce technique. It has been shown in [68,69] that at frequencies

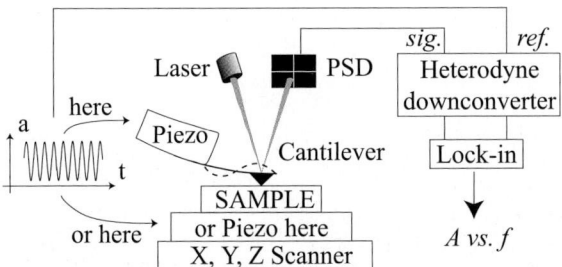

Fig. 9.7. The principles of the AFAM technique. The cantilever, in contact with a sample, is actuated by small amplitude vibrations of a few angstroms at its high-frequency contact resonance. Up to certain frequency (a few MHz) the cantilever vibrations at ultrasonic frequencies are detected with a typical photodiode and downshifted through a heterodyne downconverter

Fig. 9.8. The principles of the SLAM technique. A transducer placed below the sample actuates an AFM cantilever at the kHz range frequency. The lock-in amplifier insures precise detection of the cantilever's vibration at the excitation frequency

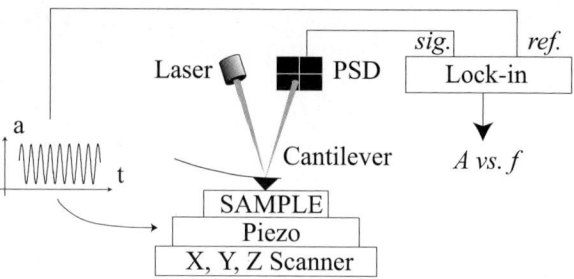

higher than the tip–sample contact resonance, and away from any spurious resonance, the ratio of d_s/z_{HF} is a sensitive measure of the contact stiffness. Therefore, subsequent SLAM imaging at different frequencies reveals the changes of contrast between areas of different elasticity, and allows calculations of the local Young's moduli.

Cuberes et al. [70] developed the HFM technique (Fig. 9.9). The HFM utilizes simultaneous excitation of both the tip and the sample. An ultrasonic transducer placed underneath actuates the sample and another ultrasonic transducer at the base of the cantilever actuates the tip. Both frequencies are in the range of MHz and differ only by a few kHz to produce beats. The cantilever's response in the beat frequency is detected by a conventional beam-bounce technique. The resulting HFM response (or image) depends on elasticity, but the quantitative interpretation remains unclear.

The UFM method was invented by Kolosov and Yamanaka [71]. Here (Fig. 9.10), an ultrasonic excitation of a few MHz is applied to a tip–sample contact via an actuator placed underneath the sample or above the AFM cantilever. The fast ultrasonic excitation is modulated by a slow frequency signal between 100 Hz up to a few kHz at which the cantilever can readily respond. The AFM feedback is usually set to be slow to maintain only the long time tip–sample position. Then, the cantilever reacts to the mean force between the tip and a sample (rectification) called the ultrasonic force and calculated over at least one period of vibrations. With increasing ultrasonic amplitudes, ultrasonic force increases only slightly up to the jump-up point when the tip–sample contact breaks. The ultrasonic force at the jump-up point

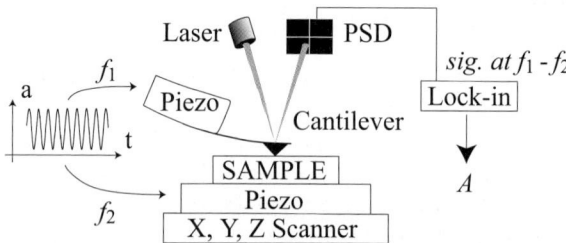

Fig. 9.9. The principles of the HFM technique. Two ultrasonic transducers (one below the sample, one above the tip) deliver ultrasonic vibrations differing by a few kHz. The lock-in amplifier insures precise detection of the cantilever's vibration at the beat frequency

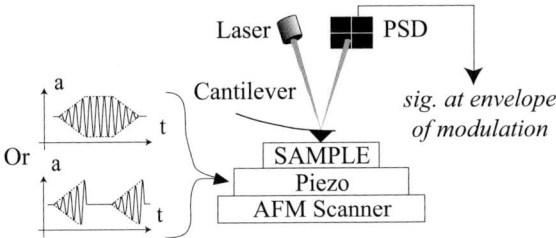

Fig. 9.10. The principles of the UFM technique. The cantilever is initially in contact with the sample, is actuated by a high-frequency ultrasonic signal at a few MHz, at which it responds only inertially. High-frequency signal is modulated by waveform of either a triangular or a trapezoidal shape. At sufficiently high ultrasonic amplitude the tip–sample contact breaks and the average cantilever's position at the modulating envelope jumps up

increases abruptly [52, 71, 72]. For larger ultrasonic amplitudes the tip stays outside the contact for a longer fraction of high-frequency cycles. Recently, it has been developed [52–54] that at least for metallic samples at ultrasonic frequencies avoiding any cantilever resonance and subharmonic [73–75], the cantilever response at the low frequency is a measure of local elasticity. In the absence of capillary forces between an AFM tip and a sample (either in dry environment or in liquids), and with the knowledge of R as well as the tip–sample adhesion energy w_{t-s}, the value of E^* is [53]

$$E^* \approx 0.2\pi^{5/2}[F_{SP}(A_2 - A_1) + \Delta\overline{F_2}A_2 - \Delta\overline{F_1}A_1]^{-3/2}R^2 w_{t-s}^{5/2} \qquad (9.7)$$

where F_{SP} is the set point, and $\Delta\overline{F_1}$, $\Delta\overline{F_2}$ are changes of the ultrasonic force for ultrasonic amplitudes A_1, A_2 measured just before (index 1) and just after (index 2) the point of jump-up.

9.4.2
Applications of Ultrasonic Techniques in Elasticity Mapping

AFM-based ultrasonic techniques have been recently applied mostly to probe elasticity of thin films and embedded structures. As a result of the AFM tip blunting due to its high-frequency tapping [72], ultrasonic methods (SLAM, AFAM, UFM) achieve the lateral resolution of 20–100 nm. In the case of soft samples (i.e., polymers), usually only qualitative information is obtained owing to the complex relationship between the contact stiffness and the dynamic Young's modulus [68]. Moreover, any sharp variations in topography of the sample correspond to variations in the tip–sample contact area and produce some artifacts [72].

SLAM measurements have been performed on a wide range of engineering materials, from Al_2O_3/Al metal-matrix composites to PVC/PB polymer blends [68, 76, 77], and 100 thin silicon oxide coatings on polyamide [69].

AFAM has measured reduced Young's moduli for 50 nm thin films Ni deposited on Si, glass-fiber/polymer composites [78], and thin diamond-like carbon coatings [62].

The UFM method was applied to probe relative elasticity of defects in highly-oriented pyrolitic graphite (HOPG) [79], Ge dots on silicon substrate [80], and heterogeneous nanostructures [81]. Also polymeric materials embedded into stiff [82] or more compliant matrices have been measured. The UFM depth sensitivity for elastic imaging was determined to be several hundred nanometers [83].

9.4.3
UFM Measurements of Adhesion Hysteresis and Their Relations to Friction at the Tip-Sample Contact

The UFM method recently has been applied to measure adhesion hysteresis and local friction in fine-polished metallic samples, HOPG, and mica [52–54].

Adhesion hysteresis (AH) is the difference between the work needed to rupture and establish the contact between two bodies [84]. AH depends on elastic, viscoelastic, plastic, adhesive, geometric properties of contacting asperities [7,57], as well as on temperature, relative humidity [85], time of contact, and any chemical reactions at the contact [86,87]. In the realm of the nanoscale AFM measurements, the elastic part of adhesion hysteresis AH_{el} is usually entangled into simultaneously occurring mechanical hysteresis and capillary forces hysteresis (Fig. 9.1). Mechanical hysteresis stems from the snap-in and snap-out behavior of the AFM cantilevers when the gradient of the tip–sample forces exceeds the cantilever spring constant. Capillary force hysteresis is due to the presence of water at the surface.

A strong correlation between friction forces and adhesion hysteresis is confirmed by theoretical predictions and experiments at the macro- and microscale. Based on macroscale measurements, Briscoe et al. [88] reported that adhesion hysteresis and friction are interrelated, which has been supported by others [89,90] through a systematic study in rolling as well as sliding motion using rubber and glass. Microscale measurements have been made with the surface force apparatus using molecularly thick polymeric surfactants sandwiched between curved mica surfaces [91].

Szoszkiewicz et al. [52–54] used a modified UFM setup to address measurements of micro- and nanoscale adhesion hysteresis and to correlate them with friction. They employed an AFM microscope as well as a nanoindenter. In their AFM-based setup, a heterodyne interferometer working in-situ is used to set the UFM high frequency (HF), which is above the first cantilever flexural resonance and does not coincide with any other cantilever's resonance or subharmonic. Trapezoidal shape of a modulating (LF) waveform was used, which allows for the tip–sample contact to be ruptured on a rising slope, and permanently reestablished on a declining slope. The mean LF cantilever deflection (UFM signal) is acquired by an AFM as a difference between instant cantilever deflection and the set point. In the nanoindenter-based setup, the nanoindenter resonance frequency (hundreds of Hz) is much lower than that of the AFM cantilever. Thus, HF has been downshifted by several decades, and an AFM scanner is used directly to vibrate a sample at high frequency with no additional transducer necessary between the AFM scanner and the tip. Besides, the nanoindenter diamond (Berkovitch) tip is more stable than any AFM tip.

Geometrical considerations of the tip–sample positions just before and just after the jump-up point on the hysteretic force-distance curves have yielded an elastic part of adhesion hysteresis in the following form [52]:

$$\text{AH}_{\text{el}} = 4\left(A_2 \overline{F}_2 - A_1 \overline{F}_1\right) = 4\left(F_{\text{SP}}(A_2 - A_1) + \Delta \overline{F}_2 A_2 - \Delta \overline{F}_1 A_1\right) . \quad (9.8)$$

Equation (9.8) can be applied as is to both AFM and nanoindenter-based measurements, because it is independent of the particular shape of $F(\delta)$ and geometry of the tip–sample contact area.

The values of AH_{el} have been theoretically and experimentally correlated with friction. Theoretical correlations were obtained based on the classical theory of adhesional friction, contact mechanics, and capillary hysteresis (CapH). It has been obtained that in case of such samples (e.g., with negligible viscoelastic and plastic contributions to adhesion hysteresis), adhesion hysteresis scales with friction at the tip–sample contact through the following relation:

$$\text{AH}_{\text{el}} = 100 \left(\frac{F_{\text{SP}} + F_{\text{cap}}}{3\pi R w_{\text{t-s}}} + 1\right)^{-2/3} (w_{\text{t-s}}/E^*)F_{\text{R}} + \text{CapH} , \quad (9.9)$$

where F_{cap} is the capillary force coming from the eventual presence of a capillary bridge between the tip and the sample, and F_{R} is the frictional force.

Considering that surfaces are often locally regarded as self-affine fractals [24,92], it has been found that adhesion hysteresis AH^{rough} and friction $F_{\text{R}}^{\text{rough}}$ on rough surfaces are reduced compared with their counterparts (AH, F_{R}) calculated on smooth surfaces

$$\text{AH}^{\text{rough}}/\text{AH} = 1 + 58.14 S_{\text{q}} R(l_{\text{c}}/l_{\text{i}})^H l_{\text{c}}^{-2} \quad (9.10)$$

$$F_{\text{R}}^{\text{rough}}/F_{\text{R}} \approx 30^{1-H} , \quad (9.11)$$

where S_{q} is the RMS roughness at the acquired AFM images length l_{i}, l_{c} is the contact length calculated through the contact mechanics (no roughness assumed), and $H = 3 - D$ is the Hurst exponent, with D being the fractal dimension.

Correlations between adhesion hysteresis and friction have been measured using the ultrasonic force microscopy (UFM) on mica, calcite, and a few metallic samples (Pt, Au, Cu, Zn, Ti and Fe) both on nano- and microscale. Adhesion hysteresis measured at the nanoscale has ranged between 4×10^{-19} J and 10^{-18} J. Adhesion hysteresis measured at the microscale has ranged between 8×10^{-17} and 14×10^{-17} J. These results are similar to the results of other groups [93–96], which have measured adhesion hysteresis by probing energy losses with an AFM cantilever vibrating with ultrasmall amplitudes or with a surface force apparatus.

At the nanoscale, the experimental data quantitatively matched the calculated values when the surface roughness of the studied materials was included. In the microscale, the experimental data qualitatively matched the calculations, but were up to three times higher than the calculated results due to the shortcomings of the model. The model in the microscale has overestimated the calculated adhesion hysteresis, because it has implicitly considered only the single asperity tip–sample contact.

As a result the UFM method can be now used for reliable friction investigations with an unprecedented high lateral resolution, which is limited only by the tip–sample contact radius. AFM-based UFM measurements are also fast because they are limited only by a few periods of the modulating UFM frequency, which is roughly 100 Hz at a present time but can be extended into the kHz range as well. Then, a typical 256×256 pixel AFM friction image is recorded within one minute, which is a few times faster than acquiring a good quality image with currently available AFM systems.

Acknowledgements. We would like to thank B. Bhushan, G. Brune, A.J. Kulik and I. Palaci for stimulating discussions. We would like as well to acknowledge the financial support of the US NSF under the grant DMR-0405319, as well as US DOE under the grant number DE-FG02-06ER46293.

References

1. Hertz HJ (1882) Reine Angew Math 92:156
2. Sneddon IN (1965) Int J Eng Sci 3:47
3. Derjaguin BV, Muller VM, Toporov YPT (1975) J Colloid Interf Sci 53:314
4. Johnson KL, Kendall K, Roberts AD (1971) Proc R Soc Lon Ser A 324:301
5. Maugis D (2000) Contact, Adhesion, and Rupture of Elastic Solids. Springer, Berlin, Heidelberg, New York
6. Burnham NA, Colton RJ, Pollock HM (1991) J Vac Sci Technol A 9:2548
7. Bhushan B (1999) Handbook of Micro/NanoTribology, 2nd edn. CRC, Boca Raton
8. Schmauder S (2002) Annu Rev Mater Res 32:437
9. Barber JR, Ciavarella M (2000) Int J Solids Struct 37:29
10. Tsakmakis C (2004) Int J Plasticity 20:167
11. Gao JP, Luedtke WD, Gourdon D, Ruths M, Israelachvili JN, Landman U (2004) J Phys Chem B 108:3410
12. Sugimura J (1998) J Jpn Soc Tribol 43:933
13. Meakin P (1993) Phys Rep 235:189
14. Kragelsky IV, Dobychin MN, Kombalov VS (1982) Friction and Wear Calculation Methods. Pergamon, New York
15. Greenwood JA, Williamson JBP (1966) Proc R Soc Lon Ser A 295:300
16. Whitehouse DJ, Archard JF (1970) Proc R Soc Lon Ser A 316:97
17. Nayak PR (1971) J Lubr Technol T Asme 93:398
18. Majumdar A, Bhushan B (1991) J Tribol T Asme 113:1
19. Blackmore D, Zhou G (1998) Int J Mach Tool Manu 38:551
20. Zahouani H, Vargiolu R, Loubet JL (1998) Math Comput Model 28:517
21. Yan W, Komvopoulos K (1998) J Appl Phys 84:3617
22. Chung JC, Lin JF (2004) J Tribol T Asme 126:646
23. Persson BNJ (2000) Sliding Friction: Physical Principles and Applications, 2nd edn. Springer, Berlin, Heidelberg, New York
24. Persson BNJ (2001) Phys Rev Lett 87:116101
25. Persson BNJ, Tosatti E (2001) J Chem Phys 115:5597
26. Buzio R, Boragno C, Valbusa U (2003) Wear 254:917
27. Luan B, Robbins M (2005) Nature 435:929
28. Miesbauer O, Gotzinger M, Peukert W (2003) Nanotechnology 14:371

29. Enachescu M, van den Oetelaar RJA, Carpick RW, Ogletree DF, Flipse CFJ, Salmeron M (1998) Phys Rev Lett 81:1877
30. Schwarz UD, Zworner O, Koster P, Wiesendanger P (1997) Phys Rev B 56:6997
31. Schwarz UD, Zworner O, Koster P, Wiesendanger R (1997) Phys Rev B 56:6987
32. Cappella B, Dietler G (1999) Surf Sci Rep 34:1
33. Sader JE, Chon JWM, Mulvaney P (1999) Rev Sci Instrum 70:3967
34. Higgins MJ, Proksch R, Sader JE, Polcik M, Mc Endoo S, Cleveland JP, Jarvis SP (2006) Rev Sci Instrum 77:013701
35. Proksch R, Schaffer TE, Cleveland JP, Callahan RC, Viani MB (2004) Nanotechnology 15:1344
36. Burnham NA, Chen X, Hodges CS, Matei GA, Thoreson EJ, Roberts CJ, Davies MC, Tendler SJB (2003) Nanotechnology 14:1
37. Gibson CT, Smith DA, Roberts CJ (2005) Nanotechnology 16:234
38. Heim LO, Kappl M, Butt HJ (2004) Langmuir 20:2760
39. Hutter JL (2005) Langmuir 21:2630
40. Manyes SG, Guell AG, Gorostiza P, Sanz F (2005) J Chem Phys 123:114711
41. Rost MJ, Crama L, Schakel P, van Tol E, van Velzen-Williams GBEM, Overgauw CF, Horst H, Dekker H, Okhuijsen B, Seynen M, Vijftigschild A, Han P, Katan AJ, Schoots K, Schumm R, van Loo W, Oosterkamp TH, Frenken JWM (2005) Rev Sci Instrum 76:053710
42. Yu MF, Kowalewski T, Ruoff RS (2000) Phys Rev Lett 85:1456
43. Salvetat JP, Briggs GAD, Bonard JM, Bacsa RW, Kulik AJ, Stockli T, Burnham NA, Forro L (1999) Phys Rev Lett 82:944
44. Palaci I, Fedrigo S, Brune H, Klinke C, Chen M, Riedo E (2005) Phys Rev Lett 94:175502
45. Carpick RW, Ogletree DF, Salmeron M (1997) Appl Phys Lett 70:1548
46. Lantz MA, O'Shea SJ, Welland ME (1997) Phys Rev B 56:15345
47. Briggs GAD (1992) Acoustic Microscopy. Oxford University Press, Oxford
48. Takata K, Hasegawa T, Hosaka S, Hosoki S, Komoda T (1989) Appl Phys Lett 55:1718
49. Guthner P, Fisher C, Dransfeld K (1989) Appl Phys Lett 48:89
50. Khuri-Yakub BT, Akamine S, Hadimioglu B, Yamada H, Quate CF (1991) Proc SPIE 1556:30
51. Szoszkiewicz R, Huey BD, Kolosov OV, Briggs GAD, Gremaud G, Kulik AJ (2003) Appl Surf Sci 210:54
52. Szoszkiewicz R, Kulik AJ, Gremaud G (2005) J Chem Phys 122:134706
53. Szoszkiewicz R, Bhushan B, Huey BD, Kulik AJ, Gremaud G (2005) J Chem Phys 122:144708
54. Szoszkiewicz R, Bhushan B, Huey BD, Kulik AJ, Gremaud G (2006) J Appl Phys 99:014310
55. Cretin B, Sthal F (1993) Appl Phys Lett 62:829
56. Vairac P, Cretin B (1999) Surf Interface Anal 27:588
57. Bhushan B (2004) Springer Handbook of Nanotechnology, 1st edn. Springer, Berlin, Heidelberg, New York
58. Kueng A, Kranz C, Lugstein A, Bertagnolli E, Mizaikoff B (2005) Angew Chem Int Ed 44:3419
59. Rabe U, Arnold W (1994) Appl Phys Lett 64:1493
60. Dupas E (2000) PhD dissertation, Ecole Polytechnique Federale de Lausanne
61. Rabe U, Amelio S, Kester E, Scherer V, Hirsekorn S, Arnold W (2000) Ultrasonics 38:430
62. Amelio S, Goldade AV, Rabe U, Scherer V, Bhushan B, Arnold W (2001) Thin Solid Films 392:75
63. Hurley DC, Shen K, Jennett NM, Turner JA (2003) J Appl Phys 94:2347
64. Hurley DC, Turner JA (2004) J Appl Phys 95:2403

65. Passeri D, Bettucci A, Germano M, Rossi M, Alippi A, Orlanducci S, Terranova ML, Ciavarella M (2005) Rev Sci Instrum 76:093904
66. Muraoka M (2005) Nanotechnology 16:542
67. Burnham NA, Kulik AJ, Gremaud G, Gallo PJ, Oulevey F (1996) J Vac Sci Technol B 14:794
68. Oulevey F (1999) PhD dissertation, Ecole Polytechnique Federale de Lausanne
69. Rochat G, Leterrier Y, Plummer CJG, Manson JAE, Szoszkiewicz R, Kulik AJ, Fayet P (2004) J Appl Phys 95:5429
70. Cuberes TM, Briggs GAD, Kolosov OV (1998) AFM Modes for Non-Linear Detection of Ultrasonic Vibration. Oxford University Press, Oxford
71. Kolosov OV, Yamanaka K (1993) Jpn J Appl Phys 32:22
72. Dinelli F, Castell MR, Ritchie DA, Mason NJ, Briggs GAD, Kolosov OV (2000) Philos Mag A 80:2299
73. Hirsekorn S, Rabe U, Arnold W (1997) Nanotechnology 8:57
74. Turner JA, Hirsekorn S, Rabe U, Arnold W (1997) J Appl Phys 82:966
75. Stark RW, Drobek T, Heckl WM (2001) Ultramicroscopy 86:207
76. Burnham NA, Gremaud G, Kulik AJ, Gallo PJ, Oulevey F (1996) J Vac Sci Technol B 14:1308
77. Dinelli F, Burnham NA, Kulik AJ, Gallo PJ, Gremaud G, Benoit W (1996) J Phys IV 6:731
78. Hurley DC, Muller MK, Kos AB, Geiss RH (2005) Adv Eng Mater 7:713
79. Yamanaka K (1996) Thin Solid Films 273:116
80. Kolosov OV, Castell MR, Marsh CD, Briggs GAD (1998) Phys Rev Lett 81:1046
81. Dinelli F, Assender HE, Takeda N (1999) Surf Interface Anal 27:562
82. Porfyrakis K, Kolosov OV, Assender HE (2001) J Appl Polym Sci 82:2790
83. Geisler H, Hoehn M, Rambach M, Meyer MA, Zschech E, Mertig M, Romanov A, Bobeth M, Pompe W, Geer RE (2001) Proc Microscopy Semiconduct Materials 2001 169:527
84. Israelachvili J (1997) Intermolecular and Surface Forces, 3rd edn. Academic, San Diego
85. Szoszkiewicz R, Kulik AJ, Gremaud G, Lekka M (2005) Appl Phys Lett 86:123901
86. Chaudhury MK, Whitesides GM (1992) Science 256:1539
87. Luengo G, Heuberger M, Israelachvili J (2000) J Phys Chem B 104:7944
88. Briscoe BJ, Evans DCB, Tabor D (1977) J Colloid Interf Sci 61:9
89. Barquins M, Courtel R (1975) Wear 32:133
90. Kendall K (1975) Wear 33:351
91. Horn RG, Israelachvili JN, Pribac F (1987) J Colloid Interf Sci 115:480
92. Wang S (2004) J Tribol T Asme 126:1
93. Yoshizawa H, Chen YL, Israelachvili J (1993) J Phys Chem 97:4128
94. Yoshizawa H, Chen YL, Israelachvili J (1993) Wear 168:161
95. Hoffmann PM, Oral A, Grimble RA, Ozer HO, Jeffery S, Pethica JB (2001) Proc R Soc Lon Ser A 457:1161
96. Oral A, Grimble RA, Ozer HO, Hoffmann PM, Pethica JB (2001) Appl Phys Lett 79:1915

10 Near-Field Raman Spectroscopy and Imaging

*Pietro Giuseppe Gucciardi · Sebastiano Trusso · Cirino Vasi ·
Salvatore Patanè · Maria Allegrini*

AFM	Atomic force microscopy
BCB	Brilliant cresyl blue
BWF	Breit–Wigner–Fano
CCD	Charge coupled device
CT	Charge transfer
CuTCNQ	Copper-tetracyanoquinodimethane
GFR	Gradient field Raman scattering
HOMO	Highest occupied molecular orbital
LUMO	Lowest unoccupied molecular orbital
KTP	Potassium titanyl phosphate
MGITC	Malachite green isothiocyanate
NA	Numerical aperture
NIR	Near-infrared
PMT	Photomultiplier tube
RBM	Radial breathing mode
RLA	Reaction limited aggregation
RRS	Resonant Raman scattering
RTP	Rubidium titanyl phosphate
SEM	Scanning electron microscopy
SERS	Surface-enhanced Raman scattering
SNOM	Scanning near-field optical microscopy
STM	Scanning tunneling microscopy
SWCNT	Single-walled carbon nanotubes
TCNQ	$7,7',8,8'$-Tetracyanoquinodimethane
TE-CARS	Tip-enhanced coherent anti-Stokes Raman scattering
TEM	Transmission electron microscopy
TERS	Tip-enhanced Raman scattering

10.1
Introduction

When light interacts with matter, the scattered photons carry with them detailed information about the structure of the sample under investigation. Among the principal light scattering phenomena providing structural information there are Rayleigh scattering, Raman scattering, hyper-Rayleigh scattering, hyper-Raman scattering, coherent anti-Stokes Raman scattering, and stimulated Raman gain or loss spectroscopy. Raman spectroscopy is, in fact, one of the simplest, quickest and most useful means to get structural, chemical and electronic information on a large variety of materials. The first observation of the Raman effect is due to Raman and Krishnan [1, 2]. They illuminated several purified gases and liquids with a sunlight beam passing through a blue-violet filter, observing that some light was scattered at lower energies, in the yellow-green region. Most important, the observed frequency shifts were not dependent on the incident radiation frequency but only on the nature of the samples, differencing this phenomenon from the more common fluorescence. When it was recognized that some of the observed frequency shifts coincided with the frequencies of infrared absorption, the effect was soon related to the vibrational properties of the system under investigation. For the discovery of this effect, Raman was awarded with the Nobel Prize in Physics in 1930.

From the experimental point of view, a typical apparatus for Raman spectroscopy consists of a monochromatic light source, a sample compartment and a section dedicated to the spectral analysis of the scattered radiation. Difficulties arise due to the low cross section of the process. Therefore, since the discovery of the Raman effect, the development of the experimental apparatuses have regarded the improvement of the sensitivity to low level signals, of the stray light rejection capabilities, and of the spectral resolution. The first notable breakthrough coincided with the introduction of the photomultiplier tube (PMT), developed in the early 1940s, in place of the photographic plate. With the development of laser sources Raman spectroscopy became a technique with a broad range of application: in 1962 Porto and Wood, first reported on the use a Rb pulsed laser in Raman spectroscopy [3]. Monochromator technology also improved, seeking for higher stray light rejection, spectral resolution and long term mechanical and thermal stability. Fitting a Raman system with two or three dispersive elements (double and triple monochromators mounting holographic gratings), together with the introduction of holographic notch filters, allowed for the reduction of elastically-scattered radiation down to 10 orders of magnitude. Therefore, spectral regions very close to the exciting laser line (less than $10 \, \text{cm}^{-1}$) became accessible for low frequency Raman spectroscopy [4] and Brillouin [5] scattering. The invention of charge coupled devices (CCD) added new capabilities to Raman systems, offering the chance to collect wide portions of the spectrum with a single acquisition. Finally, the coupling of Raman spectroscopy apparatuses with optical microscopes enabled unique Raman imaging potentialities on solid samples. Nowadays, with the development of nanotechnologies, scientists have looked forward to pushing Raman spectroscopy towards their ultimate performances in terms of molecular sensitivity and spatial resolution. Surface-enhanced Raman scattering (SERS) has been introduced as a viable technique towards single-molecule sensitivity. More recently, the development of the scanning near-field optical microscopy techniques (SNOM, or NSOM) [6–8] has tailored the optical imaging capabilities down to the nanometer scale, where the laws of near-field optics hold [9]. Near-field Raman spectroscopy, in particular, represents a challenging research field that has recently demonstrated enormous potentialities to get structural information on 10-nm length scales.

The scope of this chapter is to review the current status of near-field Raman spectroscopy. After a brief theoretical description of the Raman effect, the conventional experimental setups for Raman and micro-Raman spectroscopy are illustrated. Subsequently, the concepts of near-field optics are introduced, and the experimental setups for nano-Raman spectroscopy and imaging are described. Finally, we review the most relevant applications of this technique in materials science.

10.2
Raman Spectroscopy

When light is scattered from a molecule, most photons are elastically diffused. That is, most of the scattered photons have the same energy of the incident ones. However, a small fraction of light (approximately 1 in 10^7 photons) is scattered at energies different from, and usually lower than, the incident light energy. The process leading to this inelastic scattering is referred to as the linear Raman effect. If the incident light beam has frequency ω_L, and the scattered radiation has frequency ω_R the difference $\Delta\omega = \omega_L - \omega_R$ is called the Raman shift. The energy shift provides information on the vibrational properties of matter, but such information is somehow encoded. Therefore a theoretical model is needed.

10.2.1
Classical Description of the Raman Effect

A first and very accurate description of the Raman effect can be obtained by treating both the system and the electromagnetic fields classically. The classical treatment correctly predicts the appearance of emission at wavelengths different than the incident one of the scattered radiation, as well as some aspects of the selection rules. When a molecule interacts with an electromagnetic field E, an electric dipole moment P will be induced

$$P = \alpha \cdot E + \frac{1}{2}\beta : EE + \frac{1}{6}\gamma \vdots EEE , \tag{10.1}$$

where α, β and γ are respectively the polarizability, the first, and second hyper-polarizability tensors of the molecule. These are functions of the nuclei positions and hence of the molecule's vibrational state. The first-order term in (10.1), linear in E, is associated with Rayleigh and Raman scattering. The second-order term is associated with the hyper-Rayleigh and hyper-Raman scattering. The third-order, cubic in E, yields the second hyper-Rayleigh and second hyper-Raman scattering, describing all the four-wave mixing processes, among which the coherent anti-Stokes Raman scattering (CARS) is one of the most important. Limiting our description to the first term linear in E, the dependence of α on the nuclei positions can be accounted for, in case of small vibrations, by expanding each component of the polarizability tensor in a Taylor series near the nuclei equilibrium positions as follows:

$$\alpha_{ij} = a_{ij}^0 + \sum_k \left(\frac{\partial a_{ij}}{\partial q_k}\right)_0 q_k + \frac{1}{2}\sum_{k,l}\left(\frac{\partial^2 a_{ij}}{\partial q_k \partial q_l}\right)_0 q_k q_l + \dots , \tag{10.2}$$

where α_{ij}^0 are the polarizability tensor elements at the equilibrium positions, and q_k the vibration coordinates. The derivatives of the polarizability elements are calculated at the equilibrium position and summation is performed all over the possible vibrations. In the following we restrict our attention to the first term of (10.2), neglecting all higher-order terms. Moreover, we will consider only one of all the possible normal modes of vibration, i.e., the kth mode. The polarizability tensor for such a mode can be expressed as

$$(\alpha_{ij})_k = (\alpha_{ij})_0 + \left(\frac{\partial \alpha_{ij}}{\partial q_k}\right)_0 q_k \tag{10.3}$$

or in tensorial form

$$\boldsymbol{\alpha}_k = \boldsymbol{\alpha}_0 + q_k \boldsymbol{\alpha}'_k , \tag{10.4}$$

where $\boldsymbol{\alpha}'_k$ is a tensor whose components are the derivatives of the polarizability tensor with respect to the kth mode's coordinates.

Let us assume for q_k a simple harmonic behaviour

$$q_k = q_{k0} \cos(\omega_k t) , \tag{10.5}$$

where q_{k0} and ω_k are the amplitude and the frequency of the motion, respectively.

Assuming the incident electromagnetic field as

$$\boldsymbol{E} = \boldsymbol{E}_0 \cos(\omega_L t) \tag{10.6}$$

the induced electric dipole due to the coupling of the electromagnetic wave with the polarizability tensor related to the kth vibration mode turns out to be

$$\boldsymbol{P} = \boldsymbol{\alpha}_0 \boldsymbol{E}_0 \cos(\omega_L t) + \frac{1}{2}\boldsymbol{\alpha}'_k \boldsymbol{E}_0 q_{k0} \left[\cos(\omega_L + \omega_k)t + \cos(\omega_L - \omega_k)t\right] . \tag{10.7}$$

As we can see, the induced dipole has three components with different frequencies. The first one $\boldsymbol{P}_1 = \boldsymbol{\alpha}_0 \boldsymbol{E}_0 \cos(\omega_L t)$ gives rise to the Rayleigh scattering, that is, to a scattering at the same frequency of the incident field. In particular, such radiation does not depend on the molecule vibrational states, but only on its polarizability at the equilibrium configuration. The other terms in (10.7) describe two oscillating dipoles radiating light at frequencies $(\omega_L + \omega_k)$ and $(\omega_L - \omega_k)$, corresponding to the anti-Stokes and to the Stokes Raman scattering, respectively. In particular from (10.3) it can be envisaged that, for the Raman scattering to take place, some of the elements of $\boldsymbol{\alpha}'_k$ must be different from zero. Therefore, the molecule must undergo a net change of its polarizability in order to change its vibrational state during the interaction with the external electric field.

As an example, in Fig. 10.1 we report the Raman spectrum of carbon tetrachloride molecules.

If the properties of the polarizability tensor of the molecule are taken into account the classical theory correctly predicts the appearance of Raman lines at ± 218, ± 314, ± 459, ± 762 and $\pm 790 \, \mathrm{cm}^{-1}$.

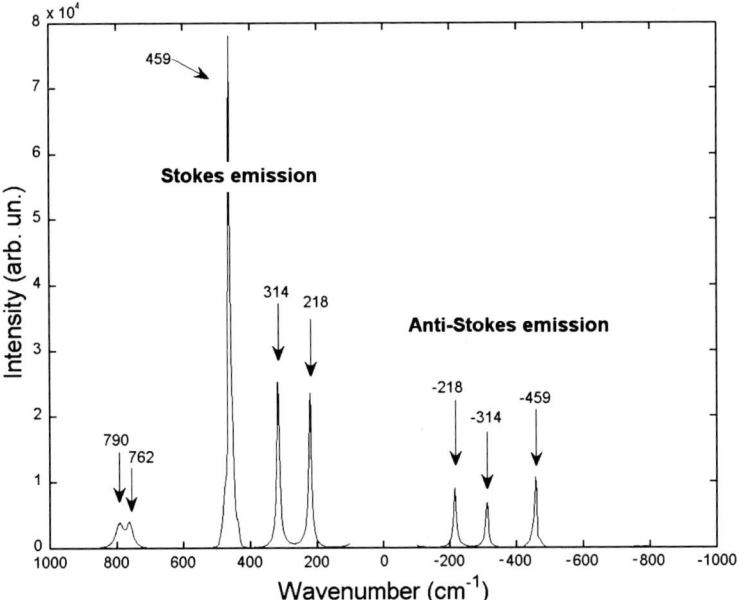

Fig. 10.1. Stokes and anti-Stokes Raman emission of CCl$_4$ at room temperature, excited by the 514.5 nm line of an Ar laser

10.2.2
Quantum Description of the Raman Effect

Although capable of predicting the appearance of the Raman peaks, the classical theory is unable to explain the clear intensity difference between the Stokes and anti-Stokes emission (see Fig. 10.1), or its dependence on the temperature. In order to correctly describe all these aspects a quantum description is needed. In this section we will introduce the semiclassical approach followed by Long [10], in which the investigated system is treated quantum mechanically and the electromagnetic field is described classically, in a time-dependent perturbation picture. In the quantum mechanical treatment the induced electric dipole of the classical theory is replaced by a transition electric dipole associated with the transition of the molecule from an initial state $|i\rangle$ to a final state $|f\rangle$, induced by an incident electric field at frequency ω_L. We assume $\widetilde{\psi}_f$ and $\widetilde{\psi}_i$ to be the perturbed time-dependent wave functions associated to the final and initial states in presence of the electric field. Denoting the electric dipole moment operator as $\hat{P}$, the molecule's transition electric dipole will be

$$P_{fi} = \langle \widetilde{\psi}_f | \hat{P} | \widetilde{\psi}_i \rangle \ . \tag{10.8}$$

The perturbed time-dependent states can be written as a series expansion

$$\widetilde{\psi}_f = \psi_f^0 + \psi_f^1 + \ldots + \psi_f^n$$
$$\widetilde{\psi}_i = \psi_i^0 + \psi_i^1 + \ldots + \psi_i^n \ , \tag{10.9}$$

where ψ_f^0 and ψ_i^0 are the unperturbed wave functions (in absence of the electric field), while ψ_f^n and ψ_i^n are the nth order modifications to ψ_f^0 and ψ_i^0 as a result of the perturbation. Each correction can be expressed as a linear combination of the unperturbed states ψ_k^0 of the system

$$\psi_i^m = \sum_k a_{ik}^m \psi_k^0$$

$$\psi_f^m = \sum_k a_{fk}^m \psi_k^0 \tag{10.10}$$

summing over all the states of the system. The coefficients a_{ik}^m and a_{fk}^m depend on the interaction Hamiltonian and are functions of the electric field $\boldsymbol{E}$; in particular, the coefficients $m = 1$ will have a linear dependence on $\boldsymbol{E}$, those with $m = 2$ will be quadratic on $\boldsymbol{E}$, and so on. Introducing the perturbative expansion of (10.9) in (10.8), the transition electric dipole results

$$\boldsymbol{P}_{fi} = \langle \psi_f^0 | \, \hat{\boldsymbol{P}} \, | \psi_i^0 \rangle + \langle \psi_f^1 | \, \hat{\boldsymbol{P}} \, | \psi_i^0 \rangle + \langle \psi_f^0 | \, \hat{\boldsymbol{P}} \, | \psi_i^1 \rangle + \langle \psi_f^2 | \, \hat{\boldsymbol{P}} \, | \psi_i^0 \rangle + \langle \psi_f^0 | \, \hat{\boldsymbol{P}} \, | \psi_i^2 \rangle$$

$$+ \langle \psi_f^3 | \, \hat{\boldsymbol{P}} \, | \psi_i^0 \rangle + \langle \psi_f^0 | \, \hat{\boldsymbol{P}} \, | \psi_i^3 \rangle + \langle \psi_f^2 | \, \hat{\boldsymbol{P}} \, | \psi_i^1 \rangle + \langle \psi_f^1 | \, \hat{\boldsymbol{P}} \, | \psi_i^2 \rangle + \dots \tag{10.11}$$

We now introduce the series expansion (10.10) into (10.11), and consider the dependence of the coefficients a_{ik}^m and a_{fk}^m on the strength of the electric field $\boldsymbol{E}$. After collecting all the terms that have the same power dependence on $\boldsymbol{E}$, we obtain the following expression

$$\boldsymbol{P}_{fi} = \boldsymbol{P}_{fi}^0 + \boldsymbol{P}_{fi}^1 + \boldsymbol{P}_{fi}^2 + \dots \tag{10.12}$$

whose first four terms are

$$\boldsymbol{P}_{fi}^0 = \langle \psi_f^0 | \, \hat{\boldsymbol{P}} \, | \psi_i^0 \rangle$$

$$\boldsymbol{P}_{fi}^1 = \langle \psi_f^0 | \, \hat{\boldsymbol{P}} \, | \psi_i^1 \rangle + \langle \psi_f^1 | \, \hat{\boldsymbol{P}} \, | \psi_i^0 \rangle$$

$$\boldsymbol{P}_{fi}^2 = \langle \psi_f^0 | \, \hat{\boldsymbol{P}} \, | \psi_i^2 \rangle + \langle \psi_f^2 | \, \hat{\boldsymbol{P}} \, | \psi_i^0 \rangle + \langle \psi_f^1 | \, \hat{\boldsymbol{P}} \, | \psi_i^1 \rangle$$

$$\boldsymbol{P}_{fi}^3 = \langle \psi_f^0 | \, \hat{\boldsymbol{P}} \, | \psi_i^3 \rangle + \langle \psi_f^3 | \, \hat{\boldsymbol{P}} \, | \psi_i^0 \rangle + \langle \psi_f^1 | \, \hat{\boldsymbol{P}} \, | \psi_i^2 \rangle + \langle \psi_f^2 | \, \hat{\boldsymbol{P}} \, | \psi_i^1 \rangle \, . \tag{10.13}$$

The first transition moment $\boldsymbol{P}_{fi}^0$ is independent of $\boldsymbol{E}$ as it involves only unperturbed wave functions. It relates to a direct transition between unperturbed initial and final states, not to a light scattering process, so it needs no further consideration for our purposes. Conversely, $\boldsymbol{P}_{fi}^1$ is linear in $\boldsymbol{E}$, $\boldsymbol{P}_{fi}^2$ is quadratic in $\boldsymbol{E}$, etc., and they describe different processes.

The linear term

$$\boldsymbol{P}_{fi}^1 = \langle \psi_f^0 | \, \hat{\boldsymbol{P}} \, | \psi_i^1 \rangle + \langle \psi_f^1 | \, \hat{\boldsymbol{P}} \, | \psi_i^0 \rangle \tag{10.14}$$

gives rise to the Rayleigh and Raman scattering. To evaluate $\boldsymbol{P}_{fi}^1$, the relation between the perturbed wave functions and the unperturbed ones has to be determined. For this purpose, the following assumptions are usually made:

– The perturbation is induced by the presence of a monochromatic electromagnetic plane wave, of frequency ω_L.
– The perturbation is of the first order.
– Interactions with the system occur through electric dipole only, while magnetic dipole and higher-order electric dipole contributions are neglected.

The above assumptions lead to the calculation of the mth component of $\boldsymbol{P}_{fi}^1$

$$
\begin{aligned}
\left(P_m^1\right)_{fi} = \frac{1}{2\hbar} \sum_k &\left(\frac{\langle \varphi_f | \hat{P}_m | \varphi_k \rangle \langle \varphi_k | \hat{P}_n | \varphi_i \rangle}{\omega_{ki} - \omega_L - i\Gamma_k} \widetilde{E}_{0n} \, \mathrm{e}^{-\mathrm{i}(\omega_L - \omega_{fi})t} \right.\\
&\left. + \frac{\langle \varphi_f | \hat{P}_m | \varphi_k \rangle \langle \varphi_k | \hat{P}_n | \varphi_i \rangle}{\omega_{ki} + \omega_L + i\Gamma_k} \widetilde{E}_{0n}^* \, \mathrm{e}^{\mathrm{i}(\omega_L + \omega_{fi})t} \right)\\
+ \frac{1}{2\hbar} \sum_k &\left(\frac{\langle \varphi_f | \hat{P}_n | \varphi_k \rangle \langle \varphi_k | \hat{P}_m | \varphi_i \rangle}{\omega_{kf} - \omega_L - i\Gamma_k} \widetilde{E}_{0n}^* \, \mathrm{e}^{\mathrm{i}(\omega_L + \omega_{fi})t} \right.\\
&\left. + \frac{\langle \varphi_f | \hat{P}_n | \varphi_k \rangle \langle \varphi_k | \hat{P}_m | \varphi_i \rangle}{\omega_{kf} + \omega_L + i\Gamma_k} \widetilde{E}_{0n} \, \mathrm{e}^{-\mathrm{i}(\omega_L - \omega_{fi})t} \right)\\
+ \,\text{c.c.} ,&
\end{aligned}
\tag{10.15}
$$

where $\omega_{ki} = \omega_k - \omega_i$, $\hat{P}_m$ and $\hat{P}_n$ are the mth and nth components of the electric dipole moment, and $\widetilde{E}_{0n}$ is the complex amplitude of the nth component of the incident field. The term Γ_k is the width of the k state ($\psi_k = \varphi_k \mathrm{e}^{-\mathrm{i}(\omega_k - i\Gamma_k)t}$), the lifetime of the state being $\tau_k = \hbar/2\Gamma_k$. The initial and the final states are considered to have an infinite lifetime. As a consequence (10.15) contains information on the line width of the k state only. The terms involving $(\omega_L + \omega_{fi})$ describe induced emission of two quanta from an initial excited level ω_i, to a lower level ω_f, if $(\omega_L + \omega_{fi}) > 0$. These terms will not be considered here. The terms with $(\omega_L - \omega_{fi})$ describe the emission of Rayleigh and Raman scattering, provided that

$$
(\omega_L - \omega_{fi}) > 0
\tag{10.16}
$$

Equation (10.15) describes the Rayleigh and Raman scattering, and can be better understood with the help of the energy level diagrams reported in Fig. 10.2a–c. In both processes the presence of the electric field induces two transitions: the absorption of a photon, bringing the system from the initial state $|i\rangle$ to a virtual (or real) level $|k\rangle$, followed by the emission of a photon, bringing the system to the final level $|f\rangle$. If $\omega_{fi} = 0$ (Fig. 10.2a) the initial and final states coincide and the scattered radiation has the same frequency of the incident one (Rayleigh scattering). When $\omega_{fi} < 0$ (Fig. 10.2b) the final state energy ($E_f = \hbar\omega_f$) is lower than the energy of the initial state ($E_i = \hbar\omega_i$), an anti-Stokes Raman photon is emitted having energy higher than the incident one. If $\omega_{fi} > 0$ the decay from the virtual state $|k\rangle$ occurs through the emission a Stokes photon having energy $\hbar\omega_S$ slightly smaller than the incident photon, plus a vibrational or rotational quantum of energy $\hbar\omega_{fi}$. Resonant Raman scattering (RRS) occurs weather the energy of the virtual state matches an allowed electronic transition.

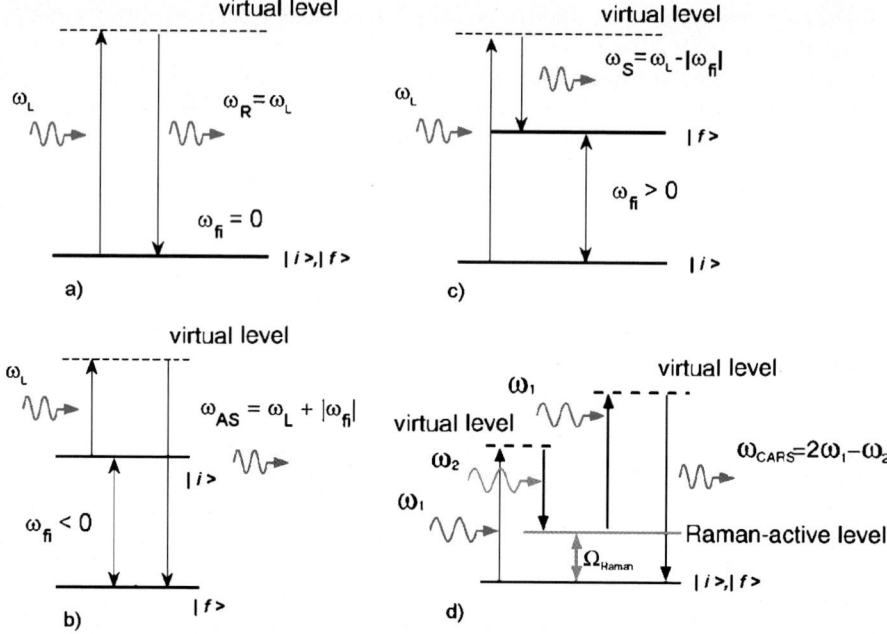

Fig. 10.2. Energy level diagram describing (**a**) Rayleigh scattering, (**b**) anti-Stokes Raman scattering, (**c**) Stokes Raman scattering, and (**d**) coherent anti-Stokes Raman scattering

In order to make a comparison with the classical description of the Raman scattering it is possible to rewrite (10.15) by introducing a generalized transition polarizability α_{fi} connected with the transition $|i\rangle \rightarrow |f\rangle$, whose components are

$$(\alpha_{mn})_{fi} = \frac{1}{\hbar} \sum_k \left(\frac{\langle \varphi_f | \hat{P}_m | \varphi_k \rangle \langle \varphi_k | \hat{P}_n | \varphi_i \rangle}{\omega_{ki} - \omega_L - i\Gamma_k} + \frac{\langle \varphi_f | \hat{P}_n | \varphi_k \rangle \langle \varphi_k | \hat{P}_m | \varphi_i \rangle}{\omega_{ki} + \omega_L + i\Gamma_k} \right).$$

(10.17)

In some cases, when the excitation energy $\hbar\omega_L$ is very small compared to any absorption energy $\hbar\omega_{ki}$ of the system, the complex terms $i\Gamma_k$ can be ignored being small with respect to ω_{ki}. Thus, the transition polarizability becomes real and inserting (10.17) in (10.15) we obtain

$$\left(P_m^1 \right)_{fi} = \frac{1}{2} (\alpha_{mn})_{fi} \left[\widetilde{E}_{0n} e^{-i\omega_s t} + \widetilde{E}_{0n}^* e^{i\omega_s t} \right]$$

(10.18)

with $\omega_s = \omega_L - \omega_{fi}$. Assuming a real electric field we obtain

$$\left(P_m^1 \right)_{fi} = \frac{1}{2} (\alpha_{mn})_{fi} \widetilde{E}_{0n} \left[e^{-i\omega_s t} + e^{i\omega_s t} \right] = (\alpha_{mn})_{fi} \widetilde{E}_{0n} \cos(\omega_s t)$$

(10.19)

which can be compared to the induced electric dipole value calculated classically. Most important is that the quantum approach provides the selection rules holding for Raman scattering. In particular, for a generic vibrational transition $|i\rangle \rightarrow |f\rangle$ to be Raman active, at least one of the components of the transition polarizability tensor $(\alpha_{mn})_{fi}$ in (10.17) must be non-zero. Finally, an expression for the intensity ratio between Stokes and anti-Stokes intensity I_S/I_{AS} of scattered radiation can be obtained from the Fermi golden rule

$$\frac{I_S}{I_{AS}} = \frac{(\omega_L - \omega_k)^4}{(\omega_L + \omega_k)^4} e^{-\left(\frac{\hbar\omega_k}{k_B T}\right)} . \tag{10.20}$$

This accounts for the observed temperature dependence of the intensity ratio.

10.2.3
Coherent Anti-Stokes Raman Scattering

At high incident laser powers the non-linear response of the sample, described by the higher-order terms in (10.1), becomes important. Coherent anti-Stokes Raman scattering (CARS) is a non-linear Raman process that combines the advantages of signal strength in stimulated-Raman spectroscopy, with the general applicability of the spontaneous-Raman spectroscopy. From a theoretical point of view, CARS is a four-wave mixing process related to the third-order susceptibility of the systems [11]. In this process, the system is coherently excited through the beating of two incoming beams having frequencies ω_1 and ω_2 and then mixed with the incoming beam at ω_1 resulting in a coherent output signal at the anti-Stokes frequency $\omega_s = 2\omega_1 - \omega_2$ (Fig. 10.2d). When the frequency difference $\omega_1 - \omega_2$ coincides with a Raman-active level Ω_{Raman} of the molecule, a CARS signal is resonantly generated. The intensity can be obtained [12] from the solution of the wave equation for the output field with the input frequency beams ω_1 and ω_2

$$\left[\nabla^2 + \frac{\omega_s^2}{c^2}\right] \varepsilon(\omega_s) E_s = -\frac{4\pi}{c^2} \frac{\partial}{\partial t} P^{(3)}(\omega_s) , \tag{10.21}$$

where $P^{(3)}(\omega_s) = \gamma(\omega_s = 2\omega_1 - \omega_2) \cdot E_1(\omega_1)E_1(\omega_1)E_2^*(\omega_2)$ is the third-order polarizability. For a slab of length l, the solution of (10.21) gives the intensity of the output field as:

$$I_S = \frac{2\pi\omega_s^2}{c\varepsilon_s} |\gamma|^2 I_1^2 I_2 \left[\frac{\sin(\Delta k \cdot l/2)}{\Delta k \cdot l/2}\right]^2 . \tag{10.22}$$

The intensity of the CARS signal is proportional to the square of the intensity of the input beam I_1 and linearly dependent on I_2. Moreover, the signal will be maximum when momentum conservation is satisfied

$$\Delta k = k_c - 2k_1 + k_2 = 0 . \tag{10.23}$$

The above relation, referred to as the phase-matching condition, is always fulfilled in co-linear experiments where beams overlap in the same direction. In this experimental scheme, the alignment is greatly simplified but rejection of the incident beams is difficult. Moreover, the spatial resolution is very poor because of the long interaction path. Since CARS originates only from a small volume at the intersection of three beams, high spatial resolution can be achieved by crossing the beams at different angles, provided care is taken to fulfil the phase-matching condition. Therefore, this configuration is specially convenient for imaging purposes [13].

10.2.4
Experimental Techniques in Raman Spectroscopy

As briefly discussed in the introduction, notable technical improvements have lead to the development of high-sensitivity apparatuses for Raman spectroscopy and imaging, now available on the market. A schematic diagram of a typical Raman scattering instrument is shown in Fig. 10.3 [14]. The most common light source is an Ar^+ laser, which provides several excitation wavelengths between 351 and 515 nm, with power outputs up to several watts. Commercial apparatuses also associate lasers at lower energies, such as HeNe (633 nm) or NIR laser diodes (785 nm), for resonant excitation or to decrease the fluorescence emission that would cover the weak Raman emission. The polarization properties of the excitation beam are controlled by polarization rotators, while plasma emission is blocked by narrow band interferential filters (not shown in Fig. 10.3). The spectrometers usually operate both in macro- and in microarrangements. In the first case, the laser is focused onto the sample by means of a lens, and the scattered radiation is collected at 90°. In the microarrangement, both the excitation and the collection of the backscattered radiation is accomplished through a microscope objective. Polarization analysis of the scattered radiation is carried out by placing an additional polarizer in the collection optical path. The light is finally focused onto the entrance slits of a double monochromator. In our setup, it is a Jobin Yvon U1000 with focal length of 1 m, provided with two holographic gratings (1800 lines/mm) and yielding a spectral resolution of $0.1 \, cm^{-1}$. The high rejection capabilities of the Rayleigh scattering makes possible the detection of Raman emission at distances on the order of a few tenths of cm^{-1} from the laser line. The system can operate both in single and multichannel configuration. In the single-channel configuration the light dispersed

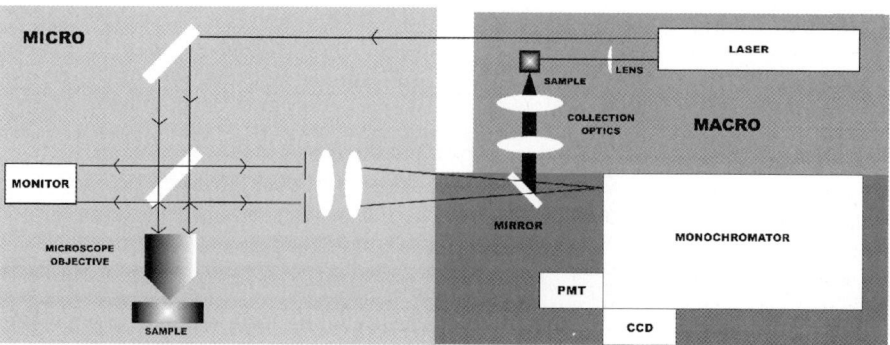

Fig. 10.3. Schematic diagram of a Raman spectroscopy apparatus operating both in micro- and macroarrangements

by the gratings is focused on the exit slits and detected by a cooled PMT. Spectra are acquired one wavelength at a time, by scanning the gratings over a defined wavelength range. In the multichannel configuration a liquid nitrogen cooled CCD is used to collect entire portions of the spectrum during a single acquisition. Such a feature is particularly useful for Raman imaging and for time-resolved Raman spectroscopy applications. This enables to follow phase transitions or aggregation phenomena as a function of time, monitoring the changes of the spectral emission. The macroconfiguration is generally used for such experiments, taking place in liquid environments.

As an example, in Fig. 10.4a we report the temporal evolution of the Raman peaks intensity at $242 \, cm^{-1}$ and at $316 \, cm^{-1}$ (Fig. 10.4b) of a system composed by an aqueous solution of the tetrakis (4-sulphonatophenyl)porphyrine (TTPS$_4$) (personal communication, Micali, Villari, 2005). Upon aggregation induced by the addition of spermine molecules, the formation of fractal J-aggregates occurs, ex-

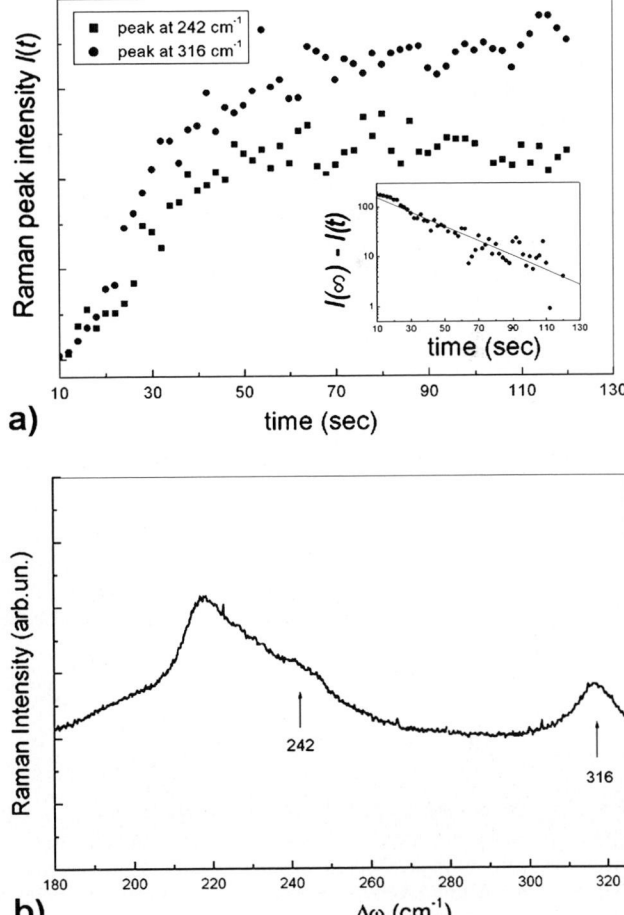

Fig. 10.4. (a) Temporal evolution of the intensitiy $I(t)$ of the Raman emission (b) at 242 and $316 \, cm^{-1}$ of a TTPS$_4$ porphyrin aqueous system. The *inset* in (a) shows an exponential fit (*solid line*) of the relation $I(\infty) - I(t)$, allowing to measure the kinetic growth rate of the aggregation phenomenon occuring in TTPS$_4$ upon the addition of spermine molecules

hibiting a large enhancement of the resonant light scattering [15, 16]. The Raman peaks, due to N–N stretching modes, identify collective motions derived from the interaction between two or more monomer entities stacked edge-to-edge (J-form) or face-to-face (H-form), respectively [15]. A detectable Raman signal appears 10 seconds after the beginning of the kinetic process. The temporal evolution of the Raman intensity $I(t)$ suggests that the aggregation process takes place within the first 40 seconds, reaching the saturation value $I(\infty)$ after such a time interval. The intensity change is fitted well by a simple exponential law (see inset of Fig. 10.4a) with a kinetic rate constant $K = 0.033\,\mathrm{s}^{-1}$. Such a value indicates the presence of a reaction limited aggregation (RLA) mechanism [17]. Static light scattering measurements indicate that at the end of the kinetic process porphyrin molecules are arranged in mass fractal aggregates, structured according to a diffusion limited aggregation (DLA) mechanism [17]. In this case, time-resolved Raman scattering measurements have been able to point out the presence of a RLA mechanism preceding the DLA one, driving the system to the final aggregates.

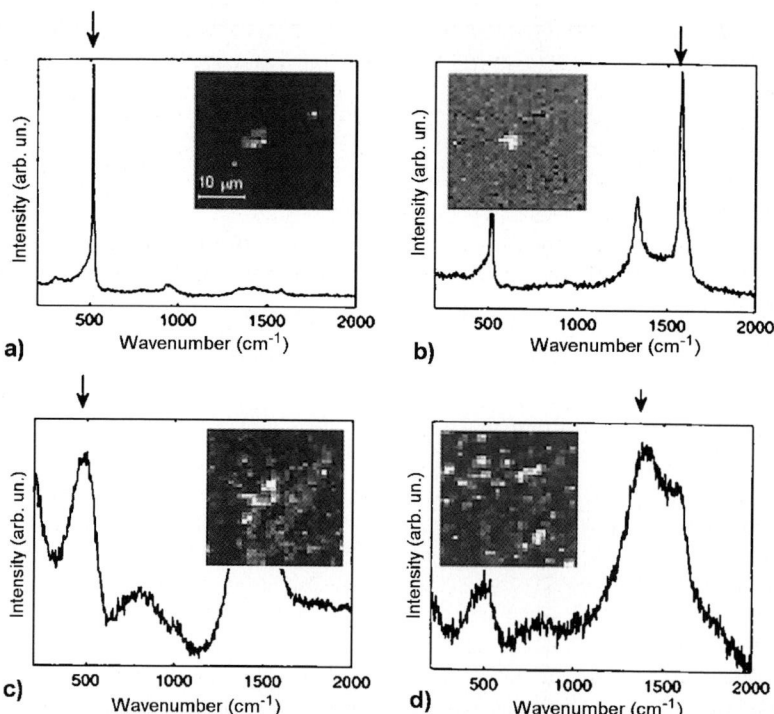

Fig. 10.5. Raman spectra collected at different positions on the surface of a-SiC thin films. The spectra are representative of four main different arrangements detected on an area of $30 \times 30\ \mu\mathrm{m}^2$. In the *inset* are shown the maps obtained by integration of a Raman band (indicated by the *arrows*) typical of the material we want to locate on the surface. Brighter spots are indicative of higher Raman emission

For operation in the micro-Raman configuration, our spectrometer is coupled to an Olympus BX100 optical microscope, equipped with a scanning motorized stage. The schematic diagram is shown on the left of Fig. 10.3. By inserting a sliding mirror in the optical path of the microscope, the surface of the sample can be imaged on a monitor. Chemical or structural imaging is accomplished by acquiring arrays of Raman spectra, and plotting some specific features (intensity, position or width) of the Raman lines characteristic of the system. The spatial resolution is theoretically limited only by the diffraction, thus depending on the numerical aperture of the objective. In practice, however, spatial resolutions below 750 nm are hardly obtainable, unless oil-immersion objectives are used.

An example of micro-Raman imaging is shown in Fig. 10.5. We have performed a study aimed to monitor the local composition of an amorphous silicon carbide thin film, deposited by pulsed laser ablation from a silicon carbide target in a vacuum, using the focused beam of a frequency-doubled Nd:YAG laser ($\lambda = 532$ nm) [18]. An array of 30×30 Raman spectra in the 150 to 3200 cm^{-1} region has been acquired, with a spectral resolution of 2.5 cm^{-1}. A 100X microscope objective (NA $= 0.9$) has been used. The scan step was 1 μm. The spectra in Fig. 10.5 highlight four different structural typologies of silicon and carbon occurring in different sites of the film. Crystalline (a) and amorphous (b) silicon have been observed, together with microcrystalline graphite (c) and amorphous carbon (d). To map the spatial location of the identified typologies, four micro-Raman maps have been carried out (insets of Fig. 10.5). At each point, the pixel intensity is obtained by integrating the area of the Raman bands associated to each species. Therefore, the Raman maps in the insets of Fig. 10.5 correspond respectively to: (a) the area of the crystalline silicon peak, (b) the peak of crystalline graphite at 1580 cm^{-1}, (c) the amorphous silicon band around 490 cm^{-1}, and (d) of the amorphous carbon band near 1450 cm^{-1}. The maps reveal the complex structure of the sample on the micrometer scale.

10.3
Near-Field Raman Spectroscopy

The main limitation of Raman scattering for applications in the domain of the nanotechnologies is its low spatial resolution. As we have seen in the previous section, even with the best microscopes, a spatial resolution better than 0.5 μm is hardly obtainable, that is much too large for nanotechnology. Therefore, the characterization of nanostructured systems is still done by electron microscopy (SEM, TEM), or by scanning probe techniques (AFM, STM), which provide information on the electronic and morphological properties of materials with nanometer scale resolution. Unfortunately, these techniques work ex-situ, often requiring special sample preparation, and can be destructive. The nano-Raman concept combines the structural, chemical and electronic information provided by Raman spectroscopy with the topographic information intrinsic to standard atomic force or shear-force microscopy, on subdiffraction spatial scales.

10.3.1
Theoretical Principles of the Near-Field Optical Microscopy

10.3.1.1
Image Formation and Spatial Resolution

The first and the easiest way to describe the formation of optical images consists in considering an aberration-free instrument and a bundle of rays obeying Fermat's principle. This simple approach is not enough to describe the limits of an intriguing instrument such as a microscope, because it does not provide information on spatial resolution. The wave theory, as proposed by Abbe [19], shows that a well-defined limit in resolution exists and cannot be overcome by the classical optical instruments based on lenses. Following Abbe's theory the smaller object that can be resolved by a standard optical microscope has a lateral dimension greater than

$$0.61 \frac{\lambda}{n \sin (\vartheta)} \ . \tag{10.24}$$

Where ϑ is the semicollection angle, n the refractive index of the medium. The expression $n \sin (\vartheta)$ is called the numerical aperture (NA) of the system. Even for the best-performing oil-immersion microscope objectives NA does not exceed 1.4, and therefore spatial resolutions better than 250 nm can not be obtained in the visible range.

10.3.1.2
Overcoming the Diffraction Limit

In 1928 Synge described a fascinating idea to overcome the diffraction limit [20]. In his speculative work, he suggested the use of a nanometric hole as the light source, scanning it at close distances from the sample surface. The modern scanning near-field optical microscopes (SNOM or NSOM) are based on this idea and, in particular, on the concept of optical tunnelling of evanescent waves. The first optical tunnelling experiment was the well-known two prisms experiment carried out by Newton [21]. Newton connected two identical prisms and illuminated the first using an incidence angle over the critical angle. He found that, while the light is completely reflected

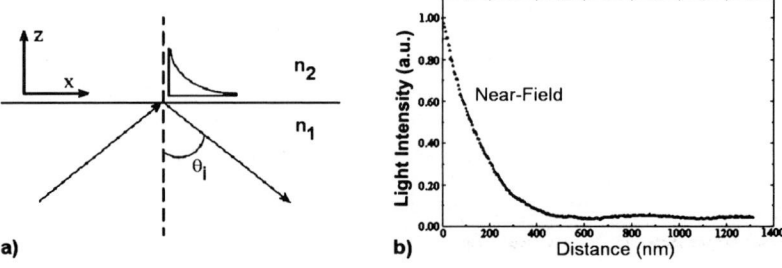

Fig. 10.6. (a) Total internal reflection scheme. **(b)** Near-field (*solid line*) exponential decay in a total internal reflection experiment

by the first prism alone, it passes through the two prisms if the "interval between the glasses is not above the ten hundred thousandth part of an inch", that is, a few tens of nanometers. Newton was not able to explain the experiment but there was clear evidence of the near-field existence. At the interface of the first prism, in fact, a non-propagating evanescent field exists. Therefore, the second prism, when placed close to the surface, is able to collect this field and to transform it in a propagating one. A simple idea of the evanescent propagation properties of the near-field is provided by the experiment sketched in Fig. 10.6a. An interface between two media with different refractive indeces, namely n_1 and n_2, is illuminated by monochromatic light in total reflection condition. Above the interface, in the medium n_2, the electromagnetic field is evanescent with an amplitude decreasing as [22, 23]

$$E(z) = E(t) \exp(-\gamma z)$$

$$\gamma = \frac{\omega}{v_2} \sqrt{\left(\frac{\sin(\vartheta_i)}{n_{21}}\right)^2 - 1} \,, \tag{10.25}$$

where v_2 is the velocity of light in the medium n_2, $n_{21} = n_2/n_1$ and, following the Snell–Descartes law, a critical angle is defined as

$$\vartheta_c = \arcsin(n_{21}) \,. \tag{10.26}$$

Newton's experiment is therefore a simple and smart method to describe the unique properties of the near-field radiation: its amplitude exponentially decays with distance, and becomes almost zero when the source-sample distance overcomes a few hundreds of nanometers (Fig. 10.6b). The actual observation of an evanescent wave in the visible range was later reported in an experiment using a laser beam as the optical source [24].

To better understand the reason why the use of the near-field provides super-resolution, it is useful to describe the nanostructure under investigation as a simple nanodipole. To image such a dipole, in terms of Fourier optics, we need either to illuminate it with evanescent components of the electromagnetic field (corresponding to high spatial frequencies), or to collect the evanescent components of the scattered light [25]. A small aperture is able to produce an evanescent electromagnetic field rich of high spatial frequencies, and spatially confined on a length scale comparable with the aperture's diameter. Synge's idea was thus to scan such a nanosource over the sample at constant velocity, collecting point by point the scattered light. Due to the evanescent character of the near-field, the probe must be kept at close distances from the surface during the scan. The picture, of course, is not exhaustive, but it is enough to understand power and problems of the near-field optics. To have a more complete point of view, it has to be kept in mind that the nanosource and sample are a system and they influence each other, resulting in a complicated distribution of the electric field between tip and sample that can hardly be treated analytically.

10.3.1.3
Intrinsic Differences Between Far and Near-Field Raman Spectroscopy

For Raman spectroscopy in the near-field, it is worth mentioning that near-field intensity in front of the fiber aperture might be much stronger than the one in a conventional far-field configuration. This difference has remarkable consequences in organic samples [26–28]. The light coming out from the small circular aperture has polarization components in all three directions x, y, and z, while the far-field radiation can only be transverse. This new situation allows excitation of different polarized Raman modes without reorienting the sample or changing the polarization of the source. Finally, a more intriguing effect arises from the exponential decay of the near-field intensity with the distance [29]. In a standard Raman experiment the electric field is usually considered as a constant, i.e., it does not change over molecular distances. In a near-field Raman spectroscopy, the field gradient effect must be taken into account. Thus, if q is the equilibrium position of some charges able to oscillate in the material, using a Taylor series it is possible to write the expansion

$$E = E_0 + \left(\frac{\partial E}{\partial q_k}\right) q_k + O(q^2) \; . \tag{10.27}$$

The polarization P becomes

$$P = \alpha_0 E_0 \cos(\omega_L t) + \frac{1}{2} q_{k0} \left(\frac{\partial \alpha}{\partial q_k}\right) E_0 \left[\cos\left\{(\omega_L - \omega_k)t\right\} + \cos\left\{(\omega_L + \omega_k)t\right\}\right]$$

$$+ \frac{1}{2} q_{k0} \left(\frac{\partial E}{\partial q_k}\right) \alpha_0 \left[\cos\left\{(\omega_L - \omega_k)t\right\} + \cos\left\{(\omega_L + \omega_k)t\right\}\right] + \dots \; , \tag{10.28}$$

where ω_L is the frequency of the incident light, and α the polarizability tensor of the material, as defined in Sect. 10.2.1. The first term gives the far-field Rayleigh scattering; the second term gives Stokes and anti-Stokes frequencies, as in the far-field Raman experiment. The third term, called the gradient field Raman (GFR) scattering, has no equivalent in the classical approach. It gives a signal carrying again Stokes and anti-Stokes contributions. However, its amplitude depends upon the field gradient and on the polarizability α, therefore different selection rules hold for Raman and GFR. While a net change of the polarizability tensor is required for a Raman transition to be active, such a requirement is no longer needed in GFR.

10.3.2
Setups for Near-Field Raman Spectroscopy

A near-field microscope is a scanning instrument, in which either the sample or the probe are raster scanned by means of a piezoelectric actuator. The images are obtained recording, point by point, the amplitude of the signal related to the physical properties of interest: the normal or the lateral force in the atomic force microscope (AFM), the current in a scanning tunnel microscope (STM), the scattered light in the scanning near-field optical microscope (SNOM). For SNOM additional care is required to control the tip to sample distance because of the exponential decay of the high spatial frequency components of the excitation radiation. A number of technical papers describing various methods to control and stabilize the distance in a SNOM setup has been reported in the literature [30, 31]. However the most used method consists in the non-optical detection of the shear force, a task accomplished by means of quartz tuning forks [32]. The rest of a standard SNOM consists in some mechanical, optical and electronic parts; the mechanical elements are devoted to the coarse positioning of the sample and of the probe. The optical parts mainly consist in the light source, usually a laser, an objective and a photodetector. Finally, the electronic parts consist in a low noise amplifier and a controller quite similar to a standard AFM or STM controller, with a feedback and an integral, proportional differential filter, but modified to detect and use the tuning fork signal to stabilize the distance. The technique supports various configurations, namely the reflection or the transmission mode, where the light is detected in a reflection or transmission geometry; the illumination or collection mode, where the near-field comes out from the probe or it is collected by the probe. The most suitable configuration depends on the sample characteristics and it has to be chosen to optimize the experiment.

10.3.2.1
Aperture Near-Field Raman Spectroscopy

Near-field optical microscopes are today produced by a number of companies. Most of them use tapered, metal-coated optical fibers ending with an apical aperture having a diameter of a few tens of nanometers, used as the nanosource. Commercial aperture probes are produced by means of the so-called heating and pulling method; unfortunately they are quite expensive, fragile and suffer of a low light throughput. Their ultimate resolution is limited to about 50 nm due to the performances of the metallic film that assure the field confinement at the apex [33, 34]. A number of alternative methods to obtain the near-field have been proposed to improve performance, reliability and cost. The cone angle of the apex plays a fundamental role in order to reduce the propagation path for the evanescent field inside the fiber as well as the losses due to the multiple internal reflection. The etching method, based on the use of hydrofluoridric acid, is proposed as an alternative to heating and pulling. This procedure is cheaper and gives cone angles at the apex as large as 45°, which increases the throughput, but usually produce a surface not suitable for the metal coating [34–37]. Tube etching and reverse tube etching are methods that produce a better surface but, up to now, they suffer from low reliability, and therefore these probes are still not commercially available [38,39]. Alternative methods come from atomic force microscopy and are based on the use of the so-called hollow cantilever. These kinds of probes are produced starting from standard AFM cantilevers drilled at the apex either by selective etching processes, or by means of focused ion beam milling [40, 41]. These processes result in a hole whose dimensions are a few tens of nanometers or less [42]. The use of these cantilevers as a nanosource requires the illumination of the probe at its back with a focused laser beam. The performances of these probes are quite promising due to their good mechanical stability that guarantees a long life performance and to their high throughput, able to preserve a well-defined polarization state [41,43]. An alternative method to generate the near-field consists in using the light scattered by a small metallic tip [44]. This approach has a number of advantages: virtually there is no limit on the ultimate resolution that depends only on the noise-background/signal ratio; these probes are able to manage higher laser power and it is possible to use plasmon resonances of the metal to improve the signal. They are also easily homemade with good results at a very low cost. The interest in nanoprobes research is key to nanospectroscopy: the improvement of the probe results in an overall advance of the instrument performances and in some cases the probe characteristics allow or deny the experiment.

At the beginning, Raman experiments have been carried out by employing an aperture-SNOM connected with a high resolution Raman spectrometer. Both illumination and collection mode operations have been demonstrated [45]. In the first scheme (Fig. 10.7a) the light from the laser beam (an air cooled Ar^+ laser in our case) is coupled to the SNOM probe. The scattered light is collected by a long working distance microscope objective (NA = 0.42) rested at 45°, and coupled to the entrance slit of a 190 mm focal length monochromator provided with a 1200 mm/line rule grating. A notch filer is used to reject the elastically-scattered light efficiently. Light is detected through a PMT operating in photon counting mode. The collection mode configuration (Fig. 10.7b) partially solves such problems. Here the sample is excited through a long working distance microscope objective. The Raman scattering is collected by the aperture at the apex of the optical fiber, which is coupled to the entrance slit of an imaging monochromator (Triax 180) for spectral analysis. Light is detected by a thermally cooled intensified CCD. In this configuration an entire Raman spectrum, between 250 and 3500 cm^{-1} with a resolution of about 20 cm^{-1}, can be acquired at a given point.

During the scan, an array of near-field Raman spectra is acquired. The comparison of these data with the topography leads to a deep and meaningful analysis of the sample characteristics, able to discriminate the chemical/Raman activity of the surface with an ultimate resolution down to 100 nm [46].

As we have seen, the simplicity of the aperture-SNOM setup has to be paid in terms of some limitations. Illumination mode Raman SNOM mainly suffers from low laser powers that can be coupled into the SNOM sensor. These values are limited to a few mW by thermal damaging of the probe. Therefore, excitation powers no larger

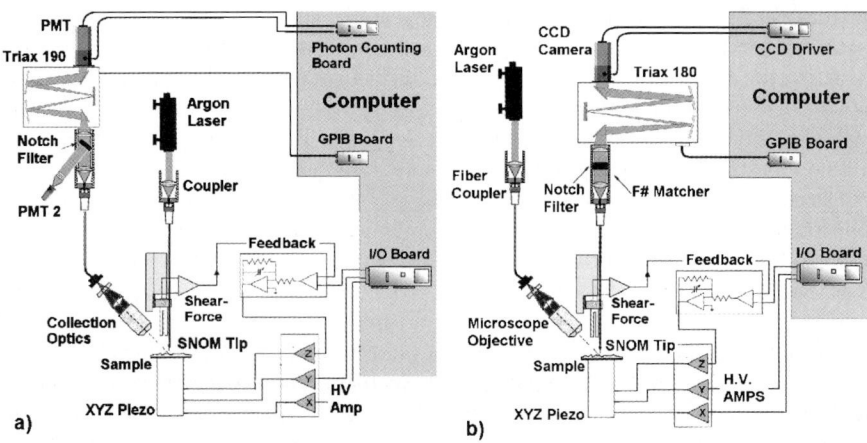

Fig. 10.7. Aperture-SNOM Raman apparatuses working in (**a**) illumination and (**b**) collection mode

than a few hundred nW can be transferred to the sample. A further problem is due to the intense Raman signal generated in the silica fiber core, coming out of the aperture, which almost completely obscures the 200 to 650 cm^{-1} region. The collection mode has an intrinsically low efficiency and resolution. As a result near-field Raman spectroscopy and mapping are suitable only to study materials with strong Raman cross sections. In spite of these serious restrictions, the technique is very interesting as it provides unique results not easily achievable by other techniques, as for example, the simultaneous mapping of the morphological and chemical properties of solid samples.

10.3.2.2
Tip-Enhanced Near-Field Raman Spectroscopy

Recently, apertureless near-field techniques [47–49] have lead to a number of interesting results both for the higher power managed by the probes and for its different working principles. The so-called tip-enhancement [50, 51] is an exceptionally promising technique able to increase the effective field exciting the sample on the local scale. This technique is based on the efficient excitation of surface plasmons at the metallic tip. This is obtained by illuminating the metal apex either with a laser light linearly-polarized along the tip axes, or through radially-polarized beams tightly-focused by means of high NA microscope objectives [52]. A typical apparatus for tip-enhanced Raman spectroscopy is depicted in Fig. 10.8. It is based on a inverted confocal microscope in combination with a probe head mounting the tip and the shear-force-based tip–sample distance control. The laser excitation is provided by a laser beam, whose output is converted into a radially-polarized beam by passing through a mode converter followed by spatial filtering. The beam is attenuated and sent into the microscope after passing a laser line filter. The beam is reflected by a dichroic beam splitter and focused by a high numerical aperture objective on the sample surface. The sample is mounted and scanned through a piezoelectric stage. The tip is positioned above the sample and centered onto the laser focus (inset of Fig. 10.8) where the longitudinal field induces the local field enhancement effect. The Raman scattered light is collected with the same objective, transmitted by the beam splitter and then detected. Light detection is accomplished either by a combination of a spectrometer and a charge coupled device (CCD) or by a photon-counting avalanche photodiode after spectral filtering using a narrow band-pass filters centered at the Raman peak energies of the system under investigation. Notch filters are used to cut the Rayleigh scattering. The metal tip is held within a few nanometers above the sample surface by using a tuning fork feedback mechanism. Sharp gold tips with a radius of curvature smaller than 50 nm are produced by electrochemical etching in hydrochloric acid [53].

Special techniques may be used to efficiently produce an incident beam having a strong longitudinal field at the center of the spot, therefore increasing the overall setup performance. A promising approach consists in realizing a pseudoradially-polarized illumination by means of four half-wave plates (Fig. 10.9), one in each quadrant, and arranged with the optical axis of each oriented segment such that the field is rotated to point in the radial direction [51, 52].

Fig. 10.8. Apertureless-SNOM Raman apparatus for tip-enhanced Raman scattering. In the *inset* the position of the tip with respect to the tightly-focused laser spot is highlighted

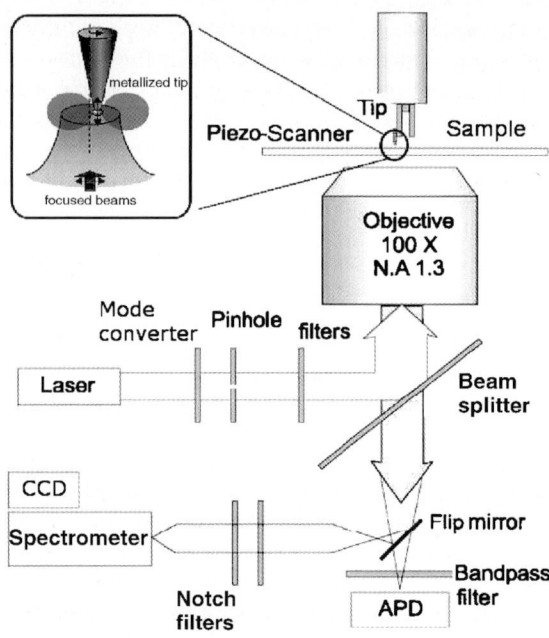

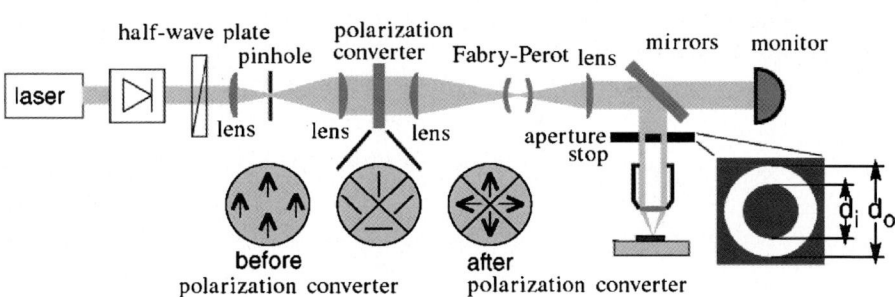

Fig. 10.9. Schematic apparatus for producing radially-polarized beams (adapted from [52])

10.4
Applications of Near-Field Raman Spectroscopy

Raman spectroscopy is a widely recognized means for chemical and structural analysis in physics, chemistry, materials science, etc. After the development of near-field optical techniques, scientists soon sensed the possibility to probe smaller and smaller volumes, pushing the spatial resolution in Raman imaging below the 500 nm barrier. Nevertheless, the intrinsic experimental difficulties due to the extremely low efficiency of the Raman scattering, combined with the limited light throughput of the optical fiber probes, has confined the applications of nano-Raman spectroscopy to a small class of molecules and solids having high Raman efficiency. Recently, new momentum has been given by the discovery of huge field-enhancement effects, provided by the local interaction of molecules with metal nanoparticles such as colloids or sharp tips. These phenomena are demonstrating their enormous potential for single-molecule recognition, single-nanostructure structural analysis, and vibrational spectroscopy on subzeptoliter (10^{-21} liters) volumes.

10.4.1
Structural Mapping

Raman spectroscopy is a well-assessed tool for the analysis of materials and devices for opto- and microelectronics ([54] and references therein). Raman spectroscopy is capable, for example, of discriminating metallic from semiconducting carbon nanotubes, and distinguishing tubes with different diameter and chirality [55]. In semiconductor technology, Raman imaging is used to map the presence of stress or local defects in real silicon devices and structures. From a qualitative point of view, in fact, a Raman frequency larger than the stress-free one indicates a compressive stress, while a Raman frequency decrease indicates tensile stress [56]. A reduction of the Raman scattering intensity can be interpreted as a fingerprint of a local phase transformation in presence of surface defects induced by the high pressures involved. These statements are however not always true [57]. For example, sophisticated models, together with a prior knowledge of the stress distribution, are required to extract quantitative information from local oxidation of silicon isolation structures, as the one shown in Fig. 10.10a, allowing to derive the stress distribution (right axis of Fig. 10.10b) from the Raman shifts (left axis of Fig. 10.10b) observed in the spectra [58].

In this context, SNOM techniques are potentially ideally suited for the application of Raman spectroscopy to the study of state-of-the-art organic and semiconductor devices on subdiffraction length scales. Furthermore, besides the improved optical

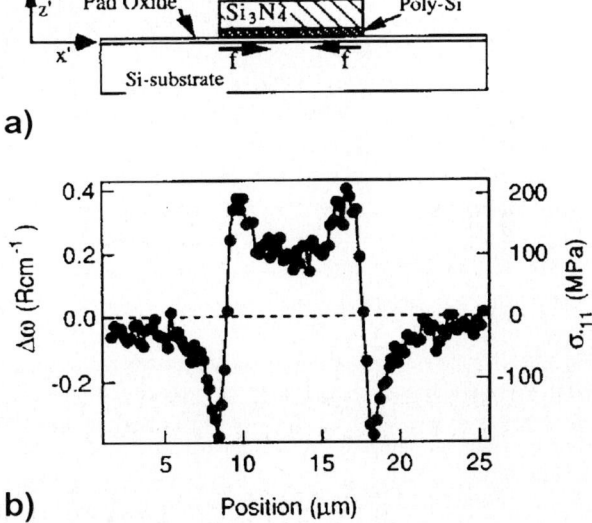

Fig. 10.10. (a) Schematic of a chemical vapor deposition Si_3N_4 film deposited on a silicon substrate. (**b**) Raman shift (*left axis*) and corresponding stress (*right axis*) in the silicon bulk, near and beneath the Si_3N_4/poly-Si line (figure reprinted from [58] with permission of the author)

resolution, an important benefit of the SNOM technique comes from the simultaneous generation of the topographic image, enabling a straightforward correlation of the Raman data with the surface morphology.

The first demonstration of such a potential was provided by Webster et al. [59,60] who mapped the stress along a defect on a silicon surface, with an aperture-SNOM working in reflection mode. Experiments were performed employing metal-coated fiber probes with ~ 150 nm apertures. The spectral analysis was carried out coupling the SNOM with a commercial Raman spectrometer (Renishaw, UK) equipped with a notch filter to reject the elastic scattering, and a cooled CCD camera as detector. Single nano-Raman spectra (Fig. 10.11a) were acquired along a scratch on the (001) surface of a silicon wafer, whose dimensions (width ~ 2 μm, depth ~ 1.6 μm) were simultaneously recovered in the topography map (Fig. 10.11b). The authors observed an increase of the peak intensity at $520 \, cm^{-1}$ (Fig. 10.11d), as well as a shift towards higher energies (Fig. 10.11e) while moving from the border of the scratch outwards,

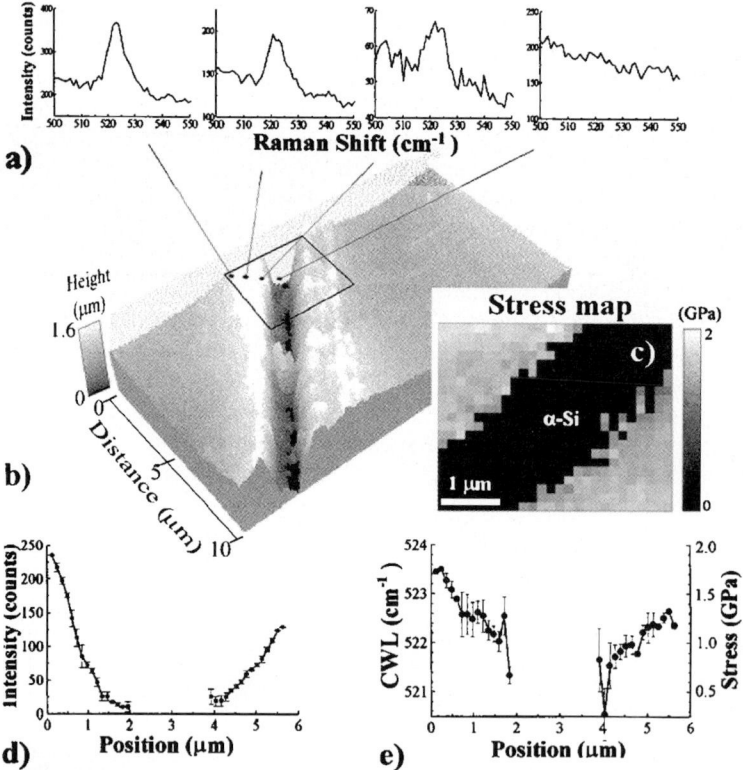

Fig. 10.11. Near-field Raman spectra acquired along a scratch produced on a silicon wafer (**a**). The scratch topography evidences a depth and width in the micrometer scale (**b**). The corresponding stress map was calculated by monitoring the fitted shift of the $520 \, cm^{-1}$ Raman peak of bulk silicon, using the uniaxial-stress approximation with a constant of $0.5 \, GPa/cm^{-1}$ (**c**). Line profiles of the $520 \, cm^{-1}$ peak intensity (**d**) and of the Raman shift (**e**) drawn along the scratch (figure reprinted from [60] with permission of the author)

in the surrounding region of plastic deformation. No Raman signal was measured inside the scratch. Using these measurements, with a uniaxial stress assumption [58] a submicrometer-resolution stress map was recovered, evidencing the occurrence of compressive stress along the two sides (Fig. 10.11c). The extinction of the 520 cm^{-1} peak inside the scratch was attributed to the formation of amorphous silicon (a-Si) due to the rapid release of the applied pressure in the scratching process. The Raman map consisted of an array of 26 × 21 spectra acquired on an area of approximately 4 × 3.5 μm^2. Each spectrum required 60 s of exposure time provided an excitation of a few tens of μW of light at 633 nm. Due to the high exposure time (1 min per spectrum), about 9 h was needed for the measurement.

Our group applied SNOM Raman techniques to image the morphological defects of 7,7′,8,8′-tetracyanoquinodimethane (TCNQ) molecular crystals on subdiffraction spatial scales [61]. Such material was proposed for applications in molecular electronic devices such as switches [62], organic light emitting sources [63], and data storage media [64]. Due to the high Raman scattering cross section of the materi-

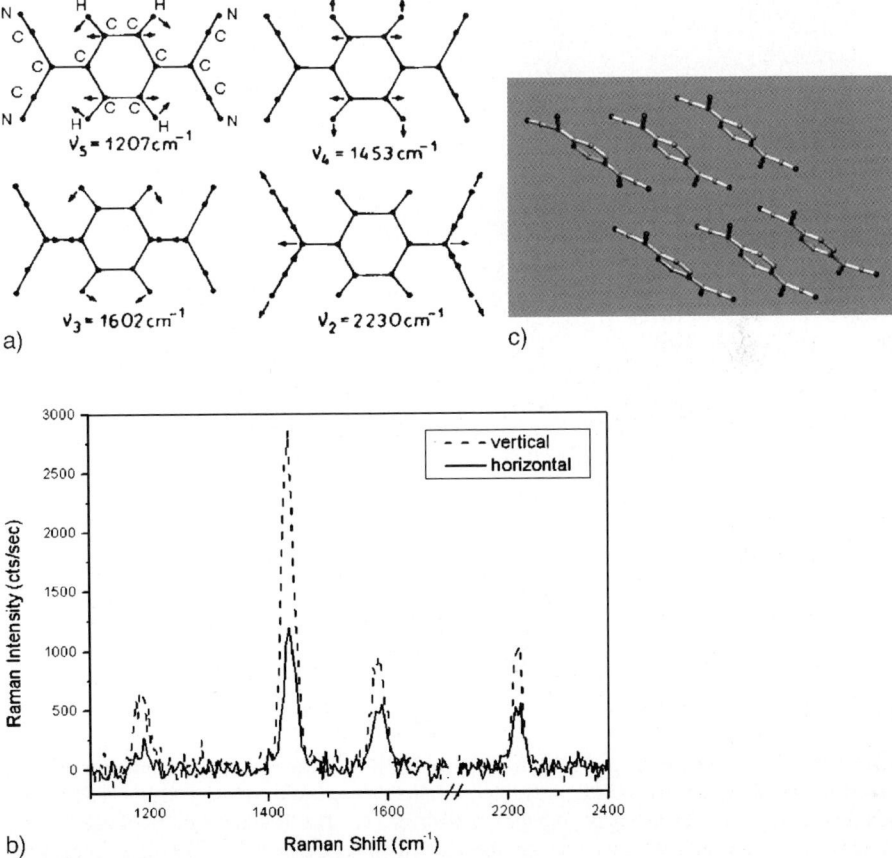

Fig. 10.12. Raman-active eigenmodes of the TCNQ molecule (**a**). Near-field Raman spectra (exposure time 100 ms) (**b**) of the TCNQ crystal along its molecular orientation directions (**c**)

als, Raman imaging with ~ 100 nm spatial resolution was achieved with very short integration times (100 ms). The specific sample was a TCNQ single crystal grown by precipitation from a solution. An aperture-SNOM working in emission mode was used (see Fig. 10.7b), mounting commercial fiber probes with 100 nm nominal apertures. A few hundreds nW of laser power ($\lambda = 514.5$ nm) were sufficient to excite the sample, providing S/N ratios better than 10. The four most intense vibrations for TCNQ [65] occur at 1207, 1453, 1602 and 2230 cm^{-1} corresponding to the eigenmodes shown in Fig. 10.12a. The two near-field Raman spectra (Fig. 10.12b, tip–sample distance of a few nanometers) reproduced the four sharp peaks mentioned above with an intensity that strongly depends on the molecular orientation of the crystal (Fig. 10.12c). Rotating the sample by 90°, we found that the Raman emission decreased by a factor of about 2 (Fig. 10.12b, solid line). Sparse defects, present on the flat sample surface, were first localized and imaged by means of a scanning micro-Raman apparatus (Fig. 10.3). The 1453 cm^{-1} vibration, being the most intense, represented the best candidate for imaging purposes. Figure 10.13a displays

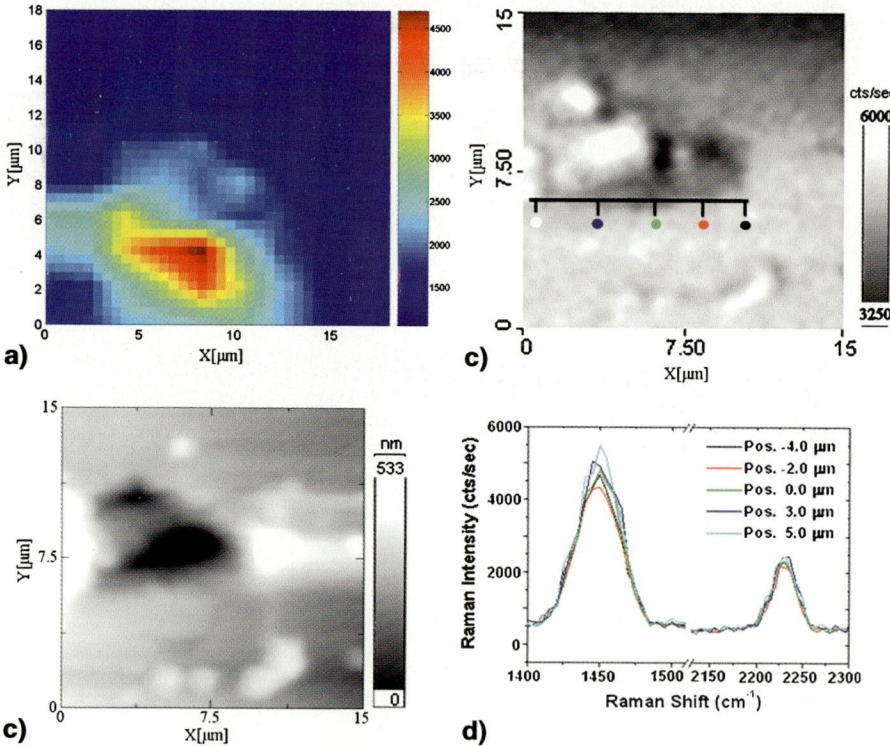

Fig. 10.13. Micro-Raman image at 1453 cm^{-1} on a $18 \times 18\ \mu m^2$ area (**a**), corresponding to a surface defect present on a TCNQ crystal. Near-field Raman map at 1453 cm^{-1} on a $15 \times 15\ \mu m^2$ area (**c**), corresponding to the topography shown in (**b**). The Raman spectra of the TCNQ molecules in (**d**) were collected along the line in (**c**) showing the fluctuation of the Raman peaks superimposed to a constant background. Exposure time 100 ms per pixel (reprinted from [61] with permission of the author)

the micro-Raman map on a $18 \times 18\,\mu m^2$ area, evidencing a spot of increased intensity. Nano-Raman investigations on similar surface portions allowed to correlate the actual topography (Fig. 10.13b) with the Raman scattering properties (Fig. 10.13c). In particular local corrugations, likely due to defects produced during the growth process, were observed on the flat sample surface. In the corresponding Raman map the corrugations appeared surrounded by a zone of depleted Raman emission, while at their center the Raman scattering increased. The depletion of the Raman activity observed at the contour of the defects could be qualitatively explained by the occurrence of TCNQ in an amorphous state. Similarly, a different molecular orientation at the level of microcrystalline clusters, or an increased material density, could account for the increase of the Raman signal in proximity of the bump apexes or nearby the

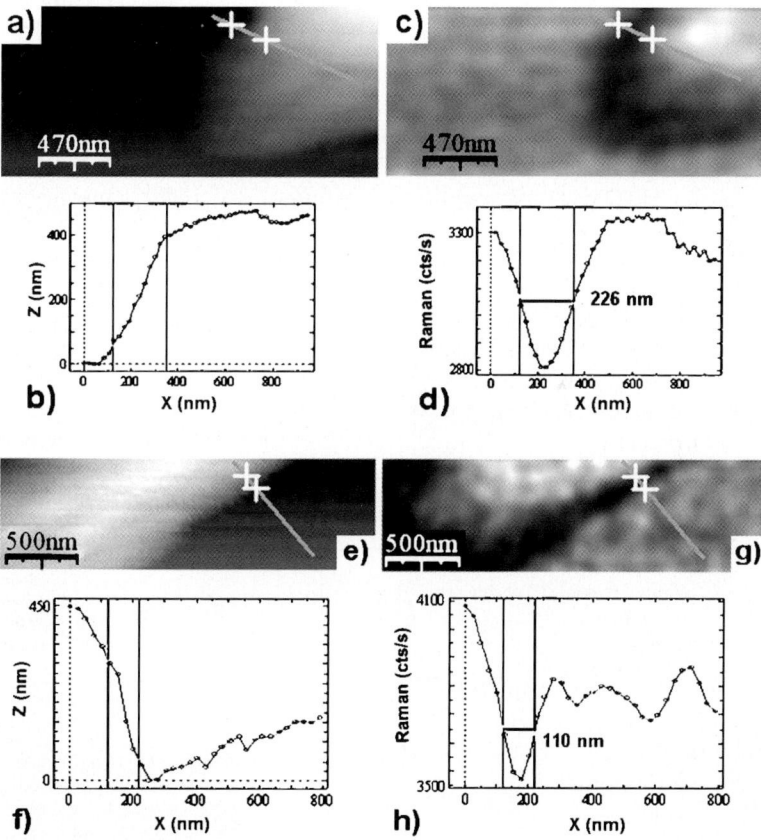

Fig. 10.14. Topography of a surface defect present on a TCNQ crystal having height of ~ 450 nm (**a**),(**b**). The corresponding Raman map at $1453\,cm^{-1}$ shows an increased scattering in the central part surrounded by an annular region of depleted scattering having a ~ 220 nm width. Scan width ($2.35 \times 1.15\,\mu m^2$). Topography (**e**) and (**g**) Raman map at $2225\,cm^{-1}$ carried out on the lateral side of the defect. From the corresponding line profiles (**f**),(**h**) we can assess the genuine nature of the optical signal, as well as a spatial resolution of the order of 100 nm. Scan width ($2.5 \times 0.74\,\mu m^2$). Exposure time 100 ms per pixel

scratches. The sparse near-field spectra in Fig. 10.13d, evidenced only changes of the 1453 cm^{-1} peak intensity and no peak shifts, within the spectrometer resolution. Due to the short integration time ($\tau_{INT} = 100$ ms), maps consisting of 128×128 points were acquired in less than 1 h. Two different zooms carried out on a zone showing a bump-like corrugation are reported in Fig. 10.14 (a,c: $2.35 \times 1.15 \,\mu m^2$; e,g: $2.5 \times 0.74 \,\mu m^2$). In both measurements the topographies (a,e) are ~ 450 nm high (line profiles b,f). In the Raman maps at 1453 cm^{-1} (c) and at 2225 cm^{-1} (g) we observed, again, a central zone of increased Raman scattering surrounded by annular regions with depleted Raman intensity. The local width of such rings can be estimated through the line profiles displayed in (d,h), and ranges from 220 to 110 nm. The latter value provides also an estimate of the spatial resolution obtained.

10.4.2
Chemical Mapping

Raman scattering allows to retrieve detailed chemical information about a large variety of samples. A large number of molecules are Raman active and their spectral signatures are sharp and well-defined. Micro- and nano-Raman imaging are thus well-suited for local mapping and discrimination of different chemical species in solid samples.

Material discrimination with near-field Raman means was demonstrated by Jahncke et al. [66], who studied a potassium titanyl phosphate (KTP) slab containing micrometric arrays of rubidium-doped zones, forming areas of rubidium titanyl phosphate (RTP). An aperture-SNOM working in emission mode was used, coupled to a double spectrometer (spectral resolution 4 cm^{-1}) mounting a cooled photomultiplier tube as the detector. Two aspheric lenses (NA = 0.55) allowed for simultaneous detection of the reflected and of the transmitted signals. The near-field Raman spectra (Fig. 10.15a) for the doped (bottom line) and the undoped (top line)

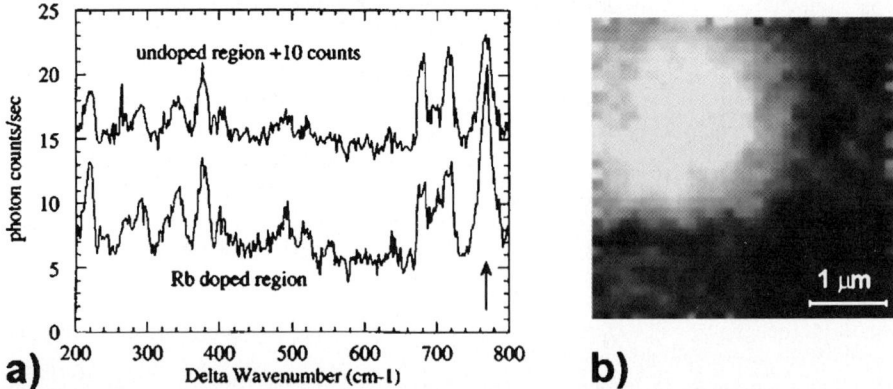

Fig. 10.15. (a) Near-field Raman spectra carried out on an undoped KTP crystal and on the Rb-doped RTP portion. (b) Near-field Raman image ($4 \times 4 \,\mu m^2$) evidencing the presence of a micrometer scale RTP-doped region on the KTP substrate. Total acquisition time 10 h (figure reprinted from [66] with permission of the author)

regions of the sample showed that the 767 cm^{-1} peak resulted almost twice as intense in the Rb-doped region with respect to the bulk region. This is mostly due to the mass difference between K and Rb [67]. As a consequence, the higher intensity area in the Raman map at this energy (Fig. 10.15b) could be attributed to a Rb-doped region. The map, an array of 32×32 Raman spectra, on a $4 \times 4 \, \mu m^2$ area, took over 10 h to be completed.

TCNQ is a strong electron acceptor and forms a variety of charge-transfer (CT) complexes with inorganic and organic donors. The CT complex formed by the interaction of a donor (D) and an acceptor (A) can be considered as a resonance hybrid of two structures: a no-bond structure (D, A), and a ionic structure (D$^+$, A$^-$) [65]. Depending on the degree of charge transfer ρ, the system contains only A$^-$ ions ($\rho = 1$), while if $0 < \rho < 1$ both A$^-$ and A^0 species exist. Copper complexes of TCNQ (CuTCNQ), in particular, exhibit unique electrical and optical field-induced switching and memory phenomena. Scanning probe techniques have provided a valuable contribution to a detailed understanding of crystallographic orientation, degree of crystallinity and chemical composition of these materials both in bulk, in thin films, and at interfaces ([68] and references therein). On the other hand, Raman imaging provides information on the chemical nature of the samples, allowing insight of the local degree of charge transfer. In the TCNQ CT complexes the electron-phonon coupling provokes both a decrease of the Raman activity and a shift of the vibration frequencies. Mapping the Raman emission at a well-defined energy is thus expected to provide chemical discrimination between TCNQ and its copper salt. We applied aperture-SNOM Raman techniques to image the formation of copper complexes of TCNQ on thin films of amorphous TCNQ [46, 61]. The apparatus used is described in Sect. 10.3.2.1.

At inspection with an optical microscope (Fig. 10.16a), the sample showed light zones, consisting of TCNQ molecules in the amorphous state, and darker areas attributable to the strong absorption of the CuTCNQ salt in the blue region of the visible spectrum [65]. The chemical nature of the two portions was confirmed by near-field Raman spectra: within the bright area the well-known TCNQ Raman peaks were observed (Fig. 10.16b), whose intensity changed from point to point; in the darker zone the chemical fingerprint of the CuTCNQ presence with given by the shift of the 1453 cm^{-1} peak towards 1380 cm^{-1} (Fig. 10.16c). In addition an overall reduction of Raman intensity by more than one order of magnitude was observed. The CuTCNQ spectrum required an integration time of 5 s per point, with respect to the 300 ms per point of the TCNQ spectra. In Fig. 10.16a we noted that, within the CuTCNQ zone, several micrometer scale brighter spots were present, probably consisting of undoped amorphous TCNQ, and originating from the inhomogeneous copper powder coverage of the surface during the oxidation process. To get insight on such phenomenon SNOM Raman was applied.

The topography (Fig. 10.17a) and the Raman emission at 1453 cm^{-1} (Fig. 10.17b) were simultaneously mapped on a $17.2 \times 14 \, \mu m^2$ wide area within the CuTCNQ zone (146×120 pts, $\tau_{INT} = 100$ ms). While the topography showed three similar doughnut shaped structures, having diameters of several micrometers and heights ranging from 200 to 500 nm, the Raman map indicated a high content of TCNQ only in two of the three structures (indicated by the white arrows). In the surrounding

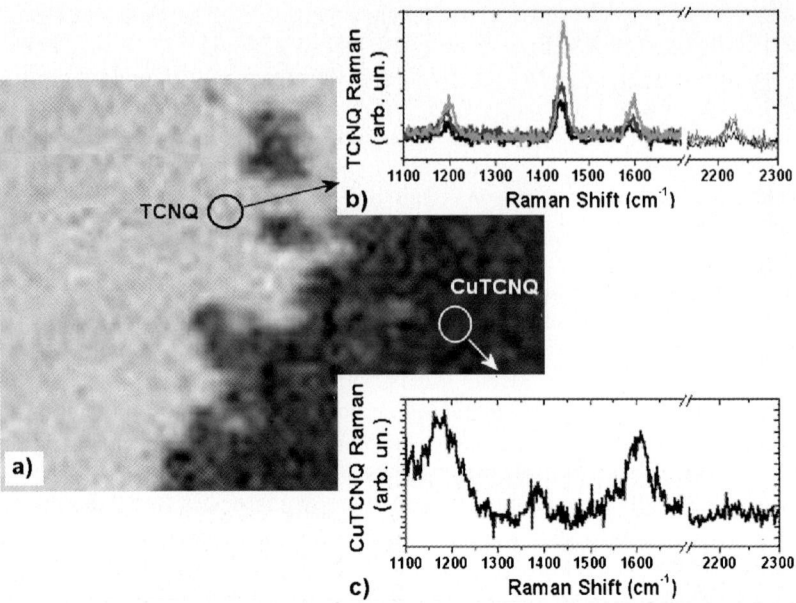

Fig. 10.16. Microphotograph of a TCNQ thin film (*light grey*) locally doped with Cu (**a**), forming CuTCNQ aggregates (*dark grey*). The chemical fingerprint of the two materials is provided by the near-field Raman spectra shown in (**b**),(**c**)

zone, the sample is mainly composed of CuTCNQ, suggesting an almost complete CT transition. A zoom was subsequently acquired on a $2.4 \times 2.0\,\mu m^2$ region. The Raman map at $1453\,cm^{-1}$ is shown in Fig. 10.17c (123×102 pts, $\tau_{INT} = 100\,ms$). On the upper left corner of the map, part of a bump-like structure appeared, characterized by a high Raman scattering, and thus suggesting the occurrence of a TCNQ-rich aggregate similar to ones observed in Fig. 10.17a. The central part of the image, instead, displayed the formation of CuTCNQ complexes, whose dimensions (235 nm, 158 nm, and 146 nm) can be determined from the line profile in Fig. 10.17d (top). On the right-hand side a high Raman intensity spot occurred (circled) indicating again, a TCNQ-rich area. The decrease of the Raman scattering, visible inside this spot, was probably due to a local copper contamination; the line profile in Fig. 10.17d (bottom) allows to assess a lateral dimension of 120 nm for such a salt cluster, as well as to observe a transition between two sites with different copper contamination occurring within 80 nm, namely the spatial resolution achieved (10 to 90% criterion).

10.4.3
Probing Single Molecules by Surface-Enhanced and Tip-Enhanced Near-Field Raman Spectroscopy

The capability to detect and identify single molecules in solids, liquids and dielectric surfaces, opens extraordinary opportunities in many disciplines such as molecular

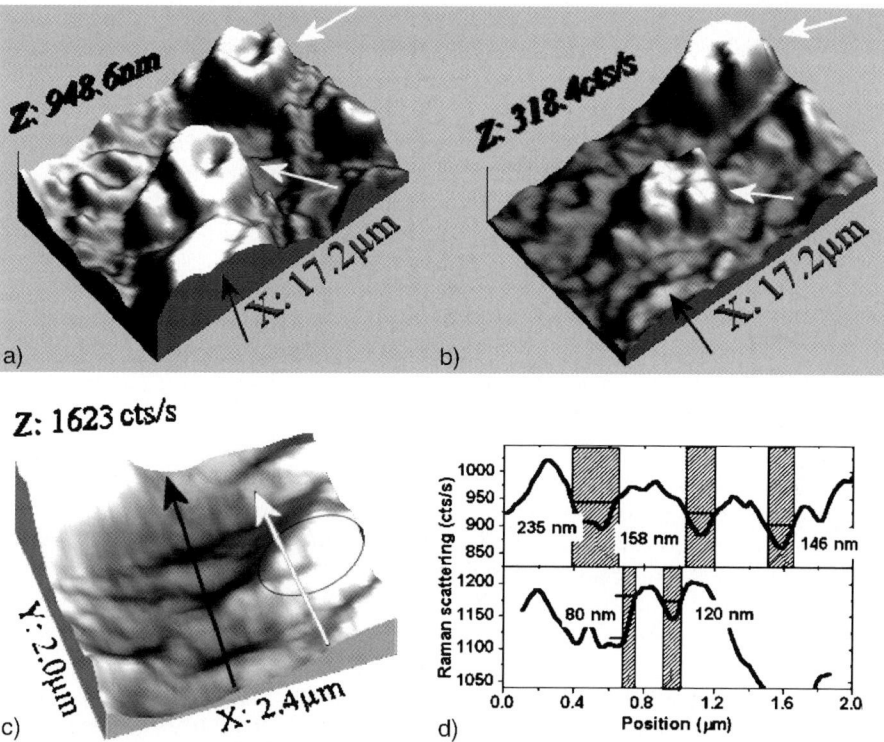

Fig. 10.17. Topography of three micrometer scale structures localized within the CuTCNQ (**a**). Scan width $17.2 \times 14\,\mu m^2$. The near-field Raman image (**b**) at the TCNQ peak emission ($1453\,cm^{-1}$) helps in assessing their chemical composition. The two structures indicated by the *white arrows*, display a high Raman scattering, have a rich content of TCNQ. The third one (*black arrows*) does not show any Raman emission at $1453\,cm^{-1}$, indicating the CuTCNQ chemical composition. Raman map at $1453\,cm^{-1}$ carried out on a $2.4 \times 2.0\,\mu m^2$ area (**c**) within the CuTCNQ zone, showing the occurrence of a TCNQ-rich spots (the one on the *upper left* corner, and the one *circled*), surrounded by CuTCNQ aggregates. Three of them, corresponding to the profile indicated by the *black arrow* (**d**, *top*) have dimensions ranging from 235 to 146 nm. Local CuTCNQ contaminations with sub-150 nm dimensions can be observed in the line profile indicated by the *black arrow* (**d**, *bottom*) (reprinted from [46] with permission of the author)

biology, analytical chemistry, and nanostructured materials. The Raman scattering, however, is an extremely inefficient process with cross sections of $\sim 10^{-30}\,cm^2$ per molecule [69], posing serious limits to its applicability to the study of molecular layers. This was the case until Fleishman et al. [70] observed unusual high Raman scattering signals from pyridine molecules adsorbed onto a roughened Ag electrode in aqueous solution. Later on, Jeanmarie et al. [71] and Albrecht et al. [72] recognized that the observed enhancement of the Raman signal could not be justified on the basis of the increased number of molecules at the surface. Two different explanations were proposed [73, 74]. First, an electromagnetic effect related to a large local field enhancement, arising from the excitation of electromagnetic resonances in the metal surface. The second mechanism, which can be defined as a chemical effect, involves new electronic states generated from the interaction between the molecules and the metallic surface, inducing resonant Raman scattering. In the electromagnetic effect, the major contribution comes from surface plasmons in metallic structures, which can enhance the electromagnetic field at the surface of the metal. As a consequence, the effect results are more evident when the molecules are adsorbed onto metallic particles having dimension smaller than the light wavelength, and when the radiation frequency is resonant with the plasmon ones. To a first approximation the enhancement of the Raman line intensity is given by [75]

$$Q_{SERS} = \left| \frac{E(r, \omega_0)}{E_0(\omega_0)} \right|^4 , \qquad (10.29)$$

where $E(r, \omega_0)$ is the total electric field at the molecule's position, and $E_0(\omega_0)$ is field associated with the incident radiation. The chemical mechanism can be better understood with the help of the energy level diagram for a molecule adsorbed on a metallic surface, shown in Fig. 10.18a. Molecules typically have their lowest lying electronic absorption bands in the near-ultraviolet, making impossible the excitation of resonant Raman scattering by means of the conventional visible laser sources. When the molecules are adsorbed on metallic substrates, however, it is not uncommon that the Fermi level of the metal could lay in the middle of the HOMO-LUMO molecular gap. In this case, charge transfer processes between the metal and the molecule, and vice versa, can occur at half of the value of the intramolecular transition energy, allowing for resonant Raman excitation visible light. The overall enhancement factor in SERS results, thus, from the combination of the two effects. While the charge-transfer mechanism can give an enhancement factor of 10 to 10^2, the electromagnetic mechanism can provide factors of 10^4 to 10^8.

The discovery of SERS opened the way to single-molecule detection by Raman means. First reports are due to Kneipp et al. [76] and Nie and Emory [69]. The first group obtained single-molecule SERS spectra [77] of cresyl violet adsorbed on silver nanoparticles [78], taking advantage of the extremely large Raman cross section of these molecules. The low molecular concentration (3.3×10^{-14} M), and the short exposure time (1 s) assured that the detected emission resulted from single molecules. Measuring the intensity of the Raman line at $1174\,cm^{-1}$ a Raman cross section of about 10^{-17} to $10^{-16}\,cm^2$ per molecule was estimated, yielding an enhancement factor of 10^{14}. In the work of Nie and Emory, Raman spectra were collected from single Rhodamine 6G molecules adsorbed on silver nanoparticles. Curiously, it was observed that only a small group of such nanoparticles, called hot particles or hot sites, showed an unusual high enhancement of the Raman signal. Comparative AFM and optical studies showed that such hot particles consisted of isolated single Ag colloids having diameters of about 110 nm. Moreover, only one out of 100 to 1000 particles resulted as hot, and out of these, only a few percent provided enhancements as high as 10^{14} to 10^{15}. The evidence that the measured spectra did refer actually to single molecules was given by polarization analysis of the scattered radiation. Bulk SERS spectra are, in fact, depolarized, while the Raman signal reported in Fig. 10.18b strongly depended on the polarization of the incident light with respect to the orientation of the nanoparticle.

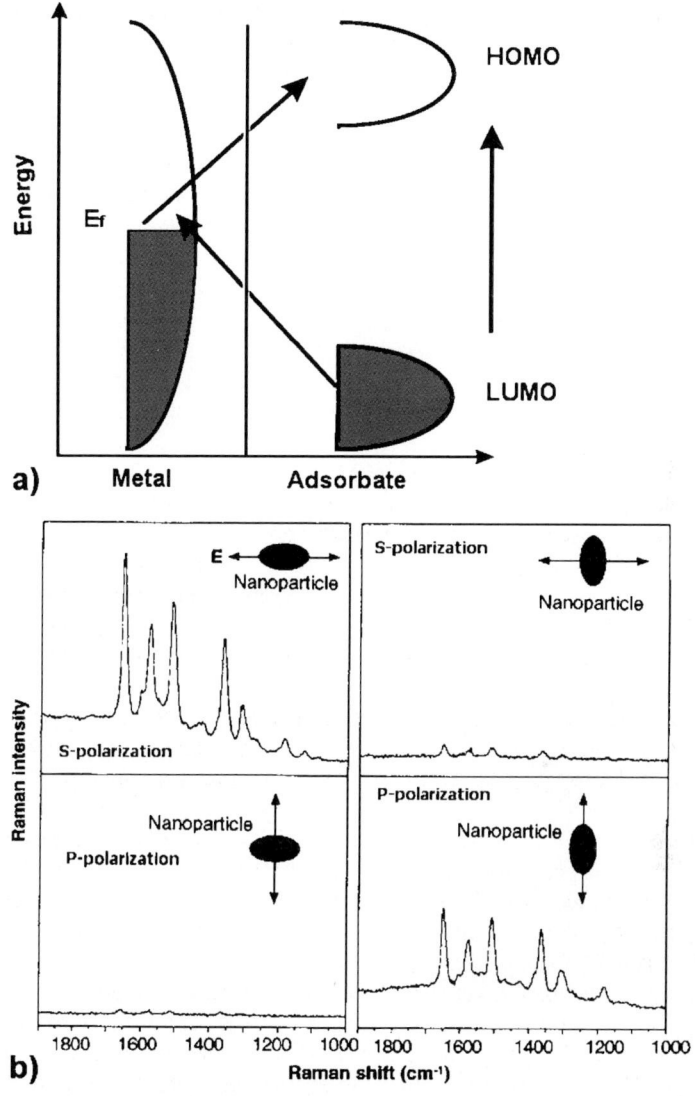

Fig. 10.18. (**a**) Schematic model of the charge transfer mechanism for a molecule adsorbed at the metal surface. (**b**) Polarization dependence of the SERS signal from R6G molecule adsorbed on the surface of a silver nanoparticle. Orientation of the nanoparticle with respect to the polarization of the incident electric are shown in the *inset* (adapted from [69])

The idea of exploiting SERS-active substrates to amplify the weak near-field Raman signal was demonstrated soon after [79, 80]. Combining surface-enhanced Raman scattering with aperture-SNOM techniques, Zenobi's group demonstrated the possibility to achieve 100 nm spatial resolution imaging dye-labeled DNA molecules [81]. The apparatus employed high throughput, chemically-etched probes [38] with apertures of $\sim$ 100 nm, delivering a laser power of $\sim$ 1 µW on the sample ($\lambda = 488$ nm). The light was collected in reflection with a microscope objective (NA = 0.65) and conveyed to a spectrometer by means of a multimode glass fiber. A brilliant cresyl blue (BCB)-labeled DNA sequence was selected as the analyte. BCB has well-known Raman bands at 1397, 1524 and 1655 cm^{-1} (the most intense vibration), assigned to the symmetrical in-plane C–C stretching. Figure 10.19a shows the Raman map at 1655 cm^{-1} on a 2×2 μm^2 area (20×20 points, $\tau_{INT} = 60$ s). The bright spot zoomed in Fig. 10.19b, whose dimensions are of 100 nm (the scan step), was attributed to a particular strong enhancement provided by the substrate in correspondence of a so-called hot site. Although there was strong SERS enhancement, the entire map took ~ 7 h to be acquired. It is interesting to observe that the enhanced Raman emission visible in correspondence of the bright spot (Fig. 10.19c, black arrow) does not coincide with the top of a silver particle. As already demonstrated by other authors, in fact, the exact location of the hot sites is not clear yet, and usually they occur in proximity of gaps or cavities among the metal colloids [69, 82].

One of the most severe limitations of SERS is that the sample must be deposited as a thin layer onto a rough film of noble metal or colloids. Moreover, due to the random distribution of the hot sites, the SERS enhancement varies across the sample

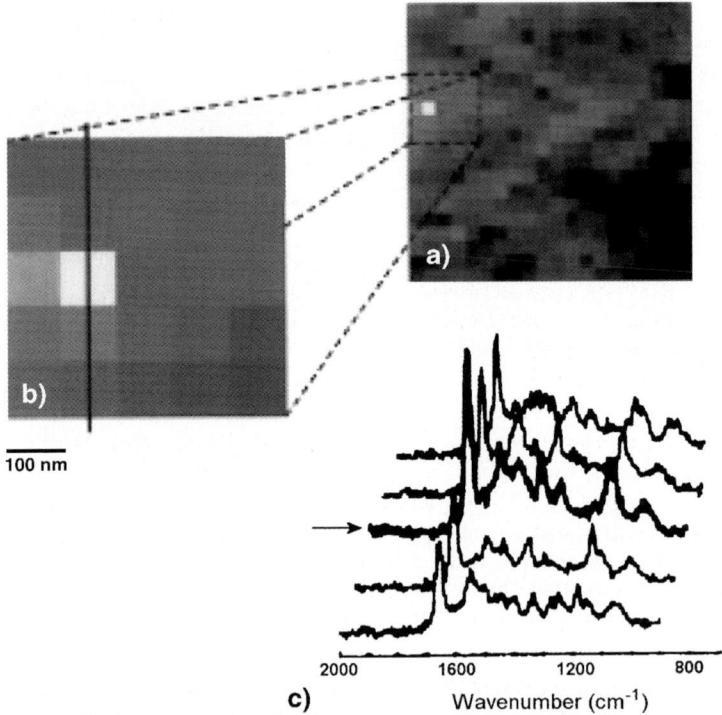

Fig. 10.19. Near-field Raman image (**a**) at 1665 cm^{-1} of dye-labeled DNA molecules deposited on a SERS-active substrate. The zoom (**b**) and the Raman spectra (**c**) acquired along the *black line* in (**b**) evidence an enhancement of the Raman emission of a spatial scale of 100 nm (reprinted from [81] with permission of the author)

and depends critically on the substrate preparation. This, in particular, limits the applicability of the method and makes quantitative measurements almost impossible. In tip-enhanced Raman scattering (TERS) the rough metal film is replaced with a sharp metal tip brought in proximity of the sample, using a scanning probe microscopy apparatus. The tip apex provides a large enhancement of the electric field [50], playing the role of the hot sites in SERS. Most of all, the enhancement arises from an interaction area comparable with the tip's radius of curvature, few tens of nanometers, well below the diffraction limited spot of the incident radiation. Therefore, the lateral optical resolution of this method is expected to be of a few nanometers. The tips are usually produced by either evaporating a noble metal (10–15 nm thick) onto an AFM tip [83, 84] or directly by electrochemical etching of a thin wire [85–87]. In the latter case the tips are glued to quartz tuning forks, and apparatuses on the detection of the shear force are used to control the tip–sample distance with sub-Angstrom precision [32, 88]. First TERS experiments were reported by the groups of Zenobi [84] and Kawata [89] on organic molecules. In both experiments the 488 nm line of an argon laser was focused on the tip from below, through an immersion oil objective (NA = 1.4), also used to collect the backscattered light. Figure 10.20a displays a Raman signal increase of a factor greater than 30 obtained when bringing a silver-coated AFM tip in contact with a thin BCB layer deposited on a glass coverslip (inset A), compared to an analogous measurement carried out with the tip retracted from the sample (inset B) [84]. Nevertheless integration times as long as 60 s were needed. Gold tips with apex diameters smaller than 50 nm, are also very effective to locally enhance the Raman scattering. Figure 10.20b displays a further example of TERS on a C_{60} thin film drop-coated on a glass substrate [84]. The Raman signal of C_{60} was easily detected when the tip was approached to the sample (inset A, integration time 200 s), whereas no signal was observed when the tip was retracted (inset B).

The origin of the TERS effect is still debated. As for SERS, it can be attributed to a combined electromagnetic and chemical mechanism. The first one is due to an increase of the electrostatic field in vicinity of the large apex curvatures or the edges of the tip (lightening rod effect). This effect can be further amplified by the excitation of plasmon-polaritons. In this case the latter represent resonant modes of electromagnetic waves and collective electron density fluctuations bound to the tip-surface assembly [50, 90–94]. Both effects have the result of enhancing the electric field at the tip apex with respect to the diffraction limited spot of the incident radiation. The chemical effect results from the increased polarizability of the adsorbates due to charge transfer or bond formation with the metal, inducing the formation of resonant Raman scattering states. As an example, the enhancement and the shift of the main peaks of C_{60} (marked with a *star* in Fig. 10.20b), have been attributed to the charge-transfer interaction between the sample and the gold tip [84]. The determination of the effective value of the field enhancement, and the consequent increase of the Raman scattering induced by metallic tips, is still a debated question. The field-enhancement is usually defined as the ratio

$$\xi = \frac{|E_{\text{tip}}|}{|E_0|} \tag{10.30}$$

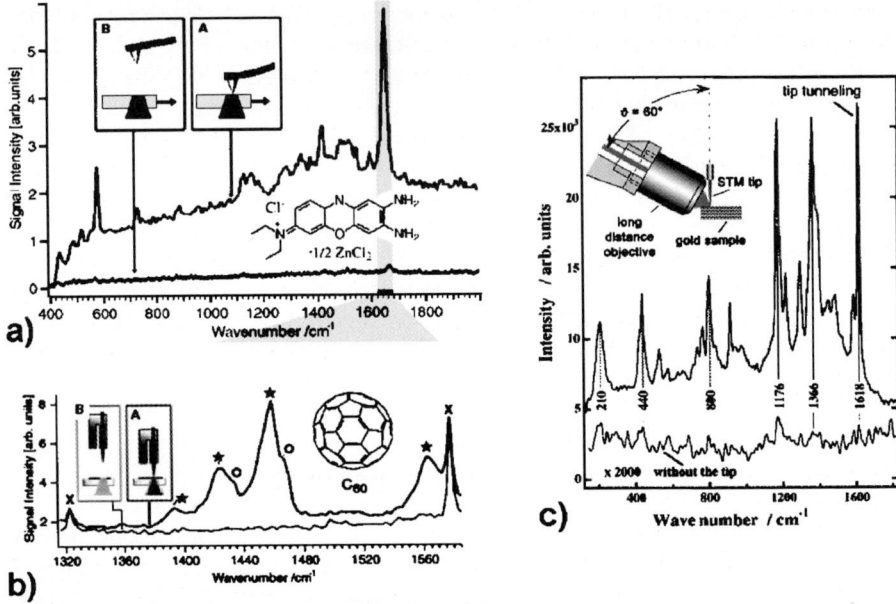

Fig. 10.20. (**a**) Tip-enhanced Raman spectra of brilliant cresyl blue dispersed on a glass support, measured with a silver-coated AFM probe. The two Raman spectra were measured (see *insets*) with the tip in contact and with the tip retracted. (**b**) Tip-enhanced Raman spectra acquired on a C_{60} thin film induced by a gold metallic tip brought in contact with the sample (see *insets*) (reprinted from [84] with permission of the author). (**c**) Comparison of RRS (without tip) and TERS (tip tunneling) spectra of malachite green isothiocyanate adsorbed at a Au(111) surface. Exposure time respective of 1 and 60 s. The *inset* shows the reflection mode configuration employed (reprinted from [100] with permission of the author)

between the actual electric field at the tip apex, E_{tip}, and the incident field E_0. Therefore, assuming that the tip acts as a nanoantenna for both the incident and the scattered photons, the Raman enhancement is quantified as the forth power of the field enhancement [95]

$$Q_{TERS} = \xi^4 . \tag{10.31}$$

Theoretical calculations [96,97] show that up to 1000-fold field enhancement might be achieved for optimum tip-surface geometry and excitation frequency excitation, leading to Raman intensity increases up to 12 orders of magnitude. Experimentally, however, the reported Raman intensity ratios $\eta = I_{TERS}/I_{RS}$, measured with and without the tip (I_{TERS}, I_{RS} respectively), range only between 1.4 and 40 [84,89,98]. Therefore, if the reduced surface area actually probed by the tip is taken into account [99], this corresponds to Raman enhancements Q up to 10^4. Much higher values have been experimentally demonstrated by Pettinger et al. [100] on malachite green isothiocyanate (MGITC) species adsorbed on a Au(111) surface. In absence of the tip (Fig. 10.20c, lower trace) MGITC exhibits intrinsic signal enhancement by resonant Raman scattering (RRS), if a suitable excitation energy is chosen (632.8 nm). The situation was observed to change radically when bringing the tip in touch with the sample (Fig. 10.20c, upper trace). For the six strongest bands, the Raman intensity increase ranged from 4000 to 14000, leading to values $Q \sim 10^6$ when considering the sample area effectively probed by the tip (radius $a \sim 90$ nm). This result is remarkable since such high values were reported using an apparatus working in reflection mode (inset of Fig. 10.20c) in which the microscope objective rests on the same side of the tip, and only long working distance objectives can be used, with a consequent loss of signal mainly due to the much smaller light gathering capabilities of the microscope objectives (NA = 0.5 compared to NA = 1.4 of oil-immersion objectives). This configuration is particularly important, since it overcomes the need of having a sample deposited on a transparent substrate, enabling for applications on real solid state devices.

10.4.4
Near-Field Raman Spectroscopy and Imaging of Carbon Nanotubes

Since their discovery in the early 1990s by Ijima [101], carbon nanotubes have become the focus of interest by many scientists. The main reason lies in their unique electrical, mechanical, thermal and optical properties (see [102] and references therein). Moreover, from their size and structure, carbon nanotubes provide a unique system for investigating one-dimensional quantum behavior [103]. One of the most promising applications of near-field Raman spectroscopy for single-walled carbon nanotubes (SWCNT), is the capability for their characterization at an individual level. Raman spectroscopy, in fact, allows for the direct determination of the nanotubes dimension, chirality, structure, and presence of defects. The quantum numbers (n, m) defining the tube structure (namely, the atomic coordinates in the 1D unit cell of the nanotube) are, in fact, associated with the frequency of the radial breathing mode (RBM, $\omega_{RBM} \sim 100-300$ cm^{-1}); metallic nanotubes can be distinguished from semiconducting ones based on the shape of the G-band ($\omega_G \sim 1592$ cm^{-1}) together with the RBM frequency. Finally, the D-band intensity is representative of the defects and other disorder-induced effects [55, 104]. In order to detail the specific structure-dependent properties of individual nanotubes, the group of Kawata [105] employed a typical transmission mode TERS apparatus working in backscattering with a high NA microscope objective (Fig. 10.21a,b). Silver-coated AFM tips, produced by thermal evaporation (thickness 40 nm), were employed to locally enhance the electric field. Commercially available SWCNT (Carbolex) samples were analyzed after spin-coating on a glass coverslip from a sonicated tetrafluoropropanol suspension, and drying in an oven for 4 h. Such a procedure usually leads to the formation of bundles of SWCNTs with some individual tube separated from the bundles. Figure 10.21c displays the tip-enhanced Raman spectrum of an individual tube. The tangential G-band appeared split into three peaks at 1593, 1570 and 1558 cm^{-1}, featuring semiconducting SWCNTs. The narrow line width of the peaks supported that the observed nanotubes consisted of only one or just a few tubes. A small D-band was observed around 1331 cm^{-1}, representing a defect mode. A tube diameter of ~ 1.5 nm was determined through measurements based on the AFM profiling and on the RBM energy ($E_{RBM} = 248/d$) [104]. Accordingly, the most probable quantum

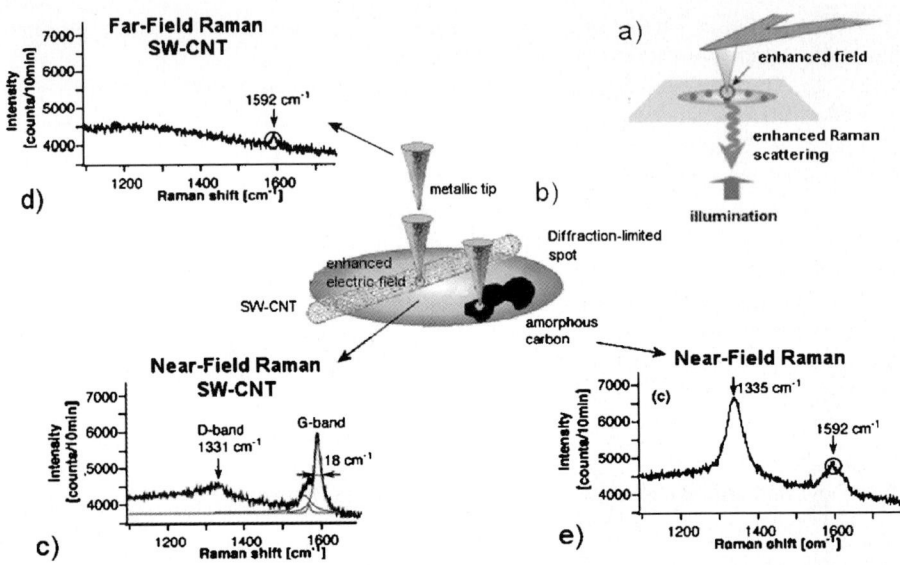

Fig. 10.21. Experimental configuration of the apparatus (**a**) used to induce an enhancement of the electric Raman scattering on single-walled carbon nanotubes (**b**). The Raman spectra with the tip in contact (**c**) and with the tip retracted (**d**) evidence the increased scattering. Tip-enhanced Raman spectrum of an amorphous carbon cluster located nearby the nanotube (**e**) (reprinted from [105] with permission of the author)

numbers associated to the semiconducting nanotube observed were (17, 4). Most important, in the far-field Raman spectrum obtained with the tip retracted by 500 nm (Fig. 10.21d), the featured lines were not recovered, due to the absence of field enhancement. Moving the tip to a different location (Fig. 10.21b) a broad D-band around $1335 \, \text{cm}^{-1}$ was detected (Fig. 10.21e), representing defects of amorphous carbon, enhanced by the tip.

Imaging the vibrational properties of SWCNTs with 10-nm resolution is an intriguing way for mapping the local composition of an individual SWCNT along its axis. The group of Novotny developed a nano-Raman setup [98, 106] for such a purpose, combining the TERS concept with scanning capabilities. The apparatus, based on an inverted microscope, was implemented with an xyz-translation stage to raster scan the sample underneath a sharp metal tip (gold or silver). Excitation at 632.8 nm, focused on a diffraction-limited spot by means of an oil-immersion objective (NA = 1.4), allowed for the resonant excitation of SWCNTs deposited on a glass coverslip. The spectral discrimination of the Raman scattering was carried out either by a monochromator coupled with a CCD or, for imaging purposes, through a set of notch and low-pass filters followed by a single photon counting avalanche photodiode. A good S/N ratio was demonstrated with very short integration times ranging from 25 to 100 ms per pixel. The capability of performing ultrahigh spatial resolution Raman imaging with such a setup is shown in Fig. 10.22 [102]. Figure 10.22a shows the confocal Raman image of an individual semiconducting SWCNT raster scanned under the focused laser spot, integrating the G′ band signal at $2600 \, \text{cm}^{-1}$. Figure 10.22b displays the corresponding nano-Raman image obtained

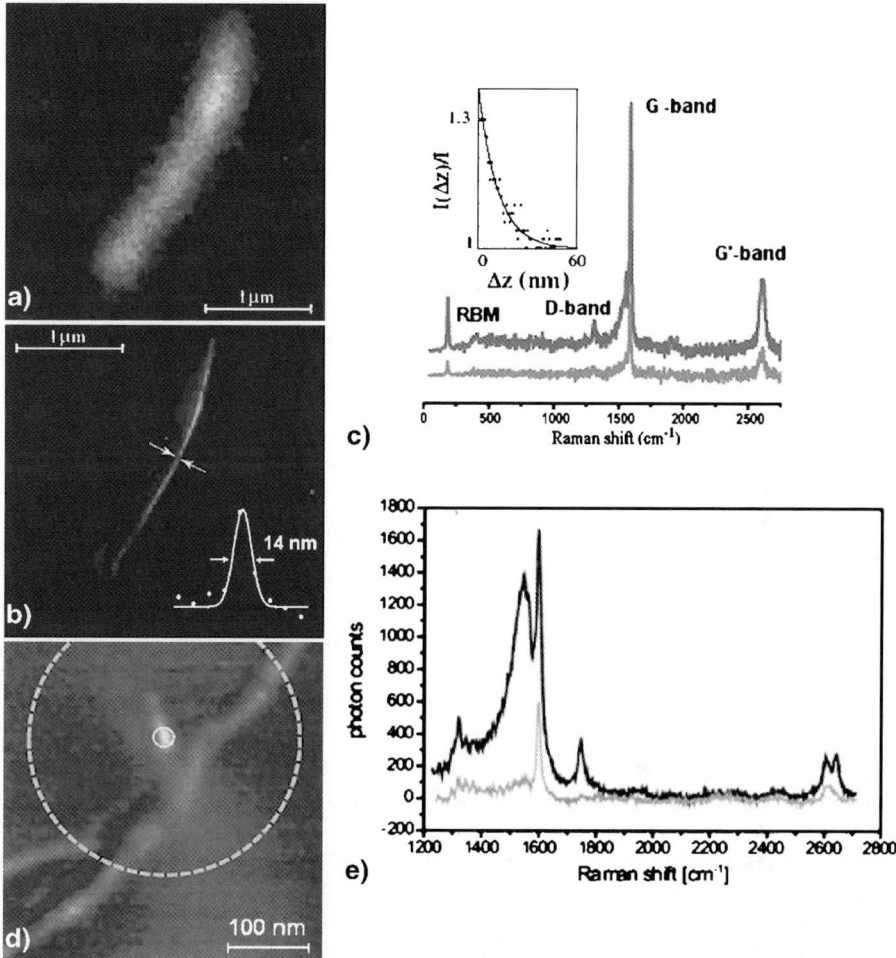

Fig. 10.22. Confocal (**a**) and near-field (**b**) Raman scattering images of a single-wall carbon nanotube. The contrast reflects the scattering intensity of the G′ band at 2640 cm^{-1}. The Raman spectra (**c**) obtained with the tip in contact (*top line*) and retracted (*bottom line*) show that an intensity enhancement takes place without any energy shift (reprinted from [102] with permission of the author). The exponential dependence of the Raman signal enhancement on the tip–sample distance is highlighted in the *inset* of (**c**) (reprinted from [98] with permission of the author). Near-field Raman image on a 400 × 400 nm^2 area of single-walled carbon nanotubes (**d**), observing the integrated scattering intensity of the G band. The near-field Raman spectrum (**e**, *black line*) acquired in correspondence of the brighter spot in (**d**) shows the metallic character of the individually probed nanotube, while the far-field spectrum (**e**, *grey line*) evidences the occurrence of several semiconducting nanotubes in the diffraction-limited probed area indicated by the *dashed circle* in (**d**) (reprinted from [107] with permission of the author)

placing the metal tip in the laser focus spot. The spatial resolution scales down from the diffraction-limit to about 14 nm. The increased Raman scattering observed when bringing the tip in proximity with the sample (black line in Fig. 10.22c), is a clear fingerprint of the TERS effect. Moreover, the exponential increase of the Raman signal as a function the tip–sample distance (inset of Fig. 10.22c) witnesses the near-field character of the scattering enhancement [98]. A more detailed analysis of the SWCNT composition is illustrated in the Raman map of Fig. 10.22d (400×400 nm^2), obtained monitoring the integrated G-band emission [107] of a few nanotubes. A very bright spot having diameter of a few tens of nanometers is evident (solid circle). The nano-Raman spectrum in Fig. 10.22e (black curve) allowed to characterize its nature. Besides the characteristic sharp G-band at 1592 cm^{-1}, in fact, a broad band having a maximum at 1550 cm^{-1} was observed, commonly attributed to a Breit–Wigner–Fano (BWF) resonance in metallic SWCNT [108]. In the Raman spectrum acquired after retracting the tip (Fig. 10.22e, gray curve), the signal was much weaker, and the BWF was nearly absent. This phenomenon was interpreted in terms of the different effective areas probed in the far and in the near-field measurement. The far-field spot (dashed circle in Fig. 10.22d) gave rise to a spectrum in which the signal was an average of all the tubes present within it. Therefore, since the spectrum does not show a pronounced BWF resonance, most of the nanotubes must be semiconducting. Conversely, the probed area in the near-field spectrum was only a few tens of nanometers (solid circle in Fig. 10.22d), and therefore the appearance of a strong BWF resonance highlighted the presence of a small metallic nanotube dispersed among other semiconducting tubes.

10.4.5
Coherent Anti-Stokes Near-Field Raman Imaging

The field amplification produced at the apex of a metallic tip has been demonstrated as useful for both linear and non-linear spectroscopy such as fluorescence [109], infrared absorption [110], Raman scattering, two-photon-excited fluorescence [111], and second harmonic generation [112–114]. In most cases, however, the far-field background, due to the diffraction-limited illumination spot, provides a noisy plateau difficult to subtract, which deteriorates the image contrast, and therefore the spatial resolution obtainable. Two-photon-excited fluorescence strongly reduces the background, but photobleaching and quenching induced by the metallic tip are unavoidable phenomena depleting the signal. Furthermore, for biological applications, the samples must be first stained with specific fluorescent probes [115]. Tip-enhanced coherent anti-Stokes Raman scattering (TE-CARS) was introduced by the group of Kawata for selective background-free vibrational imaging at the nanometer scale [116, 117]. The pump (ω_1), the Stokes ($\omega_2 < \omega_1$) and the probe (ω_1) fields were provided by two mode-locked Ti:Sa lasers, inducing a third-order non-linear polarization at frequency $\omega_{CARS} = 2\omega_1 - \omega_2$, which was detected. The beams, after being collinearly overlapped, were focused on the sample (Fig. 10.23a) by means of an oil-immersion objective, mounted on an inverted microscope. An AFM mounting silver-coated tip was used to locally induce the field enhancement. The backscattered CARS radiation, collected by the same objective, was detected by a single-photon counting avalanche photodiode. Figure 10.23b shows the spontaneous Raman spec-

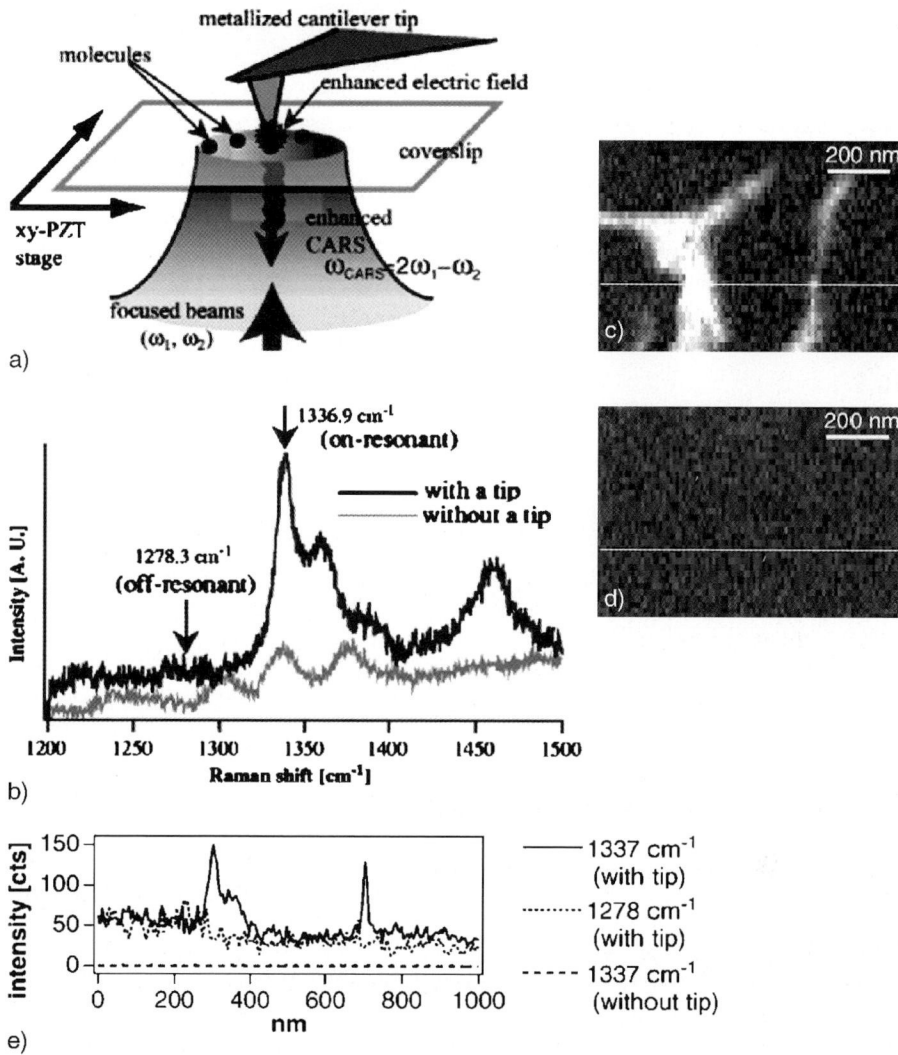

Fig. 10.23. (**a**) Schematic of tip-enhanced CARS apparatus. (**b**) Tip-enhanced (*black line*) and far-field (*grey line*) spontaneous Raman spectra of DNA nanoclusters (reprinted from [116] with permission of the author). (**c**) Tip-enhanced CARS images at resonance and (**d**) out of resonance of a DNA network. Scan area $2.4 \times 2.0\,\mu m^2$. (**e**) Line profile drawn in correspondence of the with lines at different excitation conditions (reprinted from [117] with permission of the author)

tra with (black line) and without the tip (grey line) of an aggregate of DNA molecules of poly(dA-dT). The strongest Raman peak at $1337\,cm^{-1}$ arises from the ring breathing mode of the diazole adenine molecule. TE-CARS images were acquired setting the frequency difference of the two excitation lasers to the resonance frequency at $1337\,cm^{-1}$, and compared to off-resonance images in which the frequency was set to $1278\,cm^{-1}$. The sensitivity of TE-CARS imaging was tested on DNA networks of poly(dA-dT) consisting of bundles of DNA aligned parallel on the glass substrate. The TE-CARS images in resonant and off-resonant excitation conditions are shown in Fig. 10.23c,d, respectively. The DNA filaments were clearly evident in the resonant condition, while only noise was detected off-resonance, indicating that the observed contrast is dominated by the resonant CARS signal. Figure 10.23e shows line profiles acquired along the white lines in the two optical maps, evidencing the occurrence of adenine molecules in the DNA double-helix. The estimated value of the electric field enhancement was about 100, and the estimated probed volume is of the order of around one zeptoliter. Spatial resolution well beyond the 100 nm was observed.

10.5
Conclusions

In the last decades scanning near-field optical techniques have been the basis for nanometer scale resolution spectroscopy and imaging. Single-molecule detection and identification, on the other hand, is a matter of ongoing active research. Raman spectroscopy can provide unambiguous molecular identification due to the well-defined vibrational energy peaks. However, in order to develop all its potential in nanotechnologies, a spatial resolution beyond the ~ 200 to $300\,\mathrm{nm}$ imposed by the diffraction limit is needed. Combination of Raman spectroscopy with aperture-SNOM is mainly limited by the low light outputs imposed by the fiber optic probes. More recently, the development of SERS and TERS has supplied a huge amplification of Raman scattering due to both the enhanced excitation fields and the exploitation of metal-induced resonant electronic levels. Although the physical and chemical mechanisms underlying these phenomena are not yet fully understood, single-molecule sensitivity and sub-20 nm spatial resolution Raman imaging have been demonstrated.

Acknowledgements. We kindly acknowledge D. Majolino and S. Santangelo for carefully reading the manuscript. We also acknowledge Nanotec for free software WSxM http://www.nanotec.es.

References

1. Raman CV, Krishnan KS (1928) Nature 121:501
2. Raman CV (1928) Nature 121:619
3. Porto SPS, Wood DL (1962) J Opt Soc Am 52:251
4. Bachelier G, Mlayah A, Cazayous M, Groenen J, Zwick A, Carrere H, Bedel-Pereira E, Arnoult A, Rocher A, Ponchet A (2003) Phys Rev E 67:205325
5. Maisano G, Majolino D, Mallamace F, Migliardo P, Vasi C, Vanderlingh F, Aliotta F (1986) Mol Phys 57:1083
6. Pohl DW, Denk W, Lanz M (1984) Appl Phys Lett 44:651
7. Lewis A, Isaacson M, Harootunian A, Murray A (1984) Ultramicroscopy 13:227
8. Betzig E, Trautman JK, Harris TD, Weiner JS, Kostelak RL (1991) Science 251:1468
9. Girard C, Dereux A (1996) Rep Progr Phys 59:657
10. Long DA (2002) The Raman Effect: A Unified Treatment of the Theory of Raman Scattering by Molecules. Wiley, New York

11. Maker PD, Terhune RW (1965) Phys Rev 137:A801
12. Shen YR (1984) The Principles of Non-Linear Optics. Wiley, New York
13. Duncan MD, Reintjes J, Manuccia TJ (1982) Opt Lett 7:350
14. Trusso S, Vasi C, Allegrini M, Fuso F, Pennelli G (1999) J Vac Sci Technol B 17:468
15. Monsu Scolaro L, Romeo A, Castriciano MA , Micali N (2005) Chem Comm 3018
16. Micali N, Villari V, Monsu Scolaro L, Romeo A, Catriciano MA (2005) Phys Rev E 72:050401(R)
17. Stanley HE, Ostrowsky N (1986) On Growth and Form. NATO ASI Series. Nijhoff, Dordrecht
18. Neri F, Trusso S, Vasi C, Barreca F, Valisa P (1998) Thin Solid Films 332:290
19. Abbe E (1873) Arch Mikroskop Anat 9:413
20. Synge EH (1928) Phil Mag 6:356
21. Newton I (1952) Optiks. Dover, New York, based on 4th (1730) edition
22. Courjon D (2003) Near-Field Microscopy and Near-Field Optics. Imperial College Press, London
23. Paesler MA, Moyer PJ (1996) Near-Field Optics: Theory, Instrumentation and Applications. Wiley, New York
24. Allegrini M, Ascoli C, Gozzini A (1971) Opt Commun 2:435
25. Van Labeke D, Barchiesi D, Baida F (1995) J Opt Soc Am 12:695
26. Latini G, Downes A, Fenwick O, Ambrosio A, Allegrini M, Daniel C, Silva C, Gucciardi PG, Patanè S, Daik R, Feast WJ, Cacialli F (2005) Appl Phys Lett 86:011102
27. Gucciardi PG, Patanè S, Ambrosio A, Allegrini M, Downes AD, Latini G, Fenwick O, Cacialli F (2005) Appl Phys Lett 86:203109
28. Erickson ES, Dunn RC (2005) Appl Phys Lett 87:201102
29. Ayars EJ, Hallen HD, Jahncke CL (2000) Phys Rev Lett 85:4180
30. Toledo-Crow R, Yang PC, Chen Y, Vaez-Iravani M (1992) Appl Phys Lett 60:2957
31. Betzig E, Finn PL, Weiner JS (1992) Appl Phys Lett 60:2484
32. Karrai K, Grober RD (1995) Appl Phys Lett 66:1842
33. Hecht B, Sick B, Wild UP, Deckert V, Zenobi R, Martin OJF, Pohl DW (2000) J Chem Phys 112:7761
34. Suh YD, Zenobi R (2000) Adv Mater 12:1139
35. Sayah A, Philipona C, Lambelet P, Pfeffer M, Marquis-Weible F (1998) Ultramicroscopy 71:59
36. Zeisel D, Nettesheim S, Dutoit B, Zenobi R (1996) Appl Phys Lett 68:2491
37. Wolf JF, Hillner PE, Bilewicz R, Kolsch P, Rabe JP (1999) Rev Sci Instrum 70:2751
38. Stöckle RM, Fokas C, Deckert V, Zenobi R, Sick B, Hecht B, Wild UP (1999) Appl Phys Lett 75:160
39. Patanè S, Cefalì E, Arena A, Gucciardi PG, Allegrini M (2006) Ultramicroscopy 106:475
40. Oesterschulze E, Rudow O, Mihalcea C, Scholz W, Werner S (1998) Ultramicroscopy 71:99
41. Veerman JA, Otter AM, Kuipers L, van Hulst NF (1998) Appl Phys Lett 72:3115
42. Radojewski J, Grabiec P (2003) Materials Science 21:3
43. Biagioni P, Polli D, Labardi M, Pucci A, Ruggeri G, Cerullo G, Finazzi M, Duò L (2005) Appl Phys Lett 87:223112
44. H'Dhili F, Bachelot R, Lerondel G, Barchiesi D, Royer P (2001) Appl Phys Lett 79:10
45. Gucciardi PG, Trusso S, Vasi C, Patanè S, Allegrini M (2002) Phys Chem Chem Phys 4:2747
46. Gucciardi PG, Trusso S, Vasi C, Patanè S, Allegrini M (2003) Appl Opt 42:2724
47. Zenhausern F, O'Boyle MP, Wickramasinghe HK (1994) Appl Phys Lett 65:1623; Bachelot R, Gleyzes P, Boccara AC (1995) Opt Lett 20:1924
48. Labardi M, Patanè S, Allegrini M (2000) Appl Phys Lett 77:621

49. Hillenbrand R, Keilmann F (2000) Phys Rev Lett 85:3029
50. Novotny L, Bian RX, Xie XS (1997) Phys Rev Lett 79:645
51. Hayazawa N, Saito Y, Kawata S (2004) Appl Phys Lett 85:6239
52. Dorn R, Quabis S, Leuchs G (2003) Phys Rev Lett 91:233901
53. Ren B, Picardi G, Pettinger B (2004) Rev Sci Instrum 75:837
54. Bonera E, Fanciulli M, Mariani M (2005) Appl Phys Lett 87:111913; Piazza F, Grambole D, Zhou L, Talke F, Casiraghi C, Ferrari AC, Robertson J (2004) Diam Relat Mater 13:1505
55. Jorio A, Saito R, Dresselhaus G, Dresselhaus MS (2004) Phil Trans R Soc Lond A 362:2311
56. Anastassakis E, Cantarero A, Cardona M (1990) Phys Rev B 41:7529
57. Clarke DR, Kroll MC, Kirchner PD, Cook RF, Hockey BJ (1988) Phys Rev Lett 60:2156
58. De Wolf I, Maes HE, Jones SK (1996) J Appl Phys 79:7148; De Wolf I, Anastassakis E (1999) J Appl Phys 85:7484
59. Webster S, Batchelder DN, Smith DA (1998) Appl Phys Lett 72:1478
60. Webster S, Smith DA, Batchelder DN (1998) Vib Spectrosc 18:51
61. Gucciardi PG, Trusso S, Vasi C, Patanè S, Allegrini M (2003) J Microsc Oxford 209:228
62. Gao W, Kahn A (2001) Appl Phys Lett 79:4040
63. Blochwitz J, Pfeiffer M, Fritz T, Leo K (1998) Appl Phys Lett 73:729
64. Hua ZY, Chen GR (1992) Vacuum 43:1019
65. Graja A (1997) Spectroscopy of Materials for Molecular Electronics. Scientific, Poznan
66. Jahncke CL, Paesler MA, Hallen HD (1995) Appl Phys Lett 67:2483
67. Jahncke CL, Hallen HD, Paesler MA (1996) J Raman Spectrosc 27:579
68. Higo M, Lu X, Mazur U, Hipps KW (2001) Thin Solid Films 384:90
69. Nie S, Emory SR (1997) Science 275:1102
70. Fleishmann J, Hendra PJ, McQuillan AJ (1974) Chem Phys Lett 26:163
71. Jeanmarie DL, Van Duyne RP (1977) J Electroanal Chem 84:1
72. Albrecht MG, Creighton JA (1977) J Am Chem Soc 99:5215
73. Moskovits M (1985) Rev Mod Phys 57:783
74. Otto A (1984) In: Light Scattering in Solids. Springer, Berlin Heidelberg New York
75. García-Vidal FJ, Pendry JB (1996) Phys Rev Lett 77:1163
76. Kneipp K, Wang Y, Kneipp H, Perelman LT, Itzkan I, Dasari RR, Feld MS (1997) Phys Rev Lett 78:1667
77. Otto A, Mrozek I, Grabhorn H, Akemann W (1992) J Phys Condens Matter 4:1143
78. Kneipp K, Wang Y, Kneipp H, Itzkan I, Dasari RR, Feld MS (1996) Phys Rev Lett 76:2444
79. Ziesel D, Dutoit B, Deckert V, Roth T, Zenobi R (1997) Anal Chem 69:749
80. Emory SR, Nie S (1997) Anal Chem 69:2631
81. Deckert V, Zeisel D, Zenobi R, Vo-Dinh T (1998) Anal Chem 70:2646
82. Hillenbrand R, Keilmann F, Hanarp P, Sutherland DS, Aizpurua J (2003) Appl Phys Lett 83:368
83. Wang JJ, Saito Y, Batchelder DN, Kirkham J, Robinson C, Smith DA (2005) Appl Phys Lett 86:263111
84. Stöckle C, Suh YS, Deckert V, Zenobi R (2000) Chem Phys Lett 318:131
85. Melmed AJ (1991) J Vac Sci Technol B9:601
86. Ren B, Picardi G, Pettinger B (2004) Rev Sci Instrum 75:837
87. Billot L, Berguiga L, de la Chapelle ML, Gilbert Y, Bachelot R (2005) Eur Phys J Appl Phys 31:139
88. Plake T, Ramsteiner M, Grahn HT (2002) Rev Sci Instrum 73:4250
89. Hayazawa N, Inouye Y, Sekkat Z, Kawata S (2000) Opt Commun 183:2000; Hayazawa N, Inouye Y, Sekkat Z, Kawata S (2001) Chem Phys Lett 335:369
90. Hillenbrand R, Taubner T, Keilmann F (2002) Nature 418:159
91. Knoll B, Keilmann F (2000) Opt Commun 182:321
92. Neacsu CC, Steudle GA, Raschko MB (2005) Appl Phys B 80:295

93. Ashino M, Ohtsu M (1998) Appl Phys Lett 72:1299
94. Hayazawa N, Saito Y, Kawata S (2005) Appl Phys Lett 85:6239
95. Kerker M, Wang DS, Chew H (1980) Appl Opt 19:3373
96. Jersch J, Demming F, Hildenhagen LJ, Dickmann K (1998) Appl Phys A 66:29
97. Demming AL, Festy F, Richards D (2005) J Chem Phys 122:184716
98. Hartschuh A, Sanchez EJ, Xie S, Novotny L (2003) Phys Rev Lett 90:095503
99. Patanè S, Gucciardi PG, Labardi M, Allegrini M (2004) La Rivista del Nuovo Cimento 27:1
100. Pettinger B, Ren B, Picardi G, Schuster R, Ertl G (2004) Phys Rev Lett 92:096101
101. Iijima S (1991) Nature 354:56
102. Anderson N, Hartschuh A, Cronin S, Novotny L (2005) J Am Chem Soc 127:2533
103. Venema LC, Wildoer JWG, Janssen JW, Tans SJ, Tuinstra HLJT, Kouwenhoven LP, Dekker C (1999) Science 283:52
104. Jorio A, Saito R, Hafner JH, Lieber CM, Hunter M, McClure T, Dresselhaus G, Dresselhaus MS (2001) Phys Rev Lett 86:1118
105. Hayazawa N, Yano T, Watanabe H, Inouye Y, Kawata S (2003) Chem Phys Lett 376:174
106. Hartschuh A, Anderson N, Novotny L (2005) J Microscopy 210:234
107. Hartschuh A, Anderson N, Novotny L (2004) Int J Nanosc 3:371
108. Pimenta MA, Marucci A, Empedocles SA, Bawendi MG, Hanlon EB, Rao AM, Eklund PC, Smalley RE, Dresselhaus G, Dresselhaus MS (1998) Phys Rev B 58:16016
109. Gerton JM, Wade LA, Lessard GA, Ma Z, Quake SR (2004) Phys Rev Lett 93:180801
110. Knoll B, Keilmann F (1999) Nature 399:134
111. Sanchez E, Novotny L, Xie XS (1999) Phys Rev Lett 82:4014
112. Bouhelier A, Beversluis M, Hartschuh A, Novotny L (2003) Phys Rev Lett 90:013903
113. Neacsu CC, Reider GA, Raschke MB (2005) Phys Rev B 71:201402
114. Labardi M, Allegrini M, Zavelani-Rossi M, Polli D, Cerullo G, De Silvestri S, Svelto O (2004) Opt Lett 29:62
115. Gucciardi PG, Princi P, Pisani A, Favaloro A, Cutroneo G (2005) J Kor Phys Soc 47:S86
116. Ichimura T, Hayazawa N, Hashimoto M, Inouye Y, Kawata S (2004) Appl Phys Lett 84:1768
117. Ichimura T, Hayazawa N, Hashimoto M, Inouye Y, Kawata S (2004) Phys Rev Lett 92:220801

Subject Index